Mimicking the Extracellular Matrix

The Intersection of Matrix Biology and Biomaterials

T0093566

Mimicking the Extracellular Matrix

The Intersection of Matrix Biology and Biomaterials

Edited by

Gregory A. Hudalla
University of Florida, Gainesville, USA
Email: ghudalla@bme.ufl.edu

and

William L. Murphy
University of Wisconsin–Madison, USA
Email: wlmurphy@wisc.edu

ROYAL SOCIETY
OF **CHEMISTRY**

Paperback ISBN: 978-1-83916-148-3
Hardback ISBN: 978-1-84973-833-0
EPUB ISBN: 978-1-78801-815-9

A catalogue record for this book is available from the British Library

The Royal Society of Chemistry is a charity, registered in England and Wales, Number 207890, and a company incorporated in England by Royal Charter (Registered No. RC000524), registered office: Burlington House, Piccadilly, London W1J 0BA, UK, Telephone: +44 (0) 20 7437 8656.

Visit our website at www.rsc.org/books

Preface

For at least 3000 years, humans have probed nature to identify new bio-materials, ranging from gut-derived sutures to metal tooth implants. Contemporary materials scientists continue to look to nature for innovative ideas, and this concept is now referred to as "biomimicry". Examples of biomimicry can be found throughout materials science, and virtually all implants used in modern medicine mimic some critical aspect of nature's materials. For example, polymer–ceramic composites have been designed based on the structure of bones and teeth, and elastomeric polymers have been synthesized to mimic the hierarchical ultrastructure and toughness of blood vessels and ten-dons. In each case, a detailed understanding of the fundamental com-position and structure of natural materials has led to fundamentally new chemistries and fabrication techniques. In essence, the more we think of mother nature as a materials scientist—and the more we understand her design principles—the better we become at cre-ating synthetic materials that are uniquely strong, tough, viscoelas-tic, dynamic, and/or biocompatible. Understanding the composition, structure, and function of nature's materials is the essence of "matrix biology", while designing and creating new materials using principles derived from biology is the essence of "biomaterials". *The guiding pur-pose of this book is to illuminate the exciting and dynamic intersection between matrix biology and biomaterials.*

While cells play a starring role orchestrating signaling dynamics during construction of human tissues and organs, the extracellular

Mimicking the Extracellular Matrix: The Intersection of Matrix Biology and Biomaterials
Edited by Gregory A. Hudalla and William L. Murphy
© The Royal Society of Chemistry 2020
Published by the Royal Society of Chemistry, www.rsc.org

matrix (ECM) can be viewed as the quintessential material in human biology. The ECM can exhibit a diverse array of intricately controlled properties, including stiffness, elasticity, cell adhesivity, and binding affinity for biological signals. These properties are spatially patterned, and can dynamically vary over time, resulting in intriguing and important biochemical gradients. The rapid pace of discovery in matrix biology has been matched by the rate of innovation in biomaterials science. In particular, multi-scale characterization of the natural ECM has occurred in parallel with new developments in multi-scale design and fabrication of biomaterials. As a result, the intersection between matrix biology and biomaterials is an extraordinarily active, fruitful, and reciprocal scientific interface. The provenance of the macromolecules in the ECM and the way in which they are organized—from the nanometer scale to the macroscopic scale—serve as a rich feedstock of information for designing new biomaterials. It is clear that our deeper, fundamental understanding of each of the properties of the ECM results in direct applicability to biomaterials design.

A wide variety of innovative concepts has already come from the matrix biology/biomaterials interface (as detailed throughout the chapters in this book), and we anticipate that this book will have both intended and unintended outcomes for readers. One intended outcome of this book is to disseminate recent advances in synthetic biomaterials that are inspired by natural ECMs, and to inspire further innovations in biomimicry. Biomimicry has increasing relevance in health care, such as in regenerative medicine, and in development of enabling tools for life scientists. This text aims to unify the current knowledge of ECM biology with the state-of-the-art of ECM-mimicking biomaterials. In the process, the individual chapters provide instructions for development of new biomaterials that can have direct, near-term impact on health care and life science.

Our goals at the outset of this book project were to: (1) solicit chapters from the most knowledgeable and prominent matrix biologists; (2) solicit chapters from the most innovative biomaterials scientists; (3) arrange and connect the chapters in a way that provides comprehensive information; (4) ensure that each chapter illuminates the rich interface between matrix biology and biomaterials. To achieve these goals, chapters were authored by either a leading biologist or a leading bioengineer. "Matrix biology" chapters highlight a feature of native ECMs that is integral to tissue development and homeostasis. "Biomaterials" chapters then discuss the state-of-the-art of biomaterials that mimic ECM features. Ultimately, this book is intended to appeal to both biologists and bioengineers interested in the ECM, and provide a benchmark for future efforts to develop synthetic biomaterials as ECM mimics.

The editors would like to sincerely thank each of the contributors who have made this book possible. Authors were hand-picked for each individual topic area, based on their worldwide leadership in these areas and their deep appreciation for cross-disciplinary innovation. We have achieved our goal of illuminating the rich matrix biology/biomaterials intersection thanks to the brilliance and extensive efforts of the chapter authors. The resulting book comes at a critically important time in these rapidly emerging fields. We anticipate that it will provide information to those who are entering this new cross-disciplinary intersection, and serve as an inspiration for the next generation of highly innovative biomaterials scientists.

Gregory A. Hudalla – University of Florida,
William L. Murphy – University of Wisconsin

Contents

Matrix Biology

Mimicking the Extracellular Matrix: The Intersection of Matrix Biology and Biomaterials
Edited by Gregory A. Hudalla and William L. Murphy
© The Royal Society of Chemistry 2020
Published by the Royal Society of Chemistry, www.rsc.org

Biomaterials as Mimics of the ECM

A. Building Function Through Complexity

B. Gradients and Patterns Within the
Extracellular Matrix

Chapter 8 Biomaterials: Spatial Patterning of Biomolecule Presentation Using Biomaterial Culture Methods 260
Kyle A. Kyburz, Navakanth R. Gandavarapu, Malar A. Azagarsamy, and Kristi S. Anseth

C. ECM Assembly, Organization, and Dynamics

Chapter 9 Biomaterials: Controlling Properties Over Time to Mimic the Dynamic Extracellular Matrix 285
Lisa Sawick and April Kloxin

Part I
Matrix Biology

CHAPTER 1

Matrix Biology: Extracellular Matrix – Building Function Through Complexity

LINDA J. SANDELL*[a]

[a]Department of Orthopaedic Surgery, Department of Cell Biology and Physiology, Department of Biomedical Engineering, Washington University Medical School, 660 S. Euclid Ave, CB 8233, St. Louis, MO 63110, USA
*E-mail: sandelll@wustl.edu

1.1 INTRODUCTION

1.1.1 The ECM

The extracellular matrix (ECM) is the extracellular component of a multicellular organism or tissue that provides structural and bio-chemical support to the surrounding cells. Characteristic of ECM is a complex interaction of specific large and small molecules that function as a composite structure: these structures can vary in different parts of the extracellular environment from a pericellular localization (concentrated around the cell) to interterritorial, making up the bulk of the ECM. These complex networks confer the functions of the ECM that are tissue specific; they are also dynamic, changing over time and developmental stage as well as in a response to injury or disease. ECM

Mimicking the Extracellular Matrix: The Intersection of Matrix Biology and Biomaterials
Edited by Gregory A. Hudalla and William L. Murphy
© The Royal Society of Chemistry 2020
Published by the Royal Society of Chemistry, www.rsc.org

is also a storehouse for molecules that can be released at later times, including growth factors that bind to the charged glysocaminoglycan chains of proteoglycans such as fibroblast growth factors (FGFs) and growth factors that bind to protein domains, like bone morphogenetic proteins (BMPs) and transforming growth factor beta superfamily members (TGFβs). The resident cells take their cues from their developmental program or information from cell–matrix or cell–cell interactions and synthesize appropriate ECM. Each tissue is exquisitely designed for its particular function, providing structural support and functional information that feeds back to the ECM-producing cells. For example: skin requires a barrier function, elasticity and wound healing ability; cartilage requires the ability to compress, functioning as a shock absorber; muscle requires a high capacity for work; and tendons require the ability to bend, pull and tighten. The ECM, made up of specific mixtures of collagens, proteoglycans, glycoproteins and other secreted molecules, performs all of these functions. The complexity of ECM structure and function provide a distinct challenge to tissue engineers. Figure 1.1 shows a freeze-etch electron micrograph of a chondrocyte in ear cartilage surrounded by ECM[1] while Figure 1.2 illustrates the types of molecules present in the cartilage ECM.[2]

Often the major structural component of ECM is a genetically distinct collagen, such as in skin and bone (type I collagen), tendon

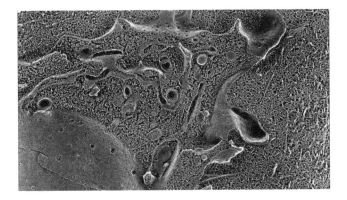

Figure 1.1   Chondrocyte with ECM. Quick-frozen, non-fixed, freeze-etched electron micrograph of a 150 day gestation fetal bovine ear chondrocyte surrounded by ECM. The cell is in the lower left area of the photograph and the matrix is in the upper right. Notice the very close association of the matrix with the cell. The plasma membrane of the cell is highly convoluted with numerous projections and infoldings. Nuclear pores are evident on the nuclear membrane, which is continuous with the cytoplasmic membrane system. The extracellular matrix consists of a finely woven meshwork that adjoins the cell membrane and is continuous throughout the extracellular space.[9]

(type III collagen), cartilage (type II collagen), cornea (type V collagen) and basement membrane (type IV collagen) often in a combination with other functional components, *e.g.* elastin for skin, proteoglycan aggrecan for cartilage, or calcium phosphate mineral for bone. Often tissues with highly functional ECM have a low cell to matrix ratio, where the cells in the tissue lay down a predictable amount of matrix during differentiation, development and growth (such as cartilage, bone, tendon). Functionally the centerpiece of connective tissues, ECM can resist compression (cartilage), make strong attachments (tendon, ligament), provide structure to the organism (bone), direct differentiation and development (mesenchyme) and guide nerve and blood vessels. Components of the ECM often assemble into large macromolecular functional units with interactions among molecules of the same or different types.

How can ECM perform so many different functions? Besides the major macromolecules the ECM harbors smaller amounts of

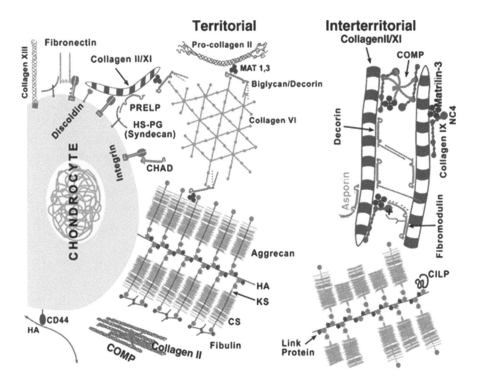

Figure 1.2   Schematic representation of cartilage ECM showing molecular constituents in cartilage and their arrangements into large multimolecular assemblies. The different compositions and organizations of macromolecules is seen at the cell surface with the number of receptors interacting with specific matrix molecules, at the interterritorial matrix closer to the cells, and the interterritorial matrix at a distance.[2]

functional molecules critical to structural functions, signaling functions, and specific degradation mechanisms. The ECM is also the home of many factors that are stored for later use, like MMPs (mobilized during injury) and growth factors (mobilized during development and regeneration).

1.1.2 Water

Water is a critical component of the ECM. The macromolecular composition of the matrix determines the matrix hydration that, in turn, determines tissue volume, creates space for molecular transport and for dynamic organization and offers compressive resistance. In the ECM, water is generally extrafibrillar (in relation to collagen fibers) and its distribution is specifically determined by the concentration of dissolved and retained proteoglycans and glycoproteins (>2 mg ml^{-1}).[3] The hydrodynamic processes controlling the water content in the ECM are those of osmosis, filtration, swelling and diffusion. In many cases, the macromolecular components influencing these hydrodynamic processes are nonideal. The nonideality (which yields parameters that vary nonlinearly with concentration) influences osmotic pressure through excluded volume polymer interactions and, for charged polymers, through the influence of active counterions on the ambient simple electrolyte concentration. Nonideal effects are manifested in dynamic parameters as well, particularly the hydrodynamic frictional terms that describe the viscous dissipation of water over the surface of the polymer chain.

The ECM functions to control many critical parameters of tissue homeostasis and response such as proliferation, apoptosis, development and morphogenesis. In this chapter, I will describe many (but not all) of the predominant ECM components and special functions required by different matrices.

1.2 MAJOR STRUCTURAL AND FUNCTIONAL COMPONENTS OF ECM

1.2.1 Collagens

The most abundant protein in all multicellular organisms is collagen and all mature collagen resides in the ECM. There are now known to be 28 different collagen types made up of at least 46 genetically distinct polypeptide chains; many other proteins contain collagenous domains. The most abundant collagens are type I (bone, skin), type II (cartilage and meniscus), type III (tendon, blood vessels), and type IV (basement membranes). Table 1.1 lists the different collagen types, their component polypeptide chains (gene products) and tissue sources.[4] The most abundant collagens (fibrillar collagen types I, II and III) are approximately 3000 nm in length and made up of uninterrupted Gly–X–Y repeats where X and Y are often proline and

Table 1.1   Vertebrate collagens.[a]

Type	Class	Composition	Distribution
I	Fibrillar	$\alpha1[I]_2\alpha2[I]$	Abundant and widespread: dermis, bone, tendon, ligament
II	Fibrillar	$\alpha1[II]_3$	Cartilage, meniscus, vitreous
III	Fibrillar	$\alpha1[III]_3$	Skin, blood vessels, intestine
IV	Network	$\alpha1[IV]_2\alpha2[IV]$ $\alpha3[IV]\alpha4[IV]\alpha5[IV]$ $\alpha5[IV]_2\alpha6[IV]$	Basement membranes
V	Fibrillar	$\alpha1[V]_3$ $\alpha1[V]_2\alpha2[V]$ $\alpha1[V]\alpha2[V]\alpha3[V]$	Widespread: bone, dermis, cornea, placenta
VI	Network	$\alpha1[VI]\alpha2[VI]\,\alpha3[VI]$ $\alpha1[VI]\alpha2[VI]\,\alpha4[VI]$	Widespread: bone, cartilage, cornea, dermis
VII	Anchoring fibrils	$\alpha1[VII]_2\alpha2[VII]$	Dermis, bladder
VIII	Network	$\alpha1[VIII]_3$ $\alpha2[VIII]_3$ $\alpha1[VIII]_2\alpha2[VIII]$	Widespread: dermis, brain, heart, kidney, cartilage
IX	FACIT[b]	$\alpha1[IX]\alpha2[IX]\alpha3[IX]$	Cartilage, cornea, vitreous
X	Network	$\alpha1[X]_3$	Cartilage hypertrophic zone
XI	Fibrillar	$\alpha1[XI]\alpha2[XI]\alpha3[XI]$	Cartilage, intervertebral disc
XII	FACIT	$\alpha1[XII]_3$	Dermis, tendon
XIII	MACIT	—	Endothelial cells, dermis, eye, heart
XIV	FACIT	$\alpha1[XIV]_3$	Widespread: bone, dermis, cartilage
XV	MULTIPLEXIN	—	Capillaries, testis, kidney, heart
XVI	FACIT	—	Dermis, kidney
XVII	MACIT	$\alpha1[XVII]_3$	Hemidesmosomes in epithelia
XVIII	MULTIPLEXIN	—	Basement membrane, liver
XIX	FACIT	—	Basement membrane
XX	FACIT	—	Cornea (chick)
XXI	FACIT	—	Stomach, kidney
XXII	FACIT	—	Tissue junctions
XXIII	MACIT	—	Heart, retina
XXIV	Fibrillar	—	Bone, cornea
XXV	MACIT	—	Brain, heart, testis
XXVI	FACIT	—	Testis, ovary
XXVII	Fibrillar	—	Cartilage
XXVIII	—	—	Dermis, sciatic nerve

[a]Modified from Shoulders and Raines, 2009.
[b]Abbreviations: FACIT, fibril-associated collagen with interrupted triple helices; MACIT, membrane-associated collagen with interrupted triple helices; MULTIPLEXIN, multiple triple-helix domains and interruptions.

hydroxy proline. In some collagens, this Gly–X–Y sequence is interrupted by globular protein sequence. Type IV collagen is about the same length, but contains many interruptions in the Gly–X–Y sequence. Three polypeptide chains of a single gene product, or multiple gene products, intertwine to form the collagen molecule. The characteristic amino acids give collagen its unique helical structure and the defining structural motif in which each of the three parallel polypeptide strands form a left-handed, polyproline II-type (PPII) helical conformation; the three chains then supercoil about each other with a one-residue stagger to form a right-handed triple helix. The individual collagenous proteins are composed of several functional domains involved in biosynthesis, fibrillogenesis, fiber structure cross-linking, cell or molecular interactions, and degradation.[5] The categories of collagen include the classical fibrillar and network-forming collagens, the fibril-associated collagens with interrupted triple helices (FACITs), membrane-associated collagens with interrupted triple helices (MACITs), and multiple triple-helix domains and interruptions (MULTIPLEXINs). In the ECM, the collagen fiber can be made up of more than one collagen type and even contain non-collagenous components.

Collagens are encoded for by large genes made up of many exons that are the functional units of the protein and regulatory apparatus. Evolutionarily, collagen arose from a single 54 base pair exon encoding 18 amino acids forming 6 Gly–X–Y triplets.[6] This exon was duplicated many times to form the fibrillar collagens and also incorporated into the other collagens. Various globular domains were added to the gene to accommodate various aspects of secretion, molecular interactions and degradation. Promoter and other regulatory elements allow specific genes to be expressed in the appropriate tissues. The study of gene structure of collagens is an intriguing example of evolution of very specific protein functions and interactions varying expression by means of the choice of gene regulation, alternative splicing and half life of the mRNA.

1.2.2 Large Chondroitin Sulfate Proteoglycans

Proteoglycans are large ECM components that are made up of a protein backbone and attached glycosaminoglycan chains (sulfated polysaccharides). The major aggregating proteoglycan of cartilage, aggrecan, has been studied as the paradigm[7] and functions to provide compressibility to cartilage: smaller amounts are found in meniscus and tendon. Aggrecan contains numerous chondroitin sulfate and keratan sulfate side chains that are distributed in the C-terminal two-thirds of the core protein, while the N-terminal one-third of the molecule carries few glycosaminoglycan chains but is relatively rich in N-linked carbohydrate, most of the cysteine residues of the protein, and behaves like a globular protein (Mr = 60 000) when isolated after proteolytic cleavage. This globular domain possesses affinity for hyaluronic acid and confers on the intact monomer the property of assembling into aggregates of many monomers bound to a central hyaluronic acid strand. This aggregate is stabilized by complementary binding activity of link protein, a Mr = 45 000 glycoprotein that also binds hyaluronic acid, having mutual affinity for the proteoglycan hyaluronic acid-binding region of aggrecan. Thus the aggrecan core protein has a molecular weight of Mr > 300 000, with over 100 glycosaminoglycan side chains containing over 100 carbohydrate residues. This "monomer" attaches *via* the link protein to a strand of hyaluronic acid, making the entire complex over 1 million kD. Large aggregating proteoglycans similar to that of cartilage have been described in several other tissues, notably tendon, bone, sclera and lung, as well as aorta smooth muscle cells, skin fibroblasts and glial cells.[8]

The core protein of aggrecan is composed of three globular domains (G1, G2 and G3) with one inter-globular domain (IGD) linking G1 and G2, and two exons for keratan sulfate (KS) chain attachment (KS domain) and for chondroitin sulfate (CS) chain attachment (CS domain) situated between G2 and G3. Attachment of these GAG chains occurs on the serine of a serine–glycine dipeptide sequence present in these regions, and one molecule of aggrecan can contain up to 100 CS chains, 30 KS chains and many O- and N-linked oligosaccharides.[9] Each CS and KS chain begins in the endoplasmic reticulum (ER) with the attachment of a xylosyl residue to a designated serine: the chains are elongated into specific glycosaminoglycan chains in the Golgi where specific glucosyl transferases, sulfatases and epimerases are resident. The G1 domain comprises the N-terminus of the core protein and has the same structural motifs as the link protein. The G3 domain, making up the C-terminus, is composed of alternatively spliced epidermal growth factor-like domains, a carbohydrate recognition domain, a complement-binding-protein-like domain and a short tail.[10,11] Other large proteoglycans in this class are versican, neurocan and brevican: each has a similar structure, but less carbohydrate than aggrecan.[9]

1.2.3 Small Proteoglycans

Smaller proteoglycans also exist in the ECM of almost all connective tissues and most belong either to the small leucine-rich proteoglycan (SLRP) family with 1-3 chondroitin or dermatan sulfate side chains, or the syndecans with a heparin sulfate side chain. The SLRPs function in hydrodynamic and structural roles, organization of the ECM, fibrillogenesis and signal transduction, particularly in development, pathophysiology and tumor growth. The members of this group are differentially expressed at the protein level in many different tissues and are highly interactive (see Table 1.2): the composition or substitution of the carbohydrate also varies, leading to specific binding affinities.[12] At the protein level, SLRPs have a variable number of tandem leucine-rich repeats comprising the major control domain. Each leucine-rich repeat has a conserved hallmark motif: LSSLxLSSNxL, where L is leucine (or isoleucine, valine or other hydrophobic amino acids), and x indicates any amino acid. They are divided into 5 subclasses encoded by 18 genes located over 7 chromosomes. Clustered genes can be regulated together during development: for example, keratocan that is downstream of lumican on chromosome 12,

Table 1.2   Interactions of SLRPs with other matrix molecules.[a]

SLRPs	Matrix assembly
Class I	
Decorin	Collagen I; collagen II and III; collagen V; collagen VI; collagen XII; collagen XIV; fibronectin; thromobospondin-1; microfibril-associated glycoprotein-1 and fibrillin-1; tenascin-X
Biglycan	Collagen I; collagen II; collagen III; collagen VI; collagen IX and biglycan; collagen II and VI complex; tropoelastin and microfibril-associated gylcoproterin-1
Asporin	Collagen I
Class II	
Fibromodulin	Collagen I; collagen II; collagen VI; collagen IX; collagen XII
Lumican	Collagen I; aggrecan; $\beta 1$ integrin; $\beta 2$ integrin; $\alpha 2\beta 1$ integrin
Osteoadherin	$\alpha v\beta 3$ integrin; noncollagenous domain 4 of collagen IX
PREPL	Perlecan and collagen
Class III	
Osteoglycin/mimecan	Collagen I
Opticin	Heparan and chondroitin sulfate proteoglycans, collagen
Class IV	
Chondroadherin	A2β1 integrin; collagen II
Class V	
Podocan	Collagen I

[a]From Chen and Birk, 2013.

is regulated by lumican.[13] The expression of SLRPs and cytokines are regulated bi-directionally through a common regulatory framework,[14] providing feedback mechanisms regulating matrix assembly and remodeling. The GAGs of SLRPs are differentially processed in development and aging, and are variable with regard to size, number, sulfation and epimerization in various tissues. For instance, lumican is predominantly a highly sulfated proteoglycan that is present in the cornea but a glycoprotein in other tissues.[15] Variations in glycosylation (both GAG and N-linked) modify the binding affinities of SLRPs at different developmental stages and in disease. For an excellent review of the functions of small proteoglycans, please see Chen (2013).[16]

The five classes of SLRPs (http://www.uniprot.org/uniprot) are: asporin, biglycan, decorin and extracellular matrix protein 2 (class I); fibromodulin, keratocan, lumican, osteomodulin and prolargin (class II); chondroadherin, nyctalopin (class IV); podocan and podocan-like protein 1 (class V). In addition to functions in collagen fibrillogenesis, these proteoglycans affect intracellular phosphorylation, a major conduit of information for cellular responses, and modulate distinct pathways, including those driven by BMP/TGF superfamily members, receptor tyrosine kinases such as ErbB family members, and IGFI receptor, and toll-like receptors.[17]

Collagen fibrillogenesis is tightly regulated by other collagens and by SLRPs to generate tissue-specific structures and therefore tissue-specific functions. In the corneal stroma, homogenous small-diameter fibrils, which are regularly packed and arranged as orthogonal lamellae, are required for corneal transparency. Lumican-deficient mice exhibit progressive corneal opacity with age. This is associated with irregularly packed, large-diameter collagen fibrils with irregular, cauliflower-like contours in the posterior stroma. The altered fibril characteristics are consistent with dysfunctional regulation of lateral fibril growth steps. The requirement of a homogeneous population of small-diameter fibrils necessary for transparency is inconsistent with lateral fibril growth in the corneal stroma. Mice that are deficient in decorin or biglycan have only a mild phenotype. However, compound mutant mice that are deficient for both decorin and biglycan demonstrate a severe phenotype with increased numbers of large-diameter fibrils and other irregularities. Fibromodulin, which is expressed during a very narrow window in development of the corneal stroma, is involved in regulation of corneal postnatal development.[18] Indeed, high degree myopia is associated with intronic variations and single-nucleotide polymorphisms in fibromodulin, proline/arginine-rich

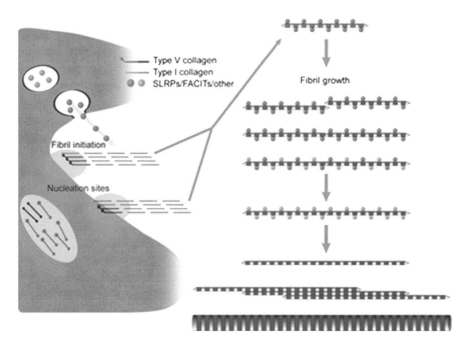

Figure 1.3   Model illustrating the involvement of SLRPs in the regulation of linear and lateral fibril growth. Collagen fibrillogenesis is a multiple-step process that is tightly regulated by the interaction of various molecules. The initial step involves heterotypic collagen I/V nucleation at the cell surface, then SLRPs bind to the protofibril surface, regulating the linear growth and lateral growth of protofibrils to mature collagen fibrils. Deficiency of SLRPs leads to dysfunctional linear and lateral fusion, with alterations in fibril structure and function.[16]

and leucine-rich repeat protein (PRELP) and opticin genes. Figure 1.3 shows the roles of small proteoglycans in collagen fibrillogenesis.

1.2.4   Biosynthesis of ECM Macromolecules

The biosynthesis of ECM components is complex due to the requirement for the extensive post-translational modifications that are vital components of almost all ECM molecules. Every ECM molecule is produced in the cell's secretory apparatus *via* the signal peptide where the proteins are transported from the cytoplasmic translational apparatus into the lumen of the ER during synthesis. The protein then undergoes specific post-translational modifications in a site-specific manner; first in the ER, then in the Golgi apparatus. The protein is exposed to the post-translational machinery that lies entirely within the lumen of the ER, where the initial direct modifications of the protein core are made as well as any N-linked oligosaccharides, and the

Golgi, where the carbohydrate chains are elongated. Certain glycosyl residues undergo epimerization and sulfation. As the protein passes through the ER, it is folded into the mature protein. The protein components of the molecule are regulated by all of the known modalities, including transcription rate, alternative promoters, splicing differences, half life of the mRNA and degradation of improperly folded protein. The addition of post-translational modifications involves hundreds of specific enzymes that modify the protein core by *O*- or *N*-glycosylation, tyrosination, hydroxylation, or sumoylation with additional modification, elongations, epimerizations and sulfations of the glycosyl residues. The secretory machinery of a connective tissue cell is specialized for a high rate of synthesis of specific ECM molecules, with the entire cell being ready to provide the necessary amino acids (large amounts of glycine and proline for collagen), with the support of the required enzymes for modifications of the amino acid and glycosyl residues. In addition, the structures of the secretory apparatus, the ER, Golgi apparatus and secretory granules must all be in place. In chondrocytes, there is a degree of compartmentalization of matrix synthesis in the cell, with aggrecan and collagen being synthesized in separate domains.[19]

Fibrillar collagens have propeptides at each end of the molecule. At some time after the fibrillar collagen types I, II, III and V leave the Golgi apparatus, the propeptides are removed by specific proteases, the ADAMTS (a disintegrin and metalloproteinase with thrombospondin motifs) proteases. The propeptides act to help align the three polypeptide chains for trimer formation and are thought to help keep the protein soluble during synthesis and may even play a role in fibrillogenesis. For type IIB collagen, the NH_2-propeptide also functions separately to inhibit the invasion of endothelial cells and osteoclasts into cartilage[20,21] and the COOH-propeptide is found as a part of the ECM in the growth plate.[22] As there is so much collagen synthesis during organogenesis and repair, collagen propeptides are found in the serum and urine and can be used as biomarkers of synthesis.[23–25]

After formation of the collagen fibrils, intermolecular cross-links are formed between hydroxylated lysine residues, forming an aldehyde. Complex cross-links are formed in collagen (pyridinolines derived from three lysine residues) and elastin (desmosines derived from four lysine residues). The cross-links confer physical and mechanical properties to the fibrils and provide the tensile strength fundamental to the structural role of collagen fibrils in connective tissues.[26] The enzyme lysyl oxidase is required for cross-link formation; it is copper dependent and is most active against native collagen fibrils.

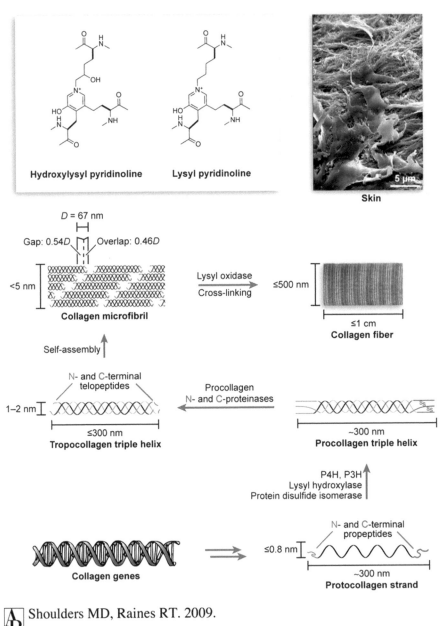

Hydroxylysyl pyridinoline **Lysyl pyridinoline**

Skin

Shoulders MD, Raines RT. 2009.
Annu. Rev. Biochem. 78:929–58

Figure 1.4 Collagen biosynthesis. Biosynthetic route from the cell to collagen fibers, for example here, the major component of skin. Size and complexity are increased by post-translational modifications: hydroxylation of lysine and proline, rearrangement of disulfide bonds, and self-assembly. Oxidation of lysine side chains leads to the spontaneous formation of hydroxylysyl pyridinoline and lysyl pyridinoline cross-links between collagen molecules in the matrix.[4]

Deficiency in copper or in lysyl cross-links results in lathyrism, characterized by poor bone formation and strength, hyperextensible skin, weak ligaments and increased occurrence of aortic aneurysms. Lysyl oxidase is also associated with cancer, being responsive to hypoxia-inducible factors, and may be involved in initiation or perpetuation of metastases.[27] Figure 1.4 is a schematic of collagen synthesis in skin.

1.3 ADDITIONAL MATRIX MOLECULES IN SPECIALIZED TISSUES

1.3.1 Heparan Sulfate Proteoglycans

Heparan sulfate proteoglycans are a diverse family of GAG-bearing protein cores that include the syndecans, the glypicans, perlecan, agrin and collagen XVIII. They play key roles during normal processes of development, tissue morphogenesis and wound healing. As key components of basement membranes in organs and tissues, they also participate in selective filtration of biological fluids, in establishing cellular barriers and in modulation of angiogenesis. Heparan sulfate is a unique polymer of *N*-acetylglucosamine and glucuronic acid, modified by specific enzymes to generate biologically active structures. The negatively charged heparan sulfate chains bind to growth factors.

1.3.1.1 Perlecan.   Perlecan is also called heparan sulfate proteoglycan 2 and is present in basement membranes, mesenchymal cells and vascular cells. However, it is also present in the musculoskeletal tissues, cartilage, meniscus and intervertebral disc, which are devoid of basement membranes and are avascular. The mature core protein is modular, containing 1 sperm protein-enterokinase-agrin (SEA) module, 4 EGF-like domains, 22 Ig-like C2 domains, 3 laminin G-like domains, 3 laminin IV type A domains, and 4 low-density lipoprotein receptor class A domains. Functions of perlecan include modulation of angiogenesis,[28] solute filtration,[29] growth factor delivery *via* heparan sulfate chains,[17] initiation of chondrogenesis,[30] regulation of cell adhesion and fibrillogenesis.[31] A knockout of perlecan in mice primarily results in developmental abnormalities of the heart, brain and cartilage.

1.3.1.2 Syndecans.   Syndecan family members classically link the functions of the cell surface heparin-binding growth factor receptors to the cytoskeleton, facilitated by their uniqueness as transmembrane heparin sulfate proteoglycans (HSPGs). They act as high capacity, low affinity co-receptors to their cognate low capacity, high affinity receptors.[32] They can be processed to release the ECM domain of the syndecan into the matrix.

1.3.2 Matricellular Proteins

Another prominent group of ECM proteins are the matricellular proteins. These are ECM proteins that modulate cell–matrix interactions and cell functions but do not seem to have a direct structural role. The family includes thrombospondin-1, thrombospondin-2, osteopontin/Spp1, osteonectin/Sparc, periostin, tenascin C and tenascin X. Expression of matricellular proteins is usually high during embryogenesis, but nearly absent during normal postnatal life. Interestingly, they reappear in response to injury.[33] The cartilage oligomeric protein (COMP) is a member of the thrombospondin gene family (Tsp5). Thrombospondin-2 is involved in collagen fibrillogenesis.[34]

Osteopontin (also called secreted phosphoprotein 1 and bone sialoprotin 1) is a good example of this group of proteins. It is expressed in many tissues and cell types and found in body fluids.[35] The secreted protein is heavily modified post-translationally by *O*-glycosylation, sulfation, and serine/threonine phosphorylation.[36] The functional domains include an integrin attachment motif (GRDGS), thrombin cleavage site, cryptic integrin attachment motif and a mineral binding polyaspartate region. While osteopontin binds tightly to hydroxyapatite and forms an integral part of mineralized bone matrix, it also acts as a cytokine involved in enhancing production of INF-gamma and IL-12 and reducing IL-10 and is essential in the pathway that leads to type I immunity.[37] It is also known to play roles in tissue repair, regulation of bone metabolism, inflammation and immunity.[38]

1.3.3 Elastic Tissues

Many tissues are elastic in nature and, while having many of the same ECM molecules as other tissues, the protein elastin and its associated proteins provide the structural and functional components that allow repetitive distending/relaxing or passive lengthening/shortening movements endowing the tissues with flexibility and rubber-like extensibility. Elastic fibers are solid, branching and unbranching, fine and thick, rod-like fibers (as in nuchal and other elastic ligaments), or they occur as concentric sheets or lamellae (in blood vessels), or arranged in three dimensional meshworks of fine fibrils, as in elastic cartilage of the ear and larynx, or they may occur as combinations of these, as seen in the skin and the lungs. The predominant molecule in elastic fibers is elastin, but other proteins are also involved, such as fibrillins[39] and a smaller glycoprotein, microfibril-associated glycoprotein (MAGP).[40] A single gene encodes elastin, albeit with alternatively spliced products; fibrillin is encoded by two genes (fibrillin-1 and fibrillin-2); MAGP is a single gene. Elastic fibers are extremely inert and difficult to extract from the ECM. While elastic fibers are fairly resistant to degradation, MMP-12 is the elastase isolated from macrophages.[41] The fibers (particularly the fibrillins) are now known to bind to TGFβ family members[42] and play an active role in tissue morphogenesis and response to injury.

1.3.4   Calcified Tissues

Calcification (or biomineralization) is the process by which hydroxy-apatite (calcium phosphate, $CaPO_4$) is deposited in the ECM, primarily on the fibrillar collagen. Physiological mineralization occurs in hard tissues, particularly the bone, the hypertrophic zone of the growth plate, and dentin, whereas pathological mineralization occurs in soft tissues such as tendon, kidney, blood vessels and synovial joints. Therefore, the deposition of mineral in the ECM of tissues must be tightly regulated.[43] In addition, the amount of mineralization must be critically regulated, as low bone mineral density leads to osteoporosis, the most prominent bone disease. The first step in mineralization is the formation of hydroxyapatite matrix vesicles[44] (or potentially related exosomes)[45] that bud from the surface membrane of hypertrophic chondrocytes, osteoblasts and odontoblasts. This is followed by prop-agation of hydroxyapatite into the ECM and its deposition between collagen fibrils. Extracellular inorganic pyrophosphate, provided by extracellular nucleotide pyrophosphatase/phosphodiesterase-1 (NPP1) and progressive ankylosis gene product (ANKH) inhibits hydroxyapa-tite formation. Tissue non-specific alkaline phosphatase (TNAP) hydro-lyzes the pyrophosphate and provides inorganic phosphate to promote mineralization. Cellular secreted proteins direct the initiation and directional growth of the mineral phase.[46] In addition, as the calcium and phosphate in body fluids are always in a metastable equilibrium, cells and tissue that are not intended to mineralize must be protected from doing so by expression of specific inhibitory proteins.

Osteoblasts drive intramembranous ossification and ossification during turnover of bone. Osteoblasts express a number of genes that code for bone matrix components and enzymes involved in bone syn-thesis, including type I collagen, tissue non-specific alkaline phospha-tase (TNAP), bone sialoprotein, osteocalcin and osteopontin: many of these proteins are under the control of the transcription factor Runx2.[46] Bone sialoprotein, among others, can nucleate the precip-itation of hydroxyapatite through a specific binding to collagen and to hydroxyapatite.[47] Osteoblasts produce bone-specific matrix onto existing ECM and, as the bone matrix continues to grow, the osteoblasts get entrapped in the matrix and terminally differentiate into osteocytes or undergo apoptosis.

Elongation of the bone requires the cartilage growth plate to allow growth by the process of endochondral bone formation. Initially, a cartilaginous template is formed made up largely of type II fibrillar collagen and the proteoglycan aggrecan. Through a very controlled process, the chondrocytes initially proliferate, then cease proliferation and begin to enlarge to become hypertrophic chondrocytes. Hypertrophic chondrocytes express a very specific complex pattern of genes that prepare the cartilage matrix to be vascularized and mineralized by means of matrix remodeling.[48] Simply, hypertrophic chondrocytes express the transcription factor Runx2 and make many of the same proteins as osteoblasts, including TNAP and bone sialoprotein and the hypertrophic-specific collagen, type X. However, in addition, hypertrophic chondrocytes produce the enzymes necessary to remodel the matrix, primarily MMP13 and MMP9. As will be discussed below, MMPs can liberate and activate growth factors from the ECM, including fibroblast growth factors (FGFs)[49] and vascular endothelial growth factor (VEGF), as well as degrade the cartilaginous matrix. Osteoblasts and osteoclasts are recruited to remodel the matrix and hypertrophic chondrocytes bud off matrix vesicles containing the proteins required for mineralization. The hypertrophic chondrocytes ultimately undergo apoptosis;[50] however, studies indicate that some hypertrophic chondrocytes may also transdifferentiate into osteoblasts.[51]

Inhibitors of mineralization are of critical importance. Most cells, unless injured or aged, do not produce the specialized ECM components and enzymes necessary to nucleate hydroxyapatite mineral formation. The serum proteins fetuin and gamma-carboxyglutamic-rich protein (MGP) help to maintain a high metastable concentration of calcium phosphate and inhibit ectopic calcification.[52] Chondrocytes (other than hypertrophic chondrocytes) and vascular smooth muscle cells synthesize high levels of MGP.

1.4   SURPRISES IN THE ECM – LATENT FUNCTIONS OF MATRIX MOLECULES

1.4.1   Degradative Enzymes in the ECM: MMPs and ADAMs

Enzymes resident in the ECM are required for normal turnover and degradation of the matrix. In some instances, the balance between anabolic and catabolic events is lost, and matrix is destroyed, as classically seen in diseases such as osteoarthritis, where cartilage is degraded. It is also well established that tumor initiation, progression and invasion are a consequence of a complex cross-talk between different cell types within the tumor microenvironment and are characterized by ability of malignant tumors to destroy matrix barriers, permitting invasion into the surrounding connective tissues, intravasation and extravasation, and metastasis to distant organs. The enzymes responsible for matrix turnover and degradation are called matrix metalloproteinses (MMPs or matrixins) and also include the adamalysins ADAMs (a disintegrin and metalloproteinase domain) and ADAMTS (with an additional thrombo-spondin domain). The MMPs are a specific group of 25 enzymes named for their dependence on metal ions for catalytic activity and their potent ability to degrade structural proteins of the ECM (Table 1.3). A typi-cal MMP has a multi-domain structure that includes a signal peptide domain (for secretion), a propeptide domain (often a latent form), and a catalytic domain. The ADAMs and ADAMTS enzymes are often mem-brane bound and include the additional domains described above. There are currently 33 members of the ADAM family and 5 members of the ADAMTS family.[53] Figure 1.5 demonstrates some of the enzymes and cleavage sites for ECM molecules used in cartilage turnover and degradation.[2]

MMPs are secreted as latent enzymes and require activation, which is tightly regulated to prevent tissue damage. The activities of most MMPs are very weak or negligible in normal steady-state tissues, but rapidly increase in response to inflammatory and oxidative stimuli. Their activity can be regulated at four levels: induction of MMP genes, vesicle trafficking and secretion, activation of latent proforms, and complexing with specific endogenous tissue inhibitors of metallopro-teinases (TIMPs). Most MMPs are activated in the ECM. The balance between production, activation and inhibition of MMPs is critical in maintaining ECM integrity. When proteolytic activity is greater than inhibition caused by TIMPs or other inhibitors, ECM breakdown occurs. Conversely, if inhibitors are too strongly expressed and prote-olysis is restricted, there is a build up of ECM proteins, with fibrosis.

1.4.2 Matrixins – Matrikines and Matricryptins

Once the ECM is established, the function and action of the matrix begins. The ECM is highly responsive and communicative with the cells and other parts of the tissue. Bioactive fragments are released from full-length proteins by limited proteolysis catalyzed by a vari-ety of enzymes such as cathepsins,[54] plasminogen activator/plasmin

Table 1.3 Human MMP members and principal biological effects.[a]

Group name	MMP	Substrates	Biological effects
Collagenase			
Collagenase-1	MMP-1	Collagens I, II, III, VII, X, gelatin, proteoglycan, link protein, entactin, tenascin	Keratinocyte migration and reepithelialization, release of bFGF, platelet aggregation, cell proliferation, pro- and anti-inflammatory, osteoclast activation, cell migration, enhanced collagen affinity, apoptosis, increased bioavailability of TGF
Collagenase-2	MMP-8	Collagens I, II, III, gelatin, proteoglycan, link protein	
Collagenase-3	MMP-13	Collagens I, II, III, IV, IX, X, XIV, proteoglycan, fibronectin, tenascin	
Gelatinase			
Gelatinase A	MMP-2	Gelatin, collagens IV, V, VII, XI, laminin, fibronectin, elastin, proteoglycan, link protein	Neurite outgrowth, generation of angiostatin-like fragment, enhanced collagen affinity, epithelial cell migration, tumor cell resistance, generation of vasoconstrictors, increased bioavailability of TGFβ pro-inflammatory, recruitment of osteoclasts
Gelatinase B	MMP-9	Gelatin, collagens III, IV, V, elastin, entactin, link protein	
Stromelysin			
Stromelysin-1	MMP-3	Proteoglycan, collagens III, IV, IX, X, laminin, fibronectin, gelatin, tenascin, link protein, elastin	Mammary epithelial cell apoptosis and alveolar formation, generation of angiostatin-like fragment, release of bFGF, increased bioavailability of IGF1 and cell proliferation, increased bioavailability of TGFβ, increased cell invasion, anti-inflammatory
Stromelysin-2	MMP-10	Collagens III, IV, V, fibronectin, laminin, proteoglycan, link protein, elastin	
Stromelysin-3	MMP-11	Fibronectin, laminin, proteoglycan, gelatin	

Membrane-type

MT1-MMP	MMP-14	Collagens I, II, III, gelatin, proteoglycan, fibronectin, laminin	Cell migration, kidney tubulogenesis, epithelial cell migration, reduced cell adhesion and spreading, embryo attachment to uterine epithelia
MT2-MMP	MMP-15	Fibronectin, tenascin, entactin, aggrecan, perlecan, laminin	
MT3-MMP	MMP-16	Collagen III, gelatin, fibronectin	
MT4-MMP	MMP-17	?	
MT5-MMP	MMP-24	Proteoglycan	
MP6-MMP	MMP-25	Gelatin	
Others			
Matrilysin	MMP-7	Proteoglycan, gelatin, fibronectin, tenascin, elastin, collagen IV, laminin, link protein	Adipocyte differentiation, generation of angiostatin-like fragment, increased bioavailability of IGF1 and cell proliferation, epithelial cell migration, increased bioavailability of TGFβ, disrupted cell aggregation and increased cell invasion, Fas-receptor mediated apoptosis, pro-inflammatory, osteocleast activation, vasoconctrictionand cell growth
Metalloelastase	MMP-12	Elastin	
Collagenase-4	MMP-18	Collagen I	
RAS I-1	MMP-19	Tenascin, gelatin, aggrecan	
Enamelysin	MMP-20	Enamel, gelatin	
XMMP	MMP-21	?	
No name	MMP-23	?	
Matrilysin 2	MMP-26	Collagen IV, fibronectin, gelatin	
No name	MMP-27	?	
Epilysin	MMP-28	Casein	

[a]From Gargiulo et al., 2014.

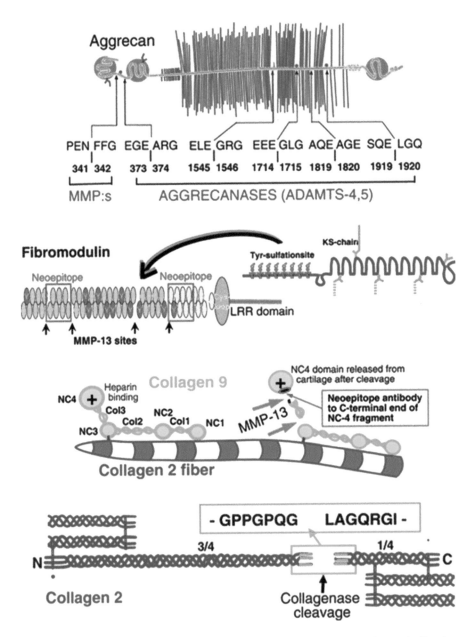

Figure 1.5   Illustrations of some specific events in cartilage breakdown, indicating specific cleavage sites and enzymes that are known to induce this cleavage at the tissue level. The degradation of aggrecan, fibromodulin, type IX and type II collagens are indicated.[2]

system[55] and MMPs. Generally these fragments provide new activities that are different from the intact molecules. MMP-2, MMP-7 and MMP-13 produce endostatin, an anti-angiogenic peptide from type XVIII,[56] MMP-9 produce tumstatin, an anti-angiogenic fragment of type IV collagen,[57] MMP-7, MMP-9, MMP-12 produce elastokines.[58] These new bioactive fragments are called matrikines and matricryptins (or elastokines if from elastin). In addition, heterotypic binding, cell-mediated mechanical forces, denaturation, multimerization, self-assembly and reactive oxygen species may also expose bioactive matricryptic sites in the ECM.[59,60] The release and activity of these new fragments adds an additional level of regulation to the ECM, modulating angiogenesis, cancer, fibrosis, inflammation, neuro-degenerative diseases and wound healing. There is a significant degree of interest in using these biologically active molecules as therapeutic agents.

Matrikines are defined by Maquart *et al.*[61] and Swindle *et al.*[62] as signaling elements that exist as subcomponents of ECM proteins and bind to cell surface receptors that belong to the cytokine, chemokine, ion channel or growth factor receptor family. These ligands (for example, tenascin EGF-like repeats) are encrypted within matrix components and modulate cellular responses mediated by growth factor receptors, and thus are considered as constrained to a surface, restricting their effect to the cells located in their vicinity. Most authors use matrikine and matricryptin interchangeably, with the definition being that they are "proteinase-generated fragments of matrix macromolecules that display cryptic bioactivities not manifested by the native, full length form of the molecule".[63] Peptides released from growth factors, chemokines and cytokines are not considered matrikines and are a separate class of bioactive fragments,[64] Recard-Blum and Salza suggest including fragments of ECM regulators (MMPs, ADAMs, and cross-linking enzymes) and ECM-affiliated proteins (*e.g.* mucins, galectins, semaphorins) as matricryptins. The ectodomains of membrane collagens (XIII, XVII, XXIII, and XXV) that are shed from the cell surface *in vivo* and released in the extracellular milieu, where they behave like cytokines, are martricryptins. The ectodomain of collagen XIII module cell adhesion, migration and proliferation and increased shedding of collagen XVII correlates with a decrease in keratinocyte motility. The ectodomains of syndecans 1-4, which are shed from the cell surface *in vivo*, contain adhesion regulatory domains and are also matricryptins.[65] Metastatin, comprised of proteolytic fragments of the link protein and the aggrecan core protein, inhibits angiogenesis and tumor growth and is a matrikine.[66] Oligosaccharides cleaved from hyaluronan[67,68] also belong to the matricryptin family and promote inflammation and improve wound healing.

The sources of matrikines vary, with each tissue having a specific set of ECM molecules and a specific set of MMPs that are capable of generating bioactive fragments. Collagens and proteoglycans,[69] elastin[58,70] and laminins[71] are major sources of matrikines. Matricellular proteins (*e.g.* SPARC, osteopontin) are additional sources. Some ECM proteins are cleaved into several fragments with either similar (*e.g.* collagen IV) or opposite activities for example SPARC, which generate both anti-angiogenic and pro-angiogenic fragments.[72] Most of the biological activities of ECM fragments are mediated by integrins, heparan sulfate proteoglycans and growth factor receptors.

Matrikines are usually produced under conditions where the producing enzyme has been activated, such as wound healing, cancer, inflammation, *etc.* Therefore the functions range from participation in angiogenesis (either pro- or anti-), ECM synthesis and remodeling and ECM assembly. In the early stages of angiogenesis, the degradation of ECM occurs in response to angiogenic stimuli. As a consequence, matrix molecules are degraded or partially modified, soluble factors are released and cryptic sites are exposed. Migration of cells through the basement membrane is a requirement for metastasis. Type IV collagens and perlecan are major components of the basement membrane: MMP cleavage of type IV collagen yields tumstatin, canstatin and arresten derived from the $\alpha 3$, $\alpha 2$ and $\alpha 1$ chains of the collagen;[73] tetrastatin, pentastatin and hexastatin are derived form the $\alpha 4$, $\alpha 5$ and $\alpha 6$ chains of type IV collagen, and endorepellin is derived from perlecan.[74] Peptides from the $\alpha 4$ and $\alpha 5$ laminin chains exhibit anti-microbial activity against *Escherichia coli* and *Staphylococcus aureus*.[75] ECM fragments also influence obesity and adipogenesis. Endostatin induces weight reduction in a murine model of obesity and regulates adipose tissue mass.[76] Elastin peptides regulate insulin resistance in mice[77] and endotrophin, a cleavage product of type IV collagen secreted by adipocytes, may play a role in obesity-related cancers.[78] Endotrophin also promotes tumor progression and fibrosis,[79] whereas endostatin has antifibrotic activity in dermal and pulmonary fibrosis.[80] Many other examples exist and are tissue specific (see Ricard-Blum and Salza, 2014 for a review and an exhaustive list of ECM bioactive fragments, their source and function).[64]

1.4.3   Growth Factors

Growth factors can be sequestered in the ECM either by binding to protein moieties or by binding to negatively charged carbohydrate chains. We have known for many years that specific cells produce growth factors at a given time and they are stored in the ECM for later use. Type IIA procollagen was one of the first ECM proteins shown to bind growth factors TGFβ, BMP-2 and BMP-7.[81] Type IIA procollagen retains the NH_2-propeptide in the ECM containing a VWF-C domain that binds to BMPs and TGFβs with high affinity (10^{-9} nM). Chordin, an inhibitor of BMPs in zebrafish[82] and decapentaplegic, which binds to BMPs in *Drosophila*, also bind *via* the same VWF-C domain.[83] These growth factors can be released by cleavage of the collagen or chordin by MMPs, releasing active BMPs.[82,84] Many of the effects of the matrix protein fibrillin are mediated through its binding to TGFβ, sequestering and then releasing the growth factor.[42] The sequestering of growth factors in the ECM allows growth factor gradients to be established and provides for a reserve supply of growth factors for later use.

In the late 1980's experiments by Comper and colleagues[85] found that the charged groups of the glycosaminoglycan chains of proteoglycans in cartilage did not participate in the water flow resistance in cartilage, as had been previously believed. Subsequently, it was discovered that these charged GAG chains bound growth factors.[86] These findings led to analysis of all GAG chains for growth factor binding and it was found that FGFs bound to HSPGs (perlecan and syndecans) at the cell surface and facilitated interaction with the actual FGF receptor.[87] Heparin in the matrix and bloodstream also binds to growth factors and has been used extensively for isolation.[88] The angiogenic growth factor VEGF is stored extracellularly, binding to the cell surface of ECM (fibronectin, HSPGs) and various MMPs.[74] Table 1.4 shows some growth factors that bind to perlecan.

Table 1.4   Interactions of perlecan with growth factors and morphogens.[a]

Growth Factor	Interacting Perlecan Moiety
FGF2	Heparan sulfate chains Domain I
FGF7	Domain III and Domain V/endorepellin
FGF10	Heparan sulfate chains, specific microdomains
FGF18	Domain III
VEGF	Heparan sulfate chains; co-localization with perlecan in tumor angiogenic vessels
PDGF	Heparan sulfate chains and Domains III and IV
Progranulin	Domain V/endorepellin
Hedgehog	Domain V/endorepellin of *Trol in Drosophilia*
TGFβ/Wnt	Protein core/UNC-52 *Trol in Drosophilia*
IL-2	Heparan sulfate chains

[a]From Whitelock *et al.*, 2008.

1.5   CONCLUSIONS

The ECM is a dynamic and response component of almost all tissues, ranging from blood vessels to cartilage. Specific combinations of molecules provide exquisite specialization of the ECM. ECM can initiate and respond to mechanical forces, developmental programs, injury and disease, and if not controlled properly can participate in progression of disease. The challenge to the tissue engineer is clearly to construct a matrix that has all of the necessary functional properties of the native matrix, and potentially to confer the complex properties necessary for the matrix to respond to the environment, whether it is normal homeostasis or a drastic insult. Many of the biologic aspects of matrix are already being used in the construction of tissue-engineered matrices, such as engineering the release of bioactive peptides. These aspects of ECM will be explained in detail in the following chapters.

REFERENCES

1. R. P. Mecham and J. E. Heuser, *Ciba Foundation Symposium 192: The Molecular Biology and Pathology of Elastic Tissues*, Plenum Press, Inc., 1991, pp. 79–109.
2. D. Heinegard, *Int. J. Exp. Pathol.*, 2009, **90**, 575–586.
3. *The Extracellular Matrix,* ed. W. D. Comper, Harwood Academic Publishers, Australia, 1996, vol. 2, pp. 1–22.
4. M. D. Shoulders and R. T. Raines, *Annu. Rev. Biochem.*, 2009, **78**, 929–958.
5. B. Siebold, R. A. Qian, R. W. Glanville, H. Hofmann, R. Deutzmann and K. Kuhn, *Eur. J. Biochem.*, 1987, **168**, 569–575.
6. L. J. Sandell and C. A. Boyd, in *Extracellular Matrix Genes*, ed. L. J. Sandell and C. A. Boyd, Academic Press, New York, 1990, pp. 1–56.
7. D. Heinegard, J. Wieslander, J. Sheehan, M. Paulsson and Y. Sommarin, *Biochem. J.*, 1985, **225**, 95–106.
8. J. H. Kimura, T. Shinomura and E. J.-M. A. Thonar, *Meth. Enzymol.*, 1987, **144**, 372–393.
9. T. N. Wight, D. K. Heinegard and V. C. Hascall, in *Cell Biology of Extracellular Matrix*, ed. E. D. Hay, Plenum Press, New York, 1991, pp. 57–59.
10. R. K. Margolis and R. U. Margolis, Exs, 1994, 70, 145–177.
11. R. A. Kosher, S. W. Gay, J. R. Kamanitz, W. M. Kulyk, B. J. Rodgers, S. Sai, T. Tanaka and M. L. Tanzer, *Dev. Biol.*, 1986, **118**, 112–117.
12. R. Plass, J. A. Last, N. C. Bartelt and G. L. Kellogg, *Nature*, 2001, **412**, 875.

13. J. H. Carlson, S. F. Porcella, G. McClarty and H. D. Caldwell, *Infect. Immun.*, 2005, **73**, 6407–6418.
14. E. S. Tasheva, A. Ke and G. W. Conrad, *Mol. Vision*, 2004, **10**, 544–554.
15. J. L. Funderburgh, M. L. Funderburgh, M. M. Mann and G. W. Conrad, *J. Biol. Chem.*, 1991, **266**, 24773–24777.
16. S. Chen and D. E. Birk, *FEBS J.*, 2013, **280**, 2120–2137.
17. J. M. Whitelock, J. Melrose and R. V. Iozzo, *Biochemistry*, 2008, **47**, 11174–11183.
18. Y. Q. Chen, A. Young, E. R. Brown, C. S. Chasela, S. A. Fiscus, I. F. Hoffman, M. Valentine, L. Emel, T. E. Taha, R. L. Goldenberg and J. S. Read, *J. Acquired Immune Defic. Syndr.*, 2010, **54**, 311–316.
19. B. M. Vertel and Y. Hitti, *Collagen Relat. Res.*, 1987, **7**, 57–75.
20. S. Hayashi, Z. Wang, J. Bryan, C. Kobayashi, R. Faccio and L. J. Sandell, *Bone*, 2011, **49**, 644–652.
21. L. J. Sandell, *Connect. Tissue Res.*, 2014, **55**, 20–25.
22. E. R. Lee, Y. Matsui and A. R. Poole, *J. Histochem. Cytochem.*, 1990, **38**, 659–673.
23. P. Garnero, *Bone*, 2014, **66C**, 46–55.
24. D. Patra, E. DeLassus, A. McAlinden and L. J. Sandell, *Matrix Biol.*, 2014, **34**, 154–160.
25. M. K. Lotz, J. Martel-Pelletier, C. Christiansen, M. L. Bandi, O. Bruyere, R. Chapurlat, J. Collette, C. Cooper, G. Giacovelli, J. A. Kanis, M. A. Karsdal, V. Kraus, W. F. Lems, I. Meulebelt, J. P. Pelletier, J. P. Raynauld, S. Reiter-Niesert, R. Rizzoli, L. J. Sandell, W. E. Van Spil and J. Y. Reginster, *Ann. Rheum. Dis.*, 2014, **72**, 1756–1763.
26. D. R. Eyre, M. A. Paz and P. M. Gallop, *Annu. Rev. Biochem.*, 1984, **53**, 717–748.
27. J. T. Erler, K. L. Bennewith, M. Nicolau, N. Dornhofer, C. Kong, Q. T. Le, J. T. Chi, S. S. Jeffrey and A. J. Giaccia, *Nature*, 2006, **440**, 1222–1226.
28. G. Bix and R. V. Iozzo, *Microsc. Res. Tech.*, 2008, **71**, 339–348.
29. S. J. Harvey and J. H. Miner, *Curr. Opin. Nephrol. Hypertens.*, 2008, **17**, 393–398.
30. M. C. Farach-Carson, J. T. Hecht and D. D. Carson, *Crit. Rev. Eukaryotic Gene Expression*, 2005, **15**, 29–48.
31. C. Kirn-Safran, M. C. Farach-Carson and D. D. Carson, *Cell. Mol. Life Sci.*, 2009, **66**, 3421–3434.
32. M. Delehedde, N. Sergeant, M. Lyon, P. S. Rudland and D. G. Fernig, *Eur. J. Biochem.*, 2001, **268**, 4423–4429.

33. M. W. Schellings, Y. M. Pinto and S. Heymans, *Cardiovasc. Res.*, 2004, **64**, 24–31.

34. T. R. Kyriakides, Y. H. Zhu, L. T. Smith, S. D. Bain, Z. Yang, M. T. Lin, K. G. Danielson, R. V. Iozzo, M. LaMarca, C. E. McKinney, E. I. Ginns and P. Bornstein, *J. Cell Biol.*, 1998, **140**, 419–430.

35. R. Zohar, W. Lee, P. Arora, S. Cheifetz, C. McCulloch and J. Sodek, *J. Cell. Physiol.*, 1997, **170**, 88–100.

36. B. Christensen, M. S. Nielsen, K. F. Haselmann, T. E. Petersen and E. S. Sorensen, *Biochem. J.*, 2005, **390**, 285–292.

37. S. Ashkar, G. F. Weber, V. Panoutsakopoulou, M. E. Sanchirico, M. Jansson, S. Zawaideh, S. R. Rittling, D. T. Denhardt, M. J. Glimcher and H. Cantor, *Science*, 2000, **287**, 860–864.

38. N. Mori, T. Majima, N. Iwasaki, S. Kon, K. Miyakawa, C. Kimura, K. Tanaka, D. T. Denhardt, S. Rittling, A. Minami and T. Uede, *Matrix Biol.*, 2007, **26**, 42–53.

39. T. Sakai, Y. Antoku, H. Iwashita, I. Goto, K. Nagamatsu and H. Shii, *Ann. Neurol.*, 1991, **29**, 664–669.

40. J. S. Kumaratilake, M. A. Gibson, J. C. Fanning and E. G. Cleary, *Eur. J. Cell Biol.*, 1989, **50**, 117–127.

41. M. J. Banda and Z. Werb, *Biochem. J.*, 1981, **193**, 589–605.

42. Z. Isogai, R. N. Ono, S. Ushiro, D. R. Keene, Y. Chen, R. Mazzieri, N. L. Charbonneau, D. P. Reinhardt, D. B. Rifkin and L. Y. Sakai, *J. Biol. Chem.*, 2003, **278**, 2750–2757.

43. C. Lian, M. Narimatsu, K. Nara and T. Hogetsu, *New Phytol.*, 2006, **171**, 825–836.

44. K. Hoshi and H. Ozawa, *Calcif. Tissue Int.*, 2000, **66**, 430–434.

45. C. Xiao, J. A. Sharp, M. Kawahara, A. R. Davalos, M. J. Difilippantonio, Y. Hu, W. Li, L. Cao, K. Buetow, T. Ried, B. P. Chadwick, C. X. Deng and B. Panning, *Cell*, 2007, **128**, 977–989.

46. H. A. Chris, A. B. Van de Lest and A. B. Vaandrager, *Curr. Opin. Orthop.*, 2007, **18**, 434–443.

47. C. E. Tye, K. R. Rattray, K. J. Warner, J. A. Gordon, J. Sodek, G. K. Hunter and H. A. Goldberg, *J. Biol. Chem.*, 2003, **278**, 7949–7955.

48. H. M. Kronenberg, *Nature*, 2003, **423**, 332–336.

49. T. Liu, R. J. Todhunter, S. Wu, W. Hou, R. Mateescu, Z. Zhang, N. I. Burton-Wurster, G. M. Acland, G. Lust and R. Wu, *Genomics*, 2007, **90**, 276–284.

50. I. M. Shapiro, C. S. Adams, T. Freeman and V. Srinivas, *Birth Defects Res., Part C*, 2005, **75**, 330–339.

51. P. Bianco, F. D. Cancedda, M. Riminucci and R. Cancedda, *Matrix Biol.*, 1998, **17**, 185–192.

52. M. Ketteler, C. Vermeer, C. Wanner, R. Westenfeld, W. Jahnen-Dechent and J. Floege, *Blood Purif.*, 2002, **20**, 473–476.
53. S. Gargiulo, P. Gamba, G. Poli and G. Leonarduzzi, *Curr. Pharm. Des.*, 2014, **20**, 2993–3018.
54. Y. Wang, C. Ding, A. E. Wluka, S. Davis, P. R. Ebeling, G. Jones and F. M. Cicuttini, *Rheumatology*, 2006, **45**, 79–84.
55. A. Bonnefoy and C. Legrand, *Thromb. Res.*, 2000, **98**, 323–332.
56. H. Fukuda, S. Mochizuki, H. Abe, H. J. Okano, C. Hara-Miyauchi, H. Okano, N. Yamaguchi, M. Nakayama, J. D'Armiento and Y. Okada, *Br. J. Cancer*, 2011, **105**, 1615–1624.
57. Y. Hamano, M. Zeisberg, H. Sugimoto, J. C. Lively, Y. Maeshima, C. Yang, R. O. Hynes, Z. Werb, A. Sudhakar and R. Kalluri, *Cancer Cell*, 2003, **3**, 589–601.
58. A. Heinz, M. C. Jung, L. Duca, W. Sippl, S. Taddese, C. Ihling, A. Rusciani, G. Jahreis, A. S. Weiss, R. H. Neubert and C. E. Schmelzer, *FEBS J.*, 2010, **277**, 1939–1956.
59. G. E. Davis, K. J. Bayless, M. J. Davis and G. A. Meininger, *Am. J. Pathol.*, 2000, **156**, 1489–1498.
60. R. Kalluri, L. G. Cantley, D. Kerjaschki and E. G. Neilson, *J. Biol. Chem.*, 2000, **275**, 20027–20032.
61. F. X. Maquart, A. Simeon, S. Pasco and J. C. Monboisse, *J. Soc. Biol.*, 1999, **193**, 423–428.
62. C. S. Swindle, K. T. Tran, T. D. Johnson, P. Banerjee, A. M. Mayes, L. Griffith and A. Wells, *J. Cell Biol.*, 2001, **154**, 459–468.
63. S. L. Schor and A. M. Schor, *Breast Cancer Res.*, 2001, **3**, 373–379.
64. S. Ricard-Blum and R. Salza, *Exp. Dermatol.*, 2014, **23**, 457–463.
65. D. Rossi, A. Pedrali, R. Gaggeri, A. Marra, L. Pignataro, E. Laurini, V. Dal Col, M. Fermeglia, S. Pricl, D. Schepmann, B. Wunsch, M. Peviani, D. Curti and S. Collina, *ChemMedChem*, 2013, **8**, 1514–1527.
66. W. Li, F. Gueyffier, G. Z. Liu, Y. Q. Zhang and L. S. Liu, *Biomed. Environ. Sci.*, 2001, **14**, 341–349.
67. S. L. Collins, K. E. Black, Y. Chan-Li, Y. H. Ahn, P. A. Cole, J. D. Powell and M. R. Horton, *Am. J. Respir. Cell Mol. Biol.*, 2011, **45**, 675–683.
68. K. Ghazi, U. Deng-Pichon, J. M. Warnet and P. Rat, *PLoS One*, 2012, **7**, e48351.
69. S. Ricard-Blum and L. Ballut, *Front. Biosci.*, 2011, **16**, 674–697.
70. F. Antonicelli, G. Bellon, L. Debelle and W. Hornebeck, *Curr. Top. Dev. Biol.*, 2007, **79**, 99–155.
71. K. Sugawara, D. Tsuruta, M. Ishii, J. C. Jones and H. Kobayashi, *Exp. Dermatol.*, 2008, **17**, 473–480.

72. C. J. Clark and E. H. Sage, *J. Cell. Biochem.*, 2008, **104**, 721–732.
73. Y. Maeshima, M. Manfredi, C. Reimer, K. A. Holthaus, H. Hopfer, B. R. Chandamuri, S. Kharbanda and R. Kalluri, *J. Biol. Chem.*, 2001, **276**, 15240–15248.
74. M. Mongiat, S. M. Sweeney, J. D. San Antonio, J. Fu and R. V. Iozzo, *J. Biol. Chem.*, 2003, **278**, 4238–4249.
75. I. Senyurek, G. Klein, H. Kalbacher, M. Deeg and B. Schittek, *Peptides*, 2010, **31**, 1468–1472.
76. M. A. Rupnick, D. Panigrahy, C. Y. Zhang, S. M. Dallabrida, B. B. Lowell, R. Langer and M. J. Folkman, *Proc. Natl. Acad. Sci. U. S. A.*, 2002, **99**, 10730–10735.
77. S. Blaise, B. Romier, C. Kawecki, M. Ghirardi, F. Rabenoelina, S. Baud, L. Duca, P. Maurice, A. Heinz, C. E. Schmelzer, M. Tarpin, L. Martiny, C. Garbar, M. Dauchez, L. Debelle and V. Durlach, *Diabetes*, 2013, **62**, 3807–3816.
78. J. Park and P. E. Scherer, *Expert Rev. Anticancer Ther.*, 2013, **13**, 111–113.
79. J. Park and P. E. Scherer, *Oncotarget*, 2012, **3**, 1487–1488.
80. Y. Yamaguchi, T. Takihara, R. A. Chambers, K. L. Veraldi, A. T. Larregina and C. A. Feghali-Bostwick, *Sci. Transl. Med.*, 2012, **4**, 136ra171.
81. Y. Zhu, A. Oganesian, D. R. Keene and L. J. Sandell, *J. Cell Biol.*, 1999, **144**, 1069–1080.
82. J. Larrain, D. Bachiller, B. Lu, E. Agius, S. Piccolo and E. M. de Robertis, *Development*, 2000, **127**, 821–830.
83. K. Yu, S. Srinivasan, O. Shimmi, B. Biehs, K. E. Rashka, D. Kimelman, M. B. O'Connor and E. Bier, *Development*, 2000, **127**, 2143–2154.
84. N. Fukui, A. McAlinden, Y. Zhu, E. Crouch, T. J. Broekelmann, R. P. Mecham and L. J. Sandell, *J. Biol. Chem.*, 2002, **277**, 2193–2201.
85. O. Zamparo and W. D. Comper, *Arch. Biochem. Biophys.*, 1989, **274**, 259–269.
86. E. Ruoslahti, *Ann. Rev. Cell Biol.*, 1988, **4**, 229–255.
87. K. Sakaguchi, M. Yanagishita, Y. Takeuchi and G. D. Aurbach, *J. Biol. Chem.*, 1991, **266**, 7270–7278.
88. T. Maciag, *Prog. Hemostasis Thromb.*, 1984, **7**, 167–182.

CHAPTER 2

Matrix Biology: Gradients and Patterns within the Extracellular Matrix

MIRIAM DOMOWICZ*[a], MAURICIO CORTES[b], AND NANCY B. SCHWARTZ[c]

[a]Departments of Pediatrics, The University of Chicago, Chicago, IL 60637, USA; [b]Departments of Pathology, Beth Israel Deaconess Medical Center, Harvard Medical School, Boston, MA 02115, USA; [c]Departments of Pediatrics and Biochemistry & Molecular Biology, Committee on Developmental Biology, The University of Chicago, Chicago, IL 60637, USA
*E-mail: mdxx@uchicago.edu

2.1 INTRODUCTION

Secreted signaling molecules produced by restricted groups of cells are known to act not only at short range on nearby cells, but in many cases, signaling pathways are activated in cells located long distances from the signaling source, eliciting differential responses by the target cells. If the secreted signaling molecules form quantitative gradients that induce distinct cellular responses in a concentration-dependent manner, such signaling molecules are referred as morphogens. The differential activation of target genes in different cells as a function of their distance from the morphogen-source cells is what underlies the formation of differentiation patterns during embryonic development,

Mimicking the Extracellular Matrix: The Intersection of Matrix Biology and Biomaterials
Edited by Gregory A. Hudalla and William L. Murphy
© The Royal Society of Chemistry 2020
Published by the Royal Society of Chemistry, www.rsc.org

which is where most of the mechanisms of morphogen movement have been studied. However, morphogens also remain active in differentiated (mature) tissues, to support their survival, maintain their stem cell niches, or to stimulate tissue regeneration and cell replacement.

The concept of morphogens was originally defined for plant biology[1,2] but became more generalized when used theoretically to explain morphogenesis during limb development.[3] Only in recent decades has experimental evidence of morphogen existence begun to accumulate by direct observation of protein gradients and demonstration of concentration-dependent cellular responses to such gradients.[4] On the basis of this evidence, many morphogen polypeptide families have emerged, including such major types as hedgehogs (HH), wingless (Wnts), epidermal growth factors (EGF), fibroblast growth factors (FGF) and transforming growth factors (TGF) as well as some metabolites such as *trans*-retinoic acid (*t*-RA). Many of these morphogens have distinct mechanisms of action (short-range or long-range) influenced by their stability and specific interactions with specific components of the extracellular matrix (ECM). Due to their varying unique properties it has been difficult to form a unified model of how morphogen gradients are formed. Nonetheless, several hypotheses for mechanisms of gradient formation have been put forward for some morphogens that may explain their long-range mechanisms of action.[2,5–11]

Signaling range is defined as the domain over which morphogens exert their effects, and it can be controlled by concentration of the signal at the source, and/or its activity, mobility, and stability.[10] The simplest mechanism whereby a gradient may be generated is passive diffusion, *i.e.*, a signal that is produced by a group of localized cells diffuses freely through the ECM, but experimental data from several biological systems indicate that many other factors may contribute to the final response. Although cellular responses are controlled by the local concentrations of the secreted signaling molecules and/or changes in receptor composition of the target cells, other factors such as morphogen recycling, degradation, post-translational modification, interaction with the ECM and receptor availability are integral in establishing the gradient shape, range and robustness.[2,5–11] Considering all these factors, the complexity of the action of morphogens as inducers of cell division, differentiation, migration, adhesion, apoptosis, and cell and tissue shape makes the study of morphogen gradients even more difficult to pursue. Regardless, it remains one of the most fascinating topics in biology, since morphogen gradients are essential for the proper patterning, diversity in multicellular organism development, and function.

This chapter will describe three models for the study of morphogen biology during vertebrate development, discuss different hypotheses regarding the movement of morphogens, present experimental evidence supporting the role of the ECM in morphogen transport, and finally, highlight the role of morphogen–ECM interactions in determining stem cell identity, niche and fate, and the design of artificial matrices for use in stem-cell-based regenerative medicine.

2.2   MODELS FOR STUDYING MORPHOGEN GRADIENTS IN DEVELOPMENT

Many examples of positional information initiated by a restrictively produced secreted molecule have been described during development, but only a few have been studied in detail in vertebrate systems. While most of the advances in understanding the formation of morphogen gradients derives from studies in *Drosophila*,[6] the level of complexity introduced by the ECM in vertebrate systems makes them more difficult to understand. This chapter concentrates on three major models (Figure 2.1): the specification of neural cell types in the vertebrate neural tube, the specification of elements in the developing limb, and the developing long-bone growth plate.

2.2.1   Cell Specification in the Neural Tube

Specification of neural cell types in the mouse spinal cord is initiated at embryonic day 10.5 (E10.5) with the production of sonic hedgehog (SHH) by the notochord and later by the neural tube floor plate (Figure 2.1A). The gradient of SHH along the dorsoventral axis of the neural tube specifies different neuronal cell fates by activating the SHH signaling pathway that controls the activity of the Gli family of transcription factors in the proliferating neural precursors. Concomitantly, cells on the dorsal surface of the neural tube start expressing members of the bone morphogenetic protein (BMP) and wingless-type MMTV integration site (Wnt) family of signaling molecules.[12,13] It is thought that a combination of these three signaling pathways gives rise to the spatial subdivision of the neural progenitor domain territory along the dorsoventral axis, with HH signaling being responsible for the patterning of the dorsal cell types and BMPs and Wnts for that of the ventral cell types.[13] The patterning of the precursor populations expands initially by proliferation under the control of FGF8 derived from the surrounding mesoderm until the progenitors exit the cell cycle and begin the processes of differentiation and migration.[12,13] For

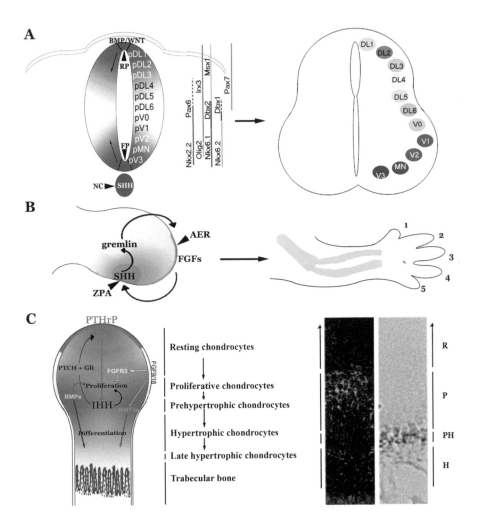

Figure 2.1 Morphogen gradient distribution and cell patterning in three developmental models. (A) Morphogens regulating neural tube development.
Gradients of SHH (blue) secreted from notochord (NC) and floor plate
(FP) and Wnt/BMP (red) secreted from roof plate (RP) induce concentration-dependent differentiation of precursor cells (pDL1-pV3) along
the dorsoventral axis (left panel) driving different progenitor cell fates
by combinatorial expression of transcription factors (color bars). After
exiting the mitotic cell cycle, cells distribute laterally generating the
distinct neuronal subtype in the dorsal spinal cord (DL1–DL6), ventral
spinal interneurons (V0–V3) and motor neurons (MN) (right panel).
(B) Morphogens regulating limb bud development. Gradients of SHH
(blue) are initiated by cells in the zone of polarizing activity (ZPA) while
FGFs (green) are expressed by cells in the apical ectodermal ridge
(AER). SHH initially activates the expression of the BMP antagonist
gremlin (yellow), which in turn maintains FGF expression in the AER by
inhibiting the action of BMPs (not shown). FGF expression maintains

generation of the ventral neural cell types, the appearance of several transcription factors (*i.e.*, Olig, Pax, Dbx, Nkx) (Figure 2.1A) controlling cell fate is dependent on increasing concentrations and duration of the SHH signal.[14–16] Furthermore, using *ex vivo* systems, exposure of neural tissue explants to greater concentrations of SHH, or longer durations of exposure, switches the identities of the cells to those of more ventral cell types.[14–16] These results suggest a mechanism for gradient sensing in which a temporal adaptation to SHH transforms the extracellular concentration of morphogen into a time-limited period of signal transduction, so that the duration of the signal is proportional to the ligand concentration.[15] Visualization of the SHH protein with antibodies or with genetically modified fully processed green fluorescent versions of the biologically active morphogen indicate that a gradient of extracellularly localized SHH is established *in vivo* that correlates with the expected biological response of the neural precursors.[17–19] Later on in development, gradients of HH, Wnt and BMP molecules participate in the axon guidance process for many of the newly generated neurons[20] together with classical guidance molecules such as netrins and slits.

2.2.2   Gradients Along the Developing Limb

The second system, the specification of elements in the developing limb (Figure 2.1B), is perhaps one of the most frequently utilized developmental models for studying patterning. This system has allowed extensive surgical manipulations that initially defined organizing centers in the limb bud and gave rise to the concept of morphogens, even before we knew their chemical identities. The limb bud is specified on the lateral flank of the embryo as mesenchymal cells from the lateral plate and somites proliferate, creating a bulge

---

proliferation by promoting SHH activity (right panel). The limb pattern (left panel) is achieved by a combination of these gradients as well as expression of TGF-β and Hox genes. (C) Morphogens regulating growth plate development. PTHrP (red) and IHH (blue) gradients coordinate chondrocyte proliferation and differentiation through a negative-feedback loop while perichondrial FGF9/18 (green) suppress chondrocyte proliferation and maturation through FGFR3 expressed by proliferative and pre-hypertrophic chondrocytes (left panel). Fluorescent immunohistochemistry (green) of a longitudinal section of postnatal day-6 mouse growth plate using an N-terminal IHH antibody, showing the extracellular distribution of IHH throughout the proliferative and hypertrophic zones; mRNA *in situ* hybridization with a IHH probe (blue) indicates that the gene is expressed only by pre-hypertrophic chondrocytes (right panel).

under the ectodermal cells. This initially undifferentiated group of mesenchymal cells is patterned to form a full limb by the action of two major, highly interactive signaling sources: the apical ectodermal ridge (AER) and the zone of polarizing activity (ZPA).[21] The AER at the distal margin of the limb bud maintains the mesenchymal cells underneath the ridge in a proliferative state, mainly by the action of FGF family members, which results in linear outgrowth of the limb, and maintains the expression of SHH in the ZPA. The ZPA is responsible for patterning the anterior–posterior (thumb–pinky) axis of the limb bud, and it is also essential for maintaining growth factor production in the AER. The BMP antagonist gremlin is expressed in a domain anterior to the ZPA and is thought to act as a signaling intermediate between SHH and FGF. BMP activity is required to induce gremlin expression,[22] which in turn maintains FGF4 expression in the AER by inhibiting the action of BMPs. FGF expression maintains proliferation by promoting SHH activity, but as the limb grows, the gremlin domain is distanced from the ZPA because its descendent cells are unable to activate gremlin transcription. Finally, cells competent to express gremlin are no longer within range of the SHH signal, leading to loss of gremlin production, and in turn loss of FGF4 and SHH synthesis, marking the termination of growth and ensuring proper limb size (Figure 2.1B). In this system, it is easy to understand how specification of the different digits is achieved based on the duration of expression and levels of SHH exposure of the limb primordium. Cells that form digit one are the only ones that have never expressed SHH or received a SHH signal, while cells that form digit five have expressed SHH continually, and the digits in between have either only received the signal, or expressed and received the signal for different periods of time[23] (Figure 2.1B). Point mutations or duplications in the limb specific regulatory enhancer region of the SHH promoter in humans lead to deregulated expression of SHH and severe forms of congenital hand deformities such as syndactyly or polydactyly.[24–26]

The limb model was also used to define extracellular signaling gradients; Sepharose beads soaked in FGF, RA or SHH proteins differentially manipulated the patterning of the limb according to where the beads were placed,[22,27,28] thus verifying that diffusion of a morphogen molecule alone was sufficient to specify location and number of skeletal elements in the limb. This system has formed the basis for designing mathematical and computational models that describe the proximo-distal generation of temporally controlled number of limb skeletal elements,[21] as well as to understand reaction–diffusion systems (Turing patterns).[29]

2.2.3   Morphogen Gradients in the Growth Plate

The developing growth plate (Figure 2.1C) is a superb model system because of the influence exerted by morphogens on the proliferation and differentiation of a single cell type, the chondrocyte. In addition, the composition of the ECM in the growth plate is well established, thereby allowing better understanding of the influence of the ECM on the gradient formation process. After condensation of the mesenchymal cells in the limb bud, cells differentiate to the chondrocyte fate and undergo rapid proliferation to elongate the skeletal elements. Cells in the center of each element exit the mitotic cell cycle and undergo differentiation to a hypertrophic phenotype. Just before differentiation is evident, cells in the central part of the element express another member of the HH family, Indian HH (IHH), which together with the parathyroid hormone related-protein (PTHrP) expressed in the distal chondrocytes of the element, are believed to initiate the formation of the growth plate[30,31] (Figure 2.1C). As cells expressing IHH progress to hypertrophy they down-regulate IHH and up-regulate characteristic matrix molecules, *e.g.*, Collagen type X. In the middle of the element, the terminal hypertrophic cells eventually die by apoptosis, and the area undergoes vascular invasion, which brings along perivascular osteoprogenitors and recruits osteoclasts, and these together with perichondrial cells start to produce bone matrix to form the bone collar and trabecular bone.[31] This process establishes two growth plates in each element, from which further growth depends on the level of morphogens expressed by the different cell types. Besides IHHs and PTHrP, other important regulators of this process are FGFs acting through FGFR3, BMP6 expressed in a pattern similar to that of IHH by the prehypertrophic chondrocytes, and Wnt5A and B expressed during the transition from the proliferative to hypertrophic stage.[31] A complex interplay of these morphogens ensures proper growth of the different skeletal elements.[26] IHH signaling in this system illustrates one of the classical cell differentiation paradigms in development: cell proliferation is partially regulated by IHH signaling, and at the highest morphogen concentration chondrocytes exit the cell cycle and start expressing IHH (prehypertrophic chondrocytes), until they continue their differentiation toward the hypertrophic phenotype by down-regulating IHH.[32] The number of cells and length of time that they expressed IHH establishes the size of the growth plate and ultimately the rate of growth.

Many other examples of morphogenic gradients have been described in vertebrate systems, such as the importance of FGF, VEGF and SHH signaling in branching formation in the lung,[33] establishment of left–right asymmetry in embryos by lefty and nodal,[34,35] and others. These systems have elucidated some characteristics of morphogen movement and gradient formation, but there are still many unanswered questions. The major challenges of using the 3-D systems of natural gradient formation have been the variety of morphogens exerting their influence on a single cell, limitations in our ability to visualize the morphogen gradients, and limitations in our ability to visualize the cellular responses triggered by the morphogens.

2.3   MOVEMENT OF SIGNALING MOLECULES

Movement of morphogens has been demonstrated experimentally in very few cases because they usually act concomitantly with more than one other morphogen, and the response by the target cells often depends on time of exposure to the morphogen, changes in behavior (receptor composition and levels, type of matrix, *etc.*) by the target cell, and whether the source of morphogen is stationary or dynamic. Thus, it is not surprising that experimental evidence supports multiple mechanisms by which morphogens move and even that the same morphogen may use more than one mode to reach target cells. In order to simplify the possibilities, we will first discuss different theoretical mechanisms of morphogen movement and then analyze the experimental observations that support them. Figure 2.2 depicts four possible models for morphogen travel.

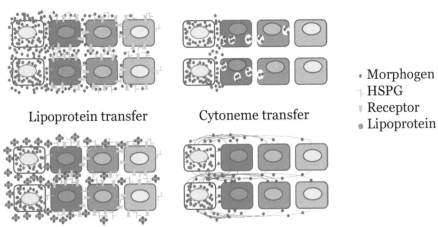

Figure 2.2   Models for morphogen transport by facilitated diffusion, transcytosis, lipoprotein transfer and cytoneme transfer. Source cells are depicted in green and target cells are depicted in blue.

Extracellular movement: Morphogen secreted by a source cell might travel through the ECM by passive diffusion, however, the shape of the gradient will not only be a result of the stiffness of the matrix but will also be influenced by components of the matrix binding and immobilizing the morphogen. Furthermore, as the morphogen interacts with the surrounding cells and ECM it could be targeted for lysosomal degradation or destroyed by extracellular proteases, leading to what is known as restricted diffusion.

Planar transcytosis: Morphogens secreted by the source may not passively travel far through the ECM but could be transported from cell to cell by an intracellular vesicle-mediated mechanism and then released to the target cells. In this model the extracellular diffusion distance is short, and the gradient is established because the population of molecules traveling intracellularly may be targeted either for degradation or secretion by each cell transited. So far, experimental support for this mechanism has only been described in *Drosophila* wing disc.[36,37]

Lipoprotein transfer: Lipid-modified morphogens could be secreted by the cell in a lipoprotein particle[38] that, through interactions with cell-surface glycerophosphatidylinositide (GPI)–anchored-heparan sulfate proteoglycans (PGs), could be distributed across several cell lengths. Lipoproteins have been shown to modulate SHH and Wnt gradients.[38,39]

Cytoneme-mediated transfer: The dispersion of morphogens in this model occurs by direct contact *via* thin cellular filopodia called cytonemes, which are implicated not only in sending but also in receiving signals.[40,41] The gradient is established either because contact with a nearby cell is more frequent than with a distant cell or because the cargo packages have to travel longer the farther away the target cell is. Even though this model represents a novel concept, supporting evidence for it in the mobility of many morphogen families (HH, Wnt and BMP) is starting to accumulate.[40,42,43]

As many of these models are mutually exclusive, it could be expected that several of these systems are in place for a particular morphogen but that one system is preferred to the others according to the cell and ECM type involved and the characteristics of the morphogen source and target cells.

Experimental evidence is available supporting all these models in different organisms and tissues,[9,44–46] but the major limitation in assessing their prevalence has been the inability to directly visualize the morphogen gradient. Antibody immunohistochemistry of many growth factors has been unreliable in sensing the extent of their gradients due to detection limitations, thus other methods of detection are being used: *e.g.*, following the activation of direct downstream targets, which are usually transcription factors, or measuring upregulation of receptors in the target-cell population. Also, expression of chimeric versions of the morphogens linked to fluorescent proteins under the control of their endogenous regulatory mechanisms has been shown to be extremely useful in assessing the extent of the gradients.[17,47] Recently, genetically encoded fluorescent detection beacons for small morphogens such as RA have been used to analyze gradients *in vivo* during embryonic development.[48] As well, kinetic interactions of ECM molecules and growth factors labeled with gold nanoparticles[49] have recently been described.

A number of theoretical/mathematical approaches have been generated to model biological morphogen gradients, which in turn may aid in refining experimental approaches addressing many of the still unanswered questions in this field.[9,50-55] These models have received a great deal of attention recently; however, they are beyond the scope of this chapter.

2.4 MORPHOGEN–ECM INTERACTIONS

Considering the three examples of morphogen-dependent pattern formation during development presented and the four postulated mechanisms of movement of morphogens, it is clear that the composition of the ECM and the interaction of ECM molecules with these secreted ligands might significantly influence gradient shape and range. Furthermore, the mechanical characteristics of the matrix, such as deformability and stiffness, could also modify the characteristics of these gradients and thus the behavior of the target cells.

For example, the skeletal dysplasias in humans, comprising a large number of disorders associated with classical structural proteins of the ECM, are often accompanied by developmental defects in pattern formation of skeletal elements or limb growth,[26] highlighting the importance of ECM molecules in modulating morphogen gradient formation.

An interesting aspect of the ECM is its diverse composition, depending on the tissue type, and its dynamic nature during development. Added to this complexity is the fact that secreted structural components of the ECM can diffuse from the site of synthesis and assemble into matrices at a distance from their biosynthetic sites,[56,57] as well as that migrating cells can modify the composition of the matrix as they migrate.[58–60] Thus, dynamic changes in matrix composition and assembly could in turn modulate the pattern of morphogen gradients.

The next section will cover the interactions of morphogens with not only their cell-surface receptors but also with molecules in the extracellular milieu, focusing on some examples of how these interactions affect the proper establishment of morphogen gradients.

2.4.1   Interaction of Morphogens with ECM Molecules

ECM molecules are characterized by multi-domain protein structures that form complex scaffolds in the extracellular space and are able to associate with plasma membrane proteins integrating transduction of extracellular signals to the cell, thus influencing many cellular processes, including differentiation, proliferation and migration. The diversity of extracellular matrix components and their function is a tightly regulated process controlled by many cellular actions, including post-translational modification and alternative splicing. Morphogens have been described to interact with both protein domains and post-translational modifications of ECM molecules to elicit distinct cellular responses.

2.4.1.1   Proteoglycans as Modulators of Morphogen Signaling.   Evidence that PGs are important in regulating the formation of gradients emerged from the study of cell-culture[61,62] and animal models.[63–66] PGs are characterized by a core protein with attached high-molecular-weight glycosaminoglycan chains (GAGs) consisting of repeating sulfated disaccharides whose chemical composition distinguishes different populations of PGs.[67] GAGs are classified as chondroitin sulfate, dermatan sulfate, heparan sulfate or keratan sulfate (CS, DS, HS or KS, respectively) based on the repeated disaccharides that form them (Table 2.1). A large variety of PG core proteins can carry varying numbers of chains from a single kind of GAG or in combination.[67] Furthermore, during gene transcript processing some core proteins have the region of GAG binding spliced out, generating what are known as part-time PGs.[67,68] From their extracellular and cell-surface localizations, PGs can modulate cell proliferation, differentiation and migration, possibly due to their ability to influence protein gradient formation and signal transduction.[69] Further diversity among the different GAG chains is accomplished by *N*- and *O*-sulfation on the disaccharide backbone, which confers unique sequences of sulfated sugars along the chain that can interact with distinct affinities with their ligands. To add to the complexity of the PGs during their

Table 2.1 Glycosaminoglycans reported to bind morphogens.

GAG	Symbol	Disaccharide	Structure	Morphogen binding
Heparan sulfate	HS	GlcA-GlcNAc IdoA-GlcNAc		FGF2;[205] FGF1; PDGF;[206] HGF;[207] VEGF;[208,209] GDNF;[210,211] SHH;[121,129] IHH;[118] Wnt;[212] BMP;[213] endostatin,[214] pleiotrophin,[215] midkine[216]
Chondroitin/dermatan sulfate	CS/DS	GlcA/IdoA-GalNAc		FGF2;[87] SHH;[121] IHH;[118] pleiotrophin,[215,217] midkine;[216] Wnt-3a;[97] HGF[218]
Keratan sulfate	KS	Gal-GlcNac		FGF2;[219] SHH[219]

biosynthesis, GAG sulfation can also be modified by extracellular sulfatases (Sulfs) in the ECM, also affecting their interactions with their ligands.[70]

FGF signaling: FGF was among the first morphogens shown to bind HS and heparin (a highly sulfated GAG closely related to HS) with strong affinity.[71] Subsequently, the list of growth factors interacting with HS in different biological models has been extended to include TGF-β, PDGF, VFGF, Wnt and HH family members, and others (Table 2.1).

FGF family members are classified as paracrine, endocrine and intracrine FGFs.[72,73] Paracrine FGFs bind HS with high affinity and thus are key in establishing their action gradients, while endocrine FGFs, like FGF19, bind HS with low affinity and thus can diffuse from their tissue of origin to the bloodstream. Paracrine FGFs act by dimerization of their cell-surface receptor, FGFR, of which there are four gene family members (FGFR1–4). Cell-surface heparan sulfate proteoglycans (HSPGs) act as co-receptors during the dimerization of FGFR by binding FGF and its receptor, thereby stabilizing the ternary complex, which then triggers tyrosine kinase activation of the receptor. The crystal structure of this complex indicates that the interaction requires HS octasaccharides[74–76] and that specific affinities vary with respect to number and position of individual sulfate groups.[77] HSPGs with transmembrane core protein domains (*e.g.*, syndecans) and the GPI-anchored types (*e.g.*, glypicans) are more likely to act as FGFs co-receptors as opposed to the ECM-secreted types (*e.g.*, perlecan). Membrane-associated HSPGs can be cleaved from the surface by specific proteases.[78] As an example, notum[79–81] can specifically cleave GPI-associated PGs. Similarly, HS-bound FGF can also be shed by the action of glycosidases such as heparanase.[82] Studies at single-cell resolution have demonstrated that cell-surface-tethered HS chains are crucial for local retention of FGF and that shedding of HS chains can spread FGF signaling to adjacent cells within a short distance during mammalian embryogenesis.[83] On the other hand, membrane-associated HSPG core proteins, such as syndecans,[84] can also modulate FGF signaling by enhancing endocytosis of the FGF–FGFR complex. Since ECM-associated HSPGs can up- or down-modulate FGF signaling depending on the cellular context, they have been proposed to either restrict ligand diffusion or protect the ligand from extracellular proteases.[85] As examples, perlecan is required for normal chondrocyte development and can form complexes with FGFR3 and FGF18[84], and the HSPG agrin potentiates the ability of FGF-2 to stimulate neurite outgrowth.[86]

Even though most studies have demonstrated interactions of FGF with HS, binding studies also have shown that FGF-2, -10, -16 and -18 can bind CS chains with lower affinity than HS does, while over-sulfated CS (CS-E) can bind with affinity comparable to that of HS.[87,88] Interestingly, studies with perlecan, a PG that contains both CS and HS chains, indicate that a perlecan variant devoid of CS can deliver bound FGF-2 to its receptor, while the perlecan form with both GAG chains acts preferentially as a sink for FGF-2 in the ECM.[89]

Elucidating the complexity of the factors involved in FGF gradient formation and the involvement of HS has been hindered by limitations in directly observing these interactions. Support for the secretion-diffusion-clearance process of FGF8 in a living embryo has been obtained by scanning fluorescence correlation spectroscopy,[90] which demonstrated an HS-bound extracellular pool of FGF8 that decreased with degradation of HS by heparatinase treatment, and increased following addition of exogenous HS.[90] Novel methodologies are being developed using gold nanoparticles to label FGF2 in order to study the dynamics of its interaction with HS by photothermal heterodyne microscopy. Using this methodology in fibroblast cell cultures, it has been possible to visualize FGF2 bound to HS on the cell surface and its diffusion from one chain to another by translocation, generating directed motion.[49] Future studies should analyze the dynamics of this phenomenon in 3-D matrices and biological tissues.

While the interactions of FGF with its receptors and HS are the most extensively studied, there is growing evidence from many biological systems indicating that HS also binds and modulates the signaling of other morphogens by similar mechanisms[91,92] (*i.e.*, PDGF/PDGFR, VEGFs/VFGFRs).

Wnt signaling: Initial studies indicate that inhibition of GAG sulfation by chlorate and addition of heparin strongly affect responses to Wnt signaling in multiple systems[93,94] and that in quail embryos and mammalian cell cultures, activation of extracellular sulfatases plays a positive role in Wnt signaling. *In vivo* data demonstrating the importance of HSPGs in Wnt signaling have been provided from *Drosophila* genetics, where removal of HSPG biosynthetic enzymes and core proteins have been shown to have dramatic effects on Wnt distribution in the ECM.[69,95,96] Specific HSPGs can bind Wnts on cell surfaces, and CS and HS can bind Wnts in the ECM.[80,94,95,97] Although the mechanism of action is not clear, it has been suggested that HSPGs help tether Wnts at the cell surface and contribute to their stability during transport in the extracellular space, allowing them more time to diffuse away from the source cells. Furthermore, desulfonation could be required for efficient signaling at the level of the target cells.[5,6] Similarly to HH family members, it has been proposed that Wnt proteins, which are lipid modified, could be secreted bound to lipoprotein particles.[98] More recent contributions using genetically engineered tethered Wnt molecules support the possibility that Wnts could control pattern and growth without forming extracellular gradients; thus future work might prompt revision of the requirement for long-range spreading of Wnt signaling.[99]

BMP signaling: Disruption of bone morphogenic protein (BMP) gradients in HSPG mutants have been also reported in *Drosophila* and vertebrate model systems,[100–103] and direct interactions with HS as well as interactions that are partly dependent on binding to a specific PG core protein (*e.g.*, betaglycan, decorin, biglycan) have also been reported.[104,105] BMP interaction with HSPG enhances internalization and degradation of BMPs in vertebrates,[102] and deletion of the HSPG-binding motif in BMP4 increases the range of action of BMP4 gradients.[106,107] On the other hand, the *Drosophila* glypican ortholog, Dally, can interact with the BMP4 ortholog, Dpp, as well as with FGFs; however, while preincubation with HS totally abolished FGF binding, BMP4 association is partially HS-resistant, suggesting the Dally protein core contributes to binding.[108] Furthermore, phenotypes generated by overexpression of Dally are partially rescued by overexpression of an HS-deficient form of Dally.[108] In vertebrates, other PG core proteins directly interact with morphogens independently of their GAG chains. For example, the glypican-3 core protein interacts with HH[109] but requires HS for optimal signaling, and the overgrowth phenotype in the glypican-3 null mouse[110,111] and the human Simpson–Golabi–Behmel syndrome[112] are in part due to over-activation of the HH signaling pathways.[109] Also, perlecan core protein binds FGF-18, and this binding can alter its mitogenic effects on chondrocytes.[113]

HH signaling: HH gradients are also modulated by CSPGs and HSPGs. The HH proteins, Sonic, Indian and Desert, activate a canonical signaling pathway that plays an important role in the regulation of cell proliferation and differentiation in embryos.[51,114,115] Secreted HH binds and antagonizes its transmembrane canonical receptor, PTCH, which in turn blocks the activity of a signaling effector called smoothened, ultimately leading to the activation of the Gli family of transcription factors and the expression of HH target genes.[114] Mutations in genes encoding HSPG biosynthetic enzymes produce HH-mutant-like phenotypes.[101,116] Reduced HS synthesis in mice carrying a hypomorphic mutation in Ext1 results in expanded IHH signaling in the growth plate.[117] Mutations in enzymes that alter levels of PG sulfation affect growth plate development by altering the distribution of IHH in the ECM;[118] as well, lack of the major CSPG, aggrecan, affects the establishment of an appropriate IHH gradient early in the formation of the growth plate.[119] The palmitoyl- and cholesterol-modified amino terminal signaling domain of HHs binds different types of GAGs[118,120,121] with stronger affinity to HS than to CS.[118] A positively charged region of SHH (aa33-38 in mouse), the Cardin–Weintraub sequence, was originally identified as a GAG binding site,[122,123] and recently an additional conserved GAG-binding site was identified by X-ray crystallography.[121] The active forms of HHs require lipid modifications and formation of multimeric lipoprotein particles,[38,124,125] and the interaction with GAGs is controlled differentially for monomeric and multimeric forms of HHs;[126] multimeric forms of HHs establish structurally different complexes with heparin or chondroitin-4-sulfate.[121] It has been proposed that, in the growth plate, proper movement of IHH secreted by source cells (pre-hypertrophic chondrocytes) to target cells (proliferative chondrocytes) requires matrix-associated CSPG (aggrecan) which either participates in the diffusion of HH or protects it from degradation, and that HSPG on the surface of the target cells, by binding with higher affinity to HH, interferes with further spreading of the gradient and presenting the morphogen to its receptor, Ptch, on the plasma membrane.[118] Recent findings[121] further support this possibility by describing low- and high-affinity binding sites for multiple HH forms. It has also been established that metalloproteases of the ADAM family mediate ectodomain shedding from the dually lipidated SHH multimers[127,128] and that this processing targets and inactivates the Cardin–Weintraub motif, influencing the SHH gradient[129] as well.

In a different system, the cerebellum, it has been proposed that a specific PG, glypican-5, acts as a co-receptor for SHH, is located adjacent to the primary cilia that act as SHH signaling organelles, and modulates cerebellar granule cell proliferation through 2-*O*-sulfo-iduronic acid residues at the nonreducing ends of their glycans.[130] Also, sulfatase 1 acts as a positive mediator of HH signaling in the ventral portion of the neural tube by modifying the sulfation state of HSPG and concentrating the amount of multimeric HH delivered to specific neural progenitor cells.[131] Recent reports suggest that contact-mediated release propagated by specialized filopodia, cytonemes, contribute to the delivery of SHH at a distance in avian limb bud[43] and *Drosophila* wing.[42,132] Although this is an intriguing novel model, it is still unclear how PGs are acting in these contexts to modulate gradient formation.

Other signaling pathways: Interactions with PGs have also been demonstrated for other important morphogens, but the actual mechanisms are less well understood. For example, left–right asymmetry in *Xenopus* embryos is established by two morphogens, nodal and lefty (members of the TGF-β family), whose gradients are dependent on a PG-rich ECM.[34,133] Also, HS and CS interact in a sulfation-dependent manner with various axon topographic guidance proteins, including slit2, netrin1, ephrinA1, ephrinA5, and semaphorin5B,[88] gradients, which are essential in the establishment of the retinotopic maps of the visual system.[134]

VEGF family members and their receptors control the expansion and remodeling of the vascular system, and neuropilins and HSPGs act as co-receptors modulating their action.[135] The heparin-binding domain of the VEGF-A isoform is thought to regulate VEGF concentration gradients; it has been demonstrated in mouse models that the heparin-binding VEGF-A$_{165}$ isoform is absolutely required for the proper development of a functional vascular network.[135–137]

2.4.1.2 Other ECM Morphogen-Interacting Proteins. The increasing number of examples of morphogens binding ECM core proteins directly, without the involvement of GAGs, implies key functions of the ECM in the presentation of morphogens. It is also important to appreciate that all morphogen gradients must coordinate their action with standard ECM receptors such as integrins and discoidin domain receptors, which interact specifically with motifs embedded in ECM proteins.[138] Despite reports of synergy or cross-talk between integrin and growth factors, it is still unclear if such observations involve cell-surface-specific interactions or cooperation at the level of the downstream signaling transduction pathways.[139] It is also possible that specific domains of ECM molecules could act on canonical morphogen receptors. For example EGF-like domains from laminins[140] and tenascin[141] can bind and activate EGF receptors, and it has been hypothesized that these domains could be released from the matrix by the action of metalloproteases.[139,140]

A typical example of morphogen regulation by ECM binding is TGF-β. Many members of this morphogen family are cleaved by a furin protease to form the mature TGF-β and a latency-associated peptide (LAP). The two peptides remain associated within the ECM as multiple latent complexes that can bind fibrillins and fibronectins. Alternatively, proTGF-β can bind an extracellular protein, emilin1, that prevents its cleavage by furin.[142] Activation of TGF-β can occur by several mechanisms, including degradation of ECM by metalloproteases, conformational changes exposing TGF-β, dissociation of the complex by thrombospondin, specific integrin activation, or mechanical strain.[139,143-145] ProBMP-7, another member of the TGF-β family, can also bind fibrillins-1 and -2.[146] Mutations in human and mouse models affecting fibrillin-1 and -2 result in syndromes affecting multiple organs, overgrowth, and limb-patterning formation defects which are linked to defects in TGF/BMP signaling.[147-149] Excess degradation of LAP and elevated TGF-β signaling is observed in knockout models of fibrilin1, suggesting that the TGF-β latent complex binding to the matrix is important for controlled release of active TGF-β from the complex.[150]

Several examples illustrate the capacity of conserved elements of ECM proteins to regulate the function of morphogens. During chondrocyte development, TGF-β1 and BMP2 bind the type-A splicing form of Col2 during early mesenchymal condensation, and as cartilage matures, Col2 is preferentially spliced to the type-B form that no longer interacts with this morphogen.[151]

In endothelial cells, hepatocyte growth factor supports cell migration by forming molecular complexes with fibronectin and vitronectin,[152] while interaction of VEGF with the type-III domain of fibronectin acts synergistically to stimulate their proliferation and migration.[153]

Wnts have several extracellular modulators, including the secreted Frizzled-related protein,[154] Cerberus,[155] and the Wnt inhibitory factor 1 (Wif-1) family.[156] In particular, Wif-1 interacts with Wnt family members through its Wif domain, while its EGF domains interact with glypicans.[157]

Several other cell-surface ECM proteins regulate HH in addition to Ptch, *i.e.*, Cdo, Boc and GAS. These bind HH *via* conserved fibronectin repeats and enhance their signaling synergistically with Ptch.[158-161] In the neural tube in particular, Cdo, Boc and Gas have been proposed to play an early role in cell fate specification of multiple neural progenitors, and later in motor neuron progenitor maintenance.[160]

2.4.2 Proteolytic Activation in the Formation of Extracellular Morphogen Gradients

Extracellular and secretory-pathway proteases such as the furin/prohormone convertases, serine proteases and matrix metalloproteases regulate extracellular remodeling and the activity and distribution of several morphogens and growth factors.[162-165] Extracellular proteases are usually involved in limited proteolysis of a few specific cleavage sites within the substrate. Their activity leads to activation or inactivation of the substrate, which involves conformational changes in the protein or separation of previously interacting domains. In some instances a proteolytic cascade may be required to provide a robust gradient providing multiple points at which positive or negative modes of regulation can occur, thereby controlling localization, timing and amplitude of the gradient.

In addition to the described examples for TGF-β (Section 2.4.1.2), other growth factors, including VEGF, HGF and nerve growth factor (NGF), require proteolytic processing for signaling and gradient formation. As was mentioned, differential splicing of VEGFA results in VEGF variants with different affinities to HSPGs (VEGF$_{165}$ and VEGF$_{189}$). In addition, the serine protease plasmin is able to proteolytically release VEGF from the matrix,[166] implying that during angiogenesis, plasminogen activation can cleave the bound forms of VEGF. Furthermore, the matrix metalloprotease MMP3 can cleave VEGF$_{164}$, generating diffusible, non-heparin-binding forms.[167,168] Sequestration of VEGF in the matrix seems to restrict VEGF gradients and guide endothelial cell migration.[91,137,169,170]

NGF is also released to the extracellular space in an activity-dependent manner in its precursor form (proNGF), and the mature NGF form is enzymatically generated by activated MMP-9. Neuronal stimulation releases the required inactive proMMP-9, which in turn is activated extracellularly by plasmin. Interestingly, the two forms (proNGF and mature NGF) play different biological roles. The mature form binds a cell-survival receptor, TrkA, while the "pro" form activates an apoptosis-promoting receptor, p75NTR. Thus, the conversion of proNGF to NGF may play a critical role in the balance of cell survival and death in the central nervous system.[171,172]

Hepatocyte growth factor (HGF) is another growth factor that is released as an inactive precursor and can be activated by several extracellular serine proteases or, during wound healing after injury, by a cascade involving thrombin and HGF activator controlling fibrosis.[173,174]

Proteolysis has also been implicated in SHH extracellular processing in which the cell-surface-tethered SHH multimers are digested by a disintegrin and metalloprotease (ADAM)-protease in concert with Scuba2 into lipidated peptide termini and truncated morphogen clusters. It has been suggested that this processing exposes PTCH-binding sites of the solubilized clusters and thereby couples SHH release with its activation.[127,129,175]

2.5 STEM CELL NICHES AND MORPHOGEN GRADIENTS

So far, we have discussed morphogen gradients in the context of controlling many patterning-type developmental events, but in adult tissue, growth factors also play vital roles in maintaining cell/tissue homeostasis, including the regulation of stem cells and their niches.

Adult tissues have the ability to regenerate and maintain a population of tissue-specific stem cells. Stem cells have the ability to produce differentiating cells by asymmetric division or simply self-renew by symmetric division, while residing in a specialized microenvironment known as the "niche". A stem cell niche provides extracellular cues for stem cell survival and identity, modulating their quiescence, self-renewal and differentiation state.

Stem cell niches have specific locations that are unique to each tissue and contain specialized supportive cells that modulate the stem cells' microenvironments through cell-surface receptors, soluble factors and their ECM.[176–179] The importance of the ECM as a reservoir of growth factors has been demonstrated in stem cell biology:[139,176] HS regulates FGF and BMP function in embryonic stem cells,[180] and GAG mimetics can potentiate the proliferation and migration of mesenchymal stem cells.[181] Furthermore, it has been suggested that the ability of human mesenchymal stem cells to differentiate to the neurogenic pathway can be modulated by HSPG.[182]

Specialized ECM structures in which the adult brain stem-cell niche is located have been described in the mouse brain lateral ventricle walls. Their composition is basement membrane-like and they contain HSPGs, in particular perlecan, conferring the ability to immobilize FGF2 and BMPs and promote growth factor activity in the neural stem cell niche.[183–186] The satellite stem cell niches in skeletal muscle are located in a compartment between the myofiber plasma membrane and the basal lamina that surrounds the myofiber, and PGs of the basal lamina can bind morphogens from satellite cells or myofiber sources such as FGF,[187] HGF[188] and others.[178,189] For instance, during skeletal muscle regeneration, HS 6-O-endosulfatases (Sulf1 and -2) reduce canonical Wnt–HS binding and regulate co-localization of the co-receptor LRP5 with caveolin3 by regulating the bioavailability of canonical Wnts for Frizzled receptors and LRP5/6 interaction in lipid rafts, which may in turn antagonize non-canonical Wnt signaling.[190]

These examples highlight the importance of ECM–morphogen interactions in defining stem cell niches and controlling the differentiation potential of the stem cells. Recently, the importance of generating suitable three-dimensional scaffolds, which mimic the native stem cell niche to enable expanding them in culture, has been high-lighted in the literature.[157,168]

2.6 ARTIFICIAL MATRICES, THE CHALLENGES

In light of the importance of morphogen–ECM interactions in modu-lating organogenesis, including the differentiation and proliferation in stem cells, it is necessary to consider the importance of the ECM when using artificial matrices to recapitulate cell behavior as it applies to regenerative therapy. The incorporation of biological principles has aided the impressive progress made in the design of artificial matri-ces or scaffolds for either regeneration *in vivo* or 3-D cell cultures with which to generate organs or tissues. Dealing with morphogens still remains a challenge because of the temporal and spatial complexity of the signal presentations and the cell specificity in responsiveness to matrix components. But as much as we would like a recapitulation of natural ECM behavior, simpler matrices bearing the appropriate growth factors may perform adequately as long as they attract stem or precursor cells, induce the expected differentiation pathways and allow cellular remodeling.

In light of the important function that PGs play in maintaining mor-phogen gradients it is not surprising that in the last decade multiple approaches have been employed to incorporate in tissue-engineered scaffolds synthetic and semi-synthetic polymers, called neoproteo-glycans, that mimic the structural and biological functions of PGs (reviewed in Weyers & Linhardt, 2013[191]). The major challenges in these efforts have to do with the dual composition of the PGs (*i.e.*, core protein and carbohydrate chains), the limited knowledge about the structural microheterogeneity of the carbohydrate structures that actively bind specific morphogens, and the limitations in performing the carbohy-drate sequential syntheses needed to generate large quantities of arti-ficial matrices for clinical purposes. Even though several designs still have many limitations, some have been successfully used for promot-ing angiogenesis,[192] osteogenic[193] and cartilage differentiation,[194] and for growth of cartilage,[195–197] cardiovascular[198] and bone[199–202] tissues.

The establishment of morphogen gradients in 2-D or 3-D artifi-cial scaffolds can be achieved using several methods, *e.g.*, microflu-idic patterning methods[171,172] or controlled deposition.[203] Numerous methods for the controlled release of morphogens from ECM-like scaffolds for use in regenerative medicine are also actively being devel-oped (review by Lee *et al.*, 2011[204]). The challenges in this area include designing methods for the stabilization of encapsulated proteins to allow extended release times and the adaptation of these approaches for clinical use.[204] Formation of gradients and patterns within artificial scaffolds is discussed in further detail in Chapters 7 and 8 by Temenoff *et al.* and Anseth *et al.*

2.7 CONCLUDING REMARKS

In the last 60 years, since the intrinsic instability of particular reaction diffusion systems suggested to Turing a possible molecular mechanism for morphogenesis,[1] stunning advances have been made in which the critical roles of the ECM in regulating cell behavior through interaction with morphogens has been well established. There are many questions still to be answered, such as: How are signaling ranges controlled? What factors regulate morphogen movement? Are different dispersal mechanisms used for a given morphogen according to tissue context? How are tissue morphogenesis and growth coordinated? How do ECM components regulate the dispersal of morphogens? Advancements in biophysical, imaging and modeling techniques will help to answer these questions in the future. This knowledge will be invaluable not only for the understanding of morphogenesis in the developmental context, but also for advancement of stem cell biology and the engineering of artificial matrices for regenerative medicine.

ABBREVIATIONS

ECM	extracellular matrix
PGs	proteoglycans
FGF	fibroblast growth factor
SHH	sonic hedgehogs
Wnts	wingless-type MMTV integration site family members
EGF	epidermal growth factor
TGF	transforming growth factor
t-RA	*trans*-retinoic acid
BMPs	bone morphogenetic proteins
VEGF	vascular endothelial growth factor
GPI	glycosylphosphatidylinisotol
GAGs	glycosaminoglycans
HS	heparan sulfate
DS	dermatan sulfate
CS	chondroitin sulfate
PTCH	patched
LAP	latency-associated peptide
NGF	nerve growth factor

REFERENCES

1. A. M. Turing, *Proc. R. Soc. B*, 1952, 37–72.
2. A. J. Zhu and M. P. Scott, *Genes Dev.*, 2004, **18**, 2985–2997.
3. L. Wolpert, *J. Theor. Biol.*, 1969, **25**, 1–47.
4. S. Urdy, *Biol. Rev. Cambridge Philos. Soc.*, 2012, **87**, 786–803.
5. J. L. Christian, *Wiley Interdiscip. Rev.: Dev. Biol.*, 2012, **1**, 3–15.
6. K. M. Cadigan, *Semin. Cell Dev. Biol.*, 2002, **13**, 83–90.
7. A. D. Lander, *Cell*, 2007, **128**, 245–256.
8. M. Strigini, *J. Neurobiol.*, 2005, **64**, 324–333.
9. K. Hironaka and Y. Morishita, *Curr. Opin. Genet. Dev.*, 2012, **22**, 553–561.
10. P. Muller and A. F. Schier, *Dev. Cell*, 2011, **21**, 145–158.
11. M. Nahmad and A. D. Lander, *Curr. Opin. Genet. Dev.*, 2011, **21**, 726–731.
12. F. Ulloa and J. Briscoe, *Cell Cycle*, 2007, **6**, 2640–2649.
13. R. Salie, V. Niederkofler and S. Arber, *Neuron*, 2005, **45**, 189–192.
14. J. Briscoe, Y. Chen, T. M. Jessell and G. Struhl, *Mol. Cell*, 2001, **7**, 1279–1291.
15. E. Dessaud, L. L. Yang, K. Hill, B. Cox, F. Ulloa, A. Ribeiro, A. Mynett, B. G. Novitch and J. Briscoe, *Nature*, 2007, **450**, 717–720.
16. J. Ericson, P. Rashbass, A. Schedl, S. Brenner-Morton, A. Kawakami, V. van Heyningen, T. M. Jessell and J. Briscoe, *Cell*, 1997, **90**, 169–180.
17. C. E. Chamberlain, J. Jeong, C. Guo, B. L. Allen and A. P. McMahon, *Development*, 2008, **135**, 1097–1106.
18. A. Gritli-Linde, P. Lewis, A. P. McMahon and A. Linde, *Dev. Biol.*, 2001, **236**, 364–386.
19. X. Huang, Y. Litingtung and C. Chiang, *Development*, 2007, **134**, 2095–2105.
20. F. Charron and M. Tessier-Lavigne, *Adv. Exp. Med. Biol.*, 2007, **621**, 116–133.
21. S. A. Newman, S. Christley, T. Glimm, H. G. Hentschel, B. Kazmierczak, Y. T. Zhang, J. Zhu and M. Alber, *Curr. Top. Dev. Biol.*, 2008, **81**, 311–340.
22. S. Nissim, S. M. Hasso, J. F. Fallon and C. J. Tabin, *Dev. Biol.*, 2006, **299**, 12–21.
23. P. W. Ingham and M. Placzek, *Nat. Rev. Genet.*, 2006, **7**, 841–850.
24. D. Furniss, L. A. Lettice, I. B. Taylor, P. S. Critchley, H. Giele, R. E. Hill and A. O. Wilkie, *Hum. Mol. Genet.*, 2008, **17**, 2417–2423.
25. M. Sun, F. Ma, X. Zeng, Q. Liu, X. L. Zhao, F. X. Wu, G. P. Wu, Z. F. Zhang, B. Gu, Y. F. Zhao, S. H. Tian, B. Lin, X. Y. Kong, X. L. Zhang, W. Yang, W. H. Lo and X. Zhang, *J. Med. Genet.*, 2008, **45**, 589–595.

26. D. Baldridge, O. Shchelochkov, B. Kelley and B. Lee, *Annu. Rev. Genomics Hum. Genet.*, 2010, **11**, 189–217.

27. S. Nissim, P. Allard, A. Bandyopadhyay, B. D. Harfe and C. J. Tabin, *Dev. Biol.*, 2007, **304**, 9–21.

28. P. J. Scherz, E. McGlinn, S. Nissim and C. J. Tabin, *Dev. Biol.*, 2007, **308**, 343–354.

29. A. Badugu, C. Kraemer, P. Germann, D. Menshykau and D. Iber, *Sci. Rep.*, 2012, **2**, 991.

30. M. B. Goldring, K. Tsuchimochi and K. Ijiri, *J. Cell. Biochem.*, 2006, **97**, 33–44.

31. F. Long and D. M. Ornitz, *Cold Spring Harbor Perspect. Biol.*, 2013, **5**, a008334.

32. T. Kobayashi, D. W. Soegiarto, Y. Yang, B. Lanske, E. Schipani, A. P. McMahon and H. M. Kronenberg, *J. Clin. Invest.*, 2005, **115**, 1734–1742.

33. R. J. Metzger, O. D. Klein, G. R. Martin and M. A. Krasnow, *Nature*, 2008, **453**, 745–750.

34. L. Marjoram and C. Wright, *Development*, 2011, **138**, 475–485.

35. Y. Tanaka, Y. Okada and N. Hirokawa, *Nature*, 2005, **435**, 172–177.

36. E. V. Entchev, A. Schwabedissen and M. Gonzalez-Gaitan, *Cell*, 2000, **103**, 981–991.

37. A. Kicheva, P. Pantazis, T. Bollenbach, Y. Kalaidzidis, T. Bittig, F. Julicher and M. Gonzalez-Gaitan, *Science*, 2007, **315**, 521–525.

38. C. Eugster, D. Panakova, A. Mahmoud and S. Eaton, *Dev. Cell*, 2007, **13**, 57–71.

39. W. Palm, M. M. Swierczynska, V. Kumari, M. Ehrhart-Bornstein, S. R. Bornstein and S. Eaton, *PLoS Biol.*, 2013, **11**, e1001505.

40. S. Roy, H. Huang, S. Liu and T. B. Kornberg, *Science*, 2014, **343**, 1244624.

41. F. Hsiung, F. A. Ramirez-Weber, D. D. Iwaki and T. B. Kornberg, *Nature*, 2005, **437**, 560–563.

42. A. Callejo, A. Bilioni, E. Mollica, N. Gorfinkiel, G. Andres, C. Ibanez, C. Torroja, L. Doglio, J. Sierra and I. Guerrero, *Proc. Natl. Acad. Sci. U. S. A.*, 2011, **108**, 12591–12598.

43. T. A. Sanders, E. Llagostera and M. Barna, *Nature*, 2013, **497**, 628–632.

44. T. B. Kornberg and A. Guha, *Curr. Opin. Genet. Dev.*, 2007, **17**, 264–271.

45. P. Muller, K. W. Rogers, S. R. Yu, M. Brand and A. F. Schier, *Development*, 2013, **140**, 1621–1638.

46. J. Ferent and E. Traiffort, *Neuroscientist*, 2015, **21**, 356–371.

47. G. Deshpande, K. Zhou, J. Y. Wan, J. Friedrich, N. Jourjine, D. Smith and P. Schedl, *PLoS Genet.*, 2013, **9**, e1003720.
48. S. Shimozono, T. Iimura, T. Kitaguchi, S. Higashijima and A. Miyawaki, *Nature*, 2013, **496**, 363–366.
49. L. Duchesne, V. Octeau, R. N. Bearon, A. Beckett, I. A. Prior, B. Lounis and D. G. Fernig, *PLoS Biol.*, 2012, **10**, e1001361.
50. H. Teimouri and A. B. Kolomeisky, *J. Chem. Phys.*, 2014, **140**, 085102.
51. E. M. Pera, H. Acosta, N. Gouignard, M. Climent and I. Arregi, *Exp. Cell Res.*, 2014, **321**, 25–31.
52. J. Lei, D. Wang, Y. Song, Q. Nie and F. Y. Wan, *Discrete Contin. Dyn. Syst. Ser B*, 2013, **18,** 721–739.
53. O. Vasieva, M. Rasolonjanahary and B. Vasiev, *Reproduction*, 2013, **145**, R175–R184.
54. S. Y. Shvartsman and R. E. Baker, *Wiley Interdiscip. Rev.: Dev. Biol.*, 2012, **1**, 715–730.
55. A. D. Lander, Q. Nie and F. Y. Wan, *Bull. Math. Biol.*, 2007, **69**, 33–54.
56. C. C. Huang, D. H. Hall, E. M. Hedgecock, G. Kao, V. Karantza, B. E. Vogel, H. Hutter, A. D. Chisholm, P. D. Yurchenco and W. G. Wadsworth, *Development*, 2003, **130**, 3343–3358.
57. C. Medioni and S. Noselli, *Development*, 2005, **132**, 3069–3077.
58. A. M. Sheppard, S. K. Hamilton and A. L. Pearlman, *J. Neurosci.*, 1991, **11**, 3928–3942.
59. A. M. Sheppard and A. L. Pearlman, *J. Comput. Neurol.*, 1997, **378**, 173–179.
60. E. C. Benesh, P. M. Miller, E. R. Pfaltzgraff, N. E. Grega-Larson, H. A. Hager, B. H. Sung, X. Qu, H. S. Baldwin, A. M. Weaver and D. M. Bader, *Mol. Biol. Cell*, 2013, **24**, 3496–3510.
61. A. C. Rapraeger, A. Krufka and B. B. Olwin, *Science*, 1991, **252**, 1705–1708.
62. A. Yayon, M. Klagsbrun, J. D. Esko, P. Leder and D. M. Ornitz, *Cell*, 1991, **64**, 841–848.
63. H. Nakato, T. A. Futch and S. B. Selleck, *Development*, 1995, **121**, 3687–3702.
64. J. Topczewski, D. S. Sepich, D. C. Myers, C. Walker, A. Amores, Z. Lele, M. Hammerschmidt, J. Postlethwait and L. Solnica-Krezel, *Dev. Cell*, 2001, **1**, 251–264.
65. J. S. Lee, S. von der Hardt, M. A. Rusch, S. E. Stringer, H. L. Stickney, W. S. Talbot, R. Geisler, C. Nusslein-Volhard, S. B. Selleck, C. B. Chien and H. Roehl, *Neuron*, 2004, **44**, 947–960.

66. D. B. Kantor, O. Chivatakarn, K. L. Peer, S. F. Oster, M. Inatani, M. J. Hansen, J. G. Flanagan, Y. Yamaguchi, D. W. Sretavan, R. J. Giger and A. L. Kolodkin, *Neuron*, 2004, **44**, 961–975.
67. N. B. Schwartz and M. S. Domowicz, *Adv. Neurobiol.*, 2014, **9**, 89–115.
68. N. B. Schwartz and M. Domowicz, *Glycoconjugate J.*, 2004, **21**, 329–341.
69. U. Hacker, K. Nybakken and N. Perrimon, *Nat. Rev. Mol. Cell Biol.*, 2005, **6**, 530–541.
70. I. Nakamura, M. G. Fernandez-Barrena, M. C. Ortiz-Ruiz, L. L. Almada, C. Hu, S. F. Elsawa, L. D. Mills, P. A. Romecin, K. H. Gulaid, C. D. Moser, J. J. Han, A. Vrabel, E. A. Hanse, N. A. Akogyeram, J. H. Albrecht, S. P. Monga, S. O. Sanderson, J. Prieto, L. R. Roberts and M. E. Fernandez-Zapico, *J. Biol. Chem.*, 2013, **288**, 21389–21398.
71. M. Asada, M. Shinomiya, M. Suzuki, E. Honda, R. Sugimoto, M. Ikekita and T. Imamura, *Biochim. Biophys. Acta*, 2009, **1790**, 40–48.
72. I. Matsuo and C. Kimura-Yoshida, *Curr. Opin. Genet. Dev.*, 2013, **23**, 399–407.
73. R. Goetz and M. Mohammadi, *Nat. Rev. Mol. Cell Biol.*, 2013, **14**, 166–180.
74. L. Pellegrini, D. F. Burke, F. von Delft, B. Mulloy and T. L. Blundell, *Nature*, 2000, **407**, 1029–1034.
75. P. Wong and W. H. Burgess, *J. Biol. Chem.*, 1998, **273**, 18617–18622.
76. S. Faham, R. E. Hileman, J. R. Fromm, R. J. Linhardt and D. C. Rees, *Science*, 1996, **271**, 1116–1120.
77. J. Kreuger, M. Salmivirta, L. Sturiale, G. Gimenez-Gallego and U. Lindahl, *J. Biol. Chem.*, 2001, **276**, 30744–30752.
78. M. L. Fitzgerald, Z. Wang, P. W. Park, G. Murphy and M. Bernfield, *J. Cell Biol.*, 2000, **148**, 811–824.
79. C. A. Kirkpatrick, B. D. Dimitroff, J. M. Rawson and S. B. Selleck, *Dev. Cell*, 2004, **7**, 513–523.
80. J. Kreuger, L. Perez, A. J. Giraldez and S. M. Cohen, *Dev. Cell*, 2004, **7**, 503–512.
81. A. Traister, W. Shi and J. Filmus, *Biochem. J.*, 2008, **410**, 503–511.
82. F. Gong, P. Jemth, M. L. Escobar Galvis, I. Vlodavsky, A. Horner, U. Lindahl and J. P. Li, *J. Biol. Chem.*, 2003, **278**, 35152–35158.
83. K. Shimokawa, C. Kimura-Yoshida, N. Nagai, K. Mukai, K. Matsubara, H. Watanabe, Y. Matsuda, K. Mochida and I. Matsuo, *Dev. Cell*, 2011, **21**, 257–272.
84. A. Elfenbein, A. Lanahan, T. X. Zhou, A. Yamasaki, E. Tkachenko, M. Matsuda and M. Simons, *Sci. Signaling*, 2012, **5**, ra36.

85. D. Yan and X. Lin, *Cold Spring Harbor Perspect. Biol.*, 2009, **1**, a002493.

86. M. J. Kim, S. L. Cotman, W. Halfter and G. J. Cole, *J. Neurobiol.*, 2003, **55**, 261–277.

87. S. S. Deepa, Y. Umehara, S. Higashiyama, N. Itoh and K. Sugahara, *J. Biol. Chem.*, 2002, **277**, 43707–43716.

88. E. L. Shipp and L. C. Hsieh-Wilson, *Chem. Biol.*, 2007, **14**, 195–208.

89. S. M. Smith, L. A. West, P. Govindraj, X. Zhang, D. M. Ornitz and J. R. Hassell, *Matrix Biol.*, 2007, **26**, 175–184.

90. S. R. Yu, M. Burkhardt, M. Nowak, J. Ries, Z. Petrasek, S. Scholpp, P. Schwille and M. Brand, *Nature*, 2009, **461**, 533–536.

91. P. Vempati, A. S. Popel and F. Mac Gabhann, *Cytokine Growth Factor Rev.*, 2014, **25**, 1–19.

92. A. Abramsson, S. Kurup, M. Busse, S. Yamada, P. Lindblom, E. Schallmeiner, D. Stenzel, D. Sauvaget, J. Ledin, M. Ringvall, U. Landegren, L. Kjellen, G. Bondjers, J. P. Li, U. Lindahl, D. Spillmann, C. Betsholtz and H. Gerhardt, *Genes Dev.*, 2007, **21**, 316–331.

93. G. K. Dhoot, M. K. Gustafsson, X. Ai, W. Sun, D. M. Standiford and C. P. Emerson, Jr., *Science*, 2001, **293**, 1663–1666.

94. F. Reichsman, L. Smith and S. Cumberledge, *J. Cell Biol.*, 1996, **135**, 819–827.

95. G. H. Baeg, X. Lin, N. Khare, S. Baumgartner and N. Perrimon, *Development*, 2001, **128**, 87–94.

96. U. Hacker, X. Lin and N. Perrimon, *Development*, 1997, **124**, 3565–3573.

97. S. Nadanaka, M. Ishida, M. Ikegami and H. Kitagawa, *J. Biol. Chem.*, 2008, **283**, 27333–27343.

98. G. Hausmann, C. Banziger and K. Basler, *Nat. Rev. Mol. Cell Biol.*, 2007, **8**, 331–336.

99. C. Alexandre, A. Baena-Lopez and J. P. Vincent, *Nature*, 2014, **505**, 180–185.

100. T. Y. Belenkaya, C. Han, D. Yan, R. J. Opoka, M. Khodoun, H. Liu and X. Lin, *Cell*, 2004, **119**, 231–244.

101. C. Han, T. Y. Belenkaya, M. Khodoun, M. Tauchi, X. Lin and X. Lin, *Development*, 2004, **131**, 1563–1575.

102. X. Jiao, P. C. Billings, M. P. O'Connell, F. S. Kaplan, E. M. Shore and D. L. Glaser, *J. Biol. Chem.*, 2007, **282**, 1080–1086.

103. G. H. Olivares, H. Carrasco, F. Aroca, L. Carvallo, F. Segovia and J. Larraín, *Dev. Biol.*, 2009, **329**, 338–349.

104. L. Schaefer and R. V. Iozzo, *J. Biol. Chem.*, 2008, **283**, 21305–21309.

105. K. C. Kirkbride, T. A. Townsend, M. W. Bruinsma, J. V. Barnett and G. C. Blobe, *J. Biol. Chem.*, 2008, **283**, 7628–7637.
106. Q. Hu, N. Ueno and R. R. Behringer, *EMBO Rep.*, 2004, **5**, 734–739.
107. B. Ohkawara, S. Iemura, P. ten Dijke and N. Ueno, *Curr. Biol.*, 2002, **12**, 205–209.
108. C. A. Kirkpatrick, S. M. Knox, W. D. Staatz, B. Fox, D. M. Lercher and S. B. Selleck, *Dev. Biol.*, 2006, **300**, 570–582.
109. M. I. Capurro, P. Xu, W. Shi, F. Li, A. Jia and J. Filmus, *Dev. Cell*, 2008, **14**, 700–711.
110. D. F. Cano-Gauci, H. H. Song, H. Yang, C. McKerlie, B. Choo, W. Shi, R. Pullano, T. D. Piscione, S. Grisaru, S. Soon, L. Sedlackova, A. K. Tanswell, T. W. Mak, H. Yeger, G. A. Lockwood, N. D. Rosenblum and J. Filmus, *J. Cell Biol.*, 1999, **146**, 255–264.
111. E. Chiao, P. Fisher, L. Crisponi, M. Deiana, I. Dragatsis, D. Schlessinger, G. Pilia and A. Efstratiadis, *Dev. Biol.*, 2002, **243**, 185–206.
112. G. Pilia, R. M. Hughes-Benzie, A. MacKenzie, P. Baybayan, E. Y. Chen, R. Huber, G. Neri, A. Cao, A. Forabosco and D. Schlessinger, *Nat. Genet.*, 1996, **12**, 241–247.
113. S. M. Smith, L. A. West and J. R. Hassell, *Arch. Biochem. Biophys.*, 2007, **468**, 244–251.
114. J. E. Hooper and M. P. Scott, *Nat. Rev. Mol. Cell Biol.*, 2005, **6**, 306–317.
115. Y. Wang, A. P. McMahon and B. L. Allen, *Curr. Opin. Cell Biol.*, 2007, **19**, 159–165.
116. Y. Bellaiche, I. The and N. Perrimon, *Nature*, 1998, **394**, 85–88.
117. L. Koziel, M. Kunath, O. G. Kelly and A. Vortkamp, *Dev. Cell*, 2004, **6**, 801–813.
118. M. Cortes, A. T. Baria and N. B. Schwartz, *Development*, 2009, **136**, 1697–1706.
119. M. S. Domowicz, M. Cortes, J. G. Henry and N. B. Schwartz, *Dev. Biol.*, 2009, **329**, 242–257.
120. F. Zhang, J. S. McLellan, A. M. Ayala, D. J. Leahy and R. J. Linhardt, *Biochemistry*, 2007, **46**, 3933–3941.
121. D. M. Whalen, T. Malinauskas, R. J. Gilbert and C. Siebold, *Proc. Natl. Acad. Sci. U. S. A.*, 2013, **110**, 16420–16425.
122. A. D. Cardin, D. A. Demeter, H. J. Weintraub and R. L. Jackson, *Methods Enzymol.*, 1991, **203**, 556–583.
123. A. D. Cardin and H. J. Weintraub, *Arteriosclerosis*, 1989, **9**, 21–32.
124. D. Panakova, H. Sprong, E. Marois, C. Thiele and S. Eaton, *Nature*, 2005, **435**, 58–65.

125. J. A. Goetz, S. Singh, L. M. Suber, F. J. Kull and D. J. Robbins, *J. Biol. Chem.*, 2006, **281**, 4087–4093.
126. P. Farshi, S. Ohlig, U. Pickhinke, S. Hoing, K. Jochmann, R. Lawrence, R. Dreier, T. Dierker and K. Grobe, *J. Biol. Chem.*, 2011, **286**, 23608–23619.
127. S. Ohlig, P. Farshi, U. Pickhinke, J. van den Boom, S. Hoing, S. Jakuschev, D. Hoffmann, R. Dreier, H. R. Scholer, T. Dierker, C. Bordych and K. Grobe, *Dev. Cell*, 2011, **20**, 764–774.
128. T. Dierker, R. Dreier, A. Petersen, C. Bordych and K. Grobe, *J. Biol. Chem.*, 2009, **284**, 8013–8022.
129. S. Ohlig, U. Pickhinke, S. Sirko, S. Bandari, D. Hoffmann, R. Dreier, P. Farshi, M. Gotz and K. Grobe, *J. Biol. Chem.*, 2012, **287**, 43708–43719.
130. R. M. Witt, M. L. Hecht, M. F. Pazyra-Murphy, S. M. Cohen, C. Noti, T. H. van Kuppevelt, M. Fuller, J. A. Chan, J. J. Hopwood, P. H. Seeberger and R. A. Segal, *J. Biol. Chem.*, 2013, **288**, 26275–26288.
131. C. Danesin, E. Agius, N. Escalas, X. Ai, C. Emerson, P. Cochard and C. Soula, *J. Neurosci.*, 2006, **26**, 5037–5048.
132. A. Bilioni, D. Sanchez-Hernandez, A. Callejo, A. C. Gradilla, C. Ibanez, E. Mollica, M. Carmen Rodriguez-Navas, E. Simon and I. Guerrero, *Dev. Biol.*, 2013, **376**, 198–212.
133. S. Oki, R. Hashimoto, Y. Okui, M. M. Shen, E. Mekada, H. Otani, Y. Saijoh and H. Hamada, *Development*, 2007, **134**, 3893–3904.
134. T. McLaughlin and D. D. M. O'Leary, *Annu. Rev. Neurosci.*, 2005, **28**, 327–355.
135. S. A. Eming and J. A. Hubbell, *Exp. Dermatol.*, 2011, **20**, 605–613.
136. P. Carmeliet, Y. S. Ng, D. Nuyens, G. Theilmeier, K. Brusselmans, I. Cornelissen, E. Ehler, V. V. Kakkar, I. Stalmans, V. Mattot, J. C. Perriard, M. Dewerchin, W. Flameng, A. Nagy, F. Lupu, L. Moons, D. Collen, P. A. D'Amore and D. T. Shima, *Nat. Med.*, 1999, **5**, 495–502.
137. C. Ruhrberg, H. Gerhardt, M. Golding, R. Watson, S. Ioannidou, H. Fujisawa, C. Betsholtz and D. T. Shima, *Genes Dev.*, 2002, **16**, 2684–2698.
138. N. Alam, H. L. Goel, M. J. Zarif, J. E. Butterfield, H. M. Perkins, B. G. Sansoucy, T. K. Sawyer and L. R. Languino, *J. Cell. Physiol.*, 2007, **213**, 649–653.
139. R. O. Hynes, *Science*, 2009, **326**, 1216–1219.
140. S. Schenk, E. Hintermann, M. Bilban, N. Koshikawa, C. Hojilla, R. Khokha and V. Quaranta, *J. Cell Biol.*, 2003, **161**, 197–209.

141. A. K. Iyer, K. T. Tran, C. W. Borysenko, M. Cascio, C. J. Camacho, H. C. Blair, I. Bahar and A. Wells, *J. Cell. Physiol.*, 2007, **211**, 748–758.
142. L. Zacchigna, C. Vecchione, A. Notte, M. Cordenonsi, S. Dupont, S. Maretto, G. Cifelli, A. Ferrari, A. Maffei, C. Fabbro, P. Braghetta, G. Marino, G. Selvetella, A. Aretini, C. Colonnese, U. Bettarini, G. Russo, S. Soligo, M. Adorno, P. Bonaldo, D. Volpin, S. Piccolo, G. Lembo and G. M. Bressan, *Cell*, 2006, **124**, 929–942.
143. D. B. Rifkin, *J. Biol. Chem.*, 2005, **280**, 7409–7412.
144. P. ten Dijke and H. M. Arthur, *Nat. Rev. Mol. Cell Biol.*, 2007, **8**, 857–869.
145. P. J. Wipff and B. Hinz, *Eur. J. Cell Biol.*, 2008, **87**, 601–615.
146. K. E. Gregory, R. N. Ono, N. L. Charbonneau, C. L. Kuo, D. R. Keene, H. P. Bachinger and L. Y. Sakai, *J. Biol. Chem.*, 2005, **280**, 27970–27980.
147. E. Arteaga-Solis, B. Gayraud, S. Y. Lee, L. Shum, L. Sakai and F. Ramirez, *J. Cell Biol.*, 2001, **154**, 275–281.
148. F. Ramirez, L. Y. Sakai, D. B. Rifkin and H. C. Dietz, *Cell. Mol. Life Sci.*, 2007, **64**, 2437–2446.
149. F. Ramirez and H. C. Dietz, *J. Biol. Chem.*, 2009, **284**, 14677–14681.
150. E. R. Neptune, P. A. Frischmeyer, D. E. Arking, L. Myers, T. E. Bunton, B. Gayraud, F. Ramirez, L. Y. Sakai and H. C. Dietz, *Nat. Genet.*, 2003, **33**, 407–411.
151. Y. Zhu, A. Oganesian, D. R. Keene and L. J. Sandell, *J. Cell Biol.*, 1999, **144**, 1069–1080.
152. S. Rahman, Y. Patel, J. Murray, K. V. Patel, R. Sumathipala, M. Sobel and E. S. Wijelath, *BMC Cell Biol.*, 2005, **6**, 8.
153. E. S. Wijelath, S. Rahman, M. Namekata, J. Murray, T. Nishimura, Z. Mostafavi-Pour, Y. Patel, Y. Suda, M. J. Humphries and M. Sobel, *Circ. Res.*, 2006, **99**, 853–860.
154. A. Uren, F. Reichsman, V. Anest, W. G. Taylor, K. Muraiso, D. P. Bottaro, S. Cumberledge and J. S. Rubin, *J. Biol. Chem.*, 2000, **275**, 4374–4382.
155. K. Willert, J. D. Brown, E. Danenberg, A. W. Duncan, I. L. Weissman, T. Reya, J. R. Yates, 3rd and R. Nusse, *Nature*, 2003, **423**, 448–452.
156. J. C. Hsieh, L. Kodjabachian, M. L. Rebbert, A. Rattner, P. M. Smallwood, C. H. Samos, R. Nusse, I. B. Dawid and J. Nathans, *Nature*, 1999, **398**, 431–436.
157. D. Sanchez-Hernandez, J. Sierra, J. R. Ortigao-Farias and I. Guerrero, *Development*, 2012, **139**, 3849–3858.

158. T. Tenzen, B. L. Allen, F. Cole, J. S. Kang, R. S. Krauss and A. P. McMahon, *Dev. Cell*, 2006, **10**, 647–656.
159. W. Zhang, J. S. Kang, F. Cole, M. J. Yi and R. S. Krauss, *Dev. Cell*, 2006, **10**, 657–665.
160. B. L. Allen, J. Y. Song, L. Izzi, I. W. Althaus, J. S. Kang, F. Charron, R. S. Krauss and A. P. McMahon, *Dev. Cell*, 2011, **20**, 775–787.
161. J. S. Kang, W. Zhang and R. S. Krauss, *Sci. Signaling*, 2007, **2007**, pe50.
162. E. K. LeMosy, *Birth Defects Res., Part C*, 2006, **78**, 243–255.
163. B. Steffensen, L. Hakkinen and H. Larjava, *Crit. Rev. Oral Biol. Med.*, 2001, **12**, 373–398.
164. D. E. Bassi, J. Fu, R. Lopez de Cicco and A. J. Klein-Szanto, *Mol. Carcinog.*, 2005, **44**, 151–161.
165. R. Roy, B. Zhang and M. A. Moses, *Exp. Cell Res.*, 2006, **312**, 608–622.
166. K. A. Houck, D. W. Leung, A. M. Rowland, J. Winer and N. Ferrara, *J. Biol. Chem.*, 1992, **267**, 26031–26037.
167. H. X. Lee, A. L. Ambrosio, B. Reversade and E. M. De Robertis, *Cell*, 2006, **124**, 147–159.
168. N. Ferrara, H. P. Gerber and J. LeCouter, *Nat. Med.*, 2003, **9**, 669–676.
169. H. Gerhardt, M. Golding, M. Fruttiger, C. Ruhrberg, A. Lundkvist, A. Abramsson, M. Jeltsch, C. Mitchell, K. Alitalo, D. Shima and C. Betsholtz, *J. Cell Biol.*, 2003, **161**, 1163–1177.
170. T. T. Chen, A. Luque, S. Lee, S. M. Anderson, T. Segura and M. L. Iruela-Arispe, *J. Cell Biol.*, 2010, **188**, 595–609.
171. M. A. Bruno and A. C. Cuello, *Proc. Natl. Acad. Sci. U. S. A.*, 2006, **103**, 6735–6740.
172. R. Lee, P. Kermani, K. K. Teng and B. L. Hempstead, *Science*, 2001, **294**, 1945–1948.
173. K. Miyazawa, T. Shimomura and N. Kitamura, *J. Biol. Chem.*, 1996, **271**, 3615–3618.
174. N. Hattori, S. Mizuno, Y. Yoshida, K. Chin, M. Mishima, T. H. Sisson, R. H. Simon, T. Nakamura and M. Miyake, *Am. J. Pathol.*, 2004, **164**, 1091–1098.
175. P. Jakobs, S. Exner, S. Schurmann, U. Pickhinke, S. Bandari, C. Ortmann, S. Kupich, P. Schulz, U. Hansen, D. G. Seidler and K. Grobe, *J. Cell Sci.*, 2014, **127**, 1726–1737.
176. M. F. Brizzi, G. Tarone and P. Defilippi, *Curr. Opin. Cell Biol.*, 2012, **24**, 645–651.

177. D. E. Discher, D. J. Mooney and P. W. Zandstra, *Science*, 2009, **324**, 1673–1677.
178. F. Gattazzo, A. Urciuolo and P. Bonaldo, *Biochim. Biophys. Acta*, 2014, **1840**, 2506–2519.
179. A. Wade, A. McKinney and J. J. Phillips, *Biochim. Biophys. Acta*, 2014, **1840**, 2520–2525.
180. D. C. Kraushaar, S. Rai, E. Condac, A. Nairn, S. Zhang, Y. Yamaguchi, K. Moremen, S. Dalton and L. Wang, *J. Biol. Chem.*, 2012, **287**, 22691–22700.
181. G. Frescaline, T. Bouderlique, M. B. Huynh, D. Papy-Garcia, J. Courty and P. Albanese, *Stem Cell Res.*, 2012, **8**, 180–192.
182. R. K. Okolicsanyi, L. R. Griffiths and L. M. Haupt, *Dev. Biol.*, 2014, **388**, 1–10.
183. A. Kerever, J. Schnack, D. Vellinga, N. Ichikawa, C. Moon, E. Arikawa-Hirasawa, J. T. Efird and F. Mercier, *Stem Cells*, 2007, **25**, 2146–2157.
184. V. Douet, A. Kerever, E. Arikawa-Hirasawa and F. Mercier, *Cell Proliferation*, 2013, **46**, 137–145.
185. A. Kerever, F. Mercier, R. Nonaka, S. de Vega, Y. Oda, B. Zalc, Y. Okada, N. Hattori, Y. Yamada and E. Arikawa-Hirasawa, *Stem Cell Res.*, 2014, **12**, 492–505.
186. V. Douet, E. Arikawa-Hirasawa and F. Mercier, *Neurosci. Lett.*, 2012, **528**, 120–125.
187. J. DiMario, N. Buffinger, S. Yamada and R. C. Strohman, *Science*, 1989, **244**, 688–690.
188. R. Tatsumi, J. E. Anderson, C. J. Nevoret, O. Halevy and R. E. Allen, *Dev. Biol.*, 1998, **194**, 114–128.
189. B. D. Cosgrove, A. Sacco, P. M. Gilbert and H. M. Blau, *Differentiation*, 2009, **78**, 185–194.
190. T. H. Tran, X. Shi, J. Zaia and X. Ai, *J. Biol. Chem.*, 2012, **287**, 32651–32664.
191. A. Weyers and R. J. Linhardt, *FEBS J.*, 2013, **280**, 2511–2522.
192. M. M. Kemp, A. Kumar, S. Mousa, E. Dyskin, M. Yalcin, P. Ajayan, R. J. Linhardt and S. A. Mousa, *Nanotechnology*, 2009, **20**, 455104.
193. U. Hempel, S. Moller, C. Noack, V. Hintze, D. Scharnweber, M. Schnabelrauch and P. Dieter, *Acta Biomater.*, 2012, **8**, 4064–4072.
194. S. Varghese, N. S. Hwang, A. C. Canver, P. Theprungsirikul, D. W. Lin and J. Elisseeff, *Matrix Biol.*, 2008, **27**, 12–21.
195. C. H. Chang, T. F. Kuo, C. C. Lin, C. H. Chou, K. H. Chen, F. H. Lin and H. C. Liu, *Biomaterials*, 2006, **27**, 1876–1888.
196. T. Deng, S. Huang, S. Zhou, L. He and Y. Jin, *J. Microencapsulation*, 2007, **24**, 163–174.

197. J. M. Coburn, M. Gibson, S. Monagle, Z. Patterson and J. H. Elis-seeff, *Proc. Natl. Acad. Sci. U. S. A.*, 2012, **109**, 10012–10017.

198. T. Nie, R. E. Akins, Jr. and K. L. Kiick, *Acta Biomater.*, 2009, **5**, 865–875.

199. B. A. Harley, A. K. Lynn, Z. Wissner-Gross, W. Bonfield, I. V. Yannas and L. J. Gibson, *J. Biomed. Mater. Res., Part A*, 2010, **92**, 1066–1077.

200. A. K. Lynn, S. M. Best, R. E. Cameron, B. A. Harley, I. V. Yannas, L. J. Gibson and W. Bonfield, *J. Biomed. Mater. Res., Part A*, 2010, **92**, 1057–1065.

201. S. N. Rath, G. Pryymachuk, O. A. Bleiziffer, C. X. Lam, A. Arku-das, S. T. Ho, J. P. Beier, R. E. Horch, D. W. Hutmacher and U. Kneser, *J. Mater. Sci.: Mater. Med.*, 2011, **22**, 1279–1291.

202. G. Bhakta, B. Rai, Z. X. Lim, J. H. Hui, G. S. Stein, A. J. van Wijnen, V. Nurcombe, G. D. Prestwich and S. M. Cool, *Biomaterials*, 2012, **33**, 6113–6122.

203. T. A. Kapur and M. S. Shoichet, *J. Biomed. Mater. Res., Part A*, 2004, **68**, 235–243.

204. K. Lee, E. A. Silva and D. J. Mooney, *J. R. Soc., Interface*, 2011, **8**, 153–170.

205. S. Guimond, M. Maccarana, B. B. Olwin, U. Lindahl and A. C. Rapraeger, *J. Biol. Chem.*, 1993, **268**, 23906–23914.

206. E. Feyzi, F. Lustig, G. Fager, D. Spillmann, U. Lindahl and M. Salmivirta, *J. Biol. Chem.*, 1997, **272**, 5518–5524.

207. M. Lyon, J. A. Deakin, K. Mizuno, T. Nakamura and J. T. Galla-gher, *J. Biol. Chem.*, 1994, **269**, 11216–11223.

208. S. Soker, D. Goldstaub, C. M. Svahn, I. Vlodavsky, B. Z. Levi and G. Neufeld, *Biochem. Biophys. Res. Commun.*, 1994, **203**, 1339–1347.

209. K. Ono, H. Hattori, S. Takeshita, A. Kurita and M. Ishihara, *Glycobiology*, 1999, **9**, 705–711.

210. M. W. Barnett, C. E. Fisher, G. Perona-Wright and J. A. Davies, *J. Cell Sci.*, 2002, **115**, 4495–4503.

211. J. A. Davies, E. A. Yates and J. E. Turnbull, *Growth Factors*, 2003, **21**, 109–119.

212. X. Ai, A. T. Do, O. Lozynska, M. Kusche-Gullberg, U. Lindahl and C. P. Emerson, Jr., *J. Cell Biol.*, 2003, **162**, 341–351.

213. R. Ruppert, E. Hoffmann and W. Sebald, *Eur. J. Biochem.*, 1996, **237**, 295–302.

214. A. Clamp, F. H. Blackhall, A. Henrioud, G. C. Jayson, K. Javaherian, J. Esko, J. T. Gallagher and C. L. Merry, *J. Biol. Chem.*, 2006, **281**, 14813–14822.

215. X. Bao, T. Mikami, S. Yamada, A. Faissner, T. Muramatsu and K. Sugahara, *J. Biol. Chem.*, 2005, **280**, 9180–9191.

216. P. Zou, K. Zou, H. Muramatsu, K. Ichihara-Tanaka, O. Habuchi, S. Ohtake, S. Ikematsu, S. Sakuma and T. Muramatsu, *Glycobiology*, 2003, **13**, 35–42.

217. N. Maeda, J. He, Y. Yajima, T. Mikami, K. Sugahara and T. Yabe, *J. Biol. Chem.*, 2003, **278**, 35805–35811.

218. K. Catlow, J. A. Deakin, M. Delehedde, D. G. Fernig, J. T. Gallagher, M. S. Pavao and M. Lyon, *Biochem. Soc. Trans.*, 2003, **31**, 352–353.

219. A. Weyers, B. Yang, K. Solakyildirim, V. Yee, L. Li, F. Zhang and R. J. Linhardt, *FEBS J.*, 2013, **280**, 2285–2293.

Matrix Biology: ECM Turnover and Temporal Fluctuation

QUYEN TRAN[a] AND BRENDA M. OGLE*[a,b]

[a]Department of Biomedical Engineering, University of Wisconsin-Madison, Madison, WI 53706, USA; [b]Department of Biomedical Engineering, University of Minnesota-Twin Cities, Minneapolis, MN 55455, USA
*E-mail: ogle@umn.edu

3.1 EXTRACELLULAR MATRIX DYNAMICS OF THE MYOCARDIUM

3.1.1 Myocardial Development

3.1.1.1 Overview. In mammals, cardiac development begins with the generation of the heart tube, comprised of an inner endothelial layer and an outer muscular layer. Following the formation of the heart tube the heart begins to loop, the first visual sign of asymmetrical development within the embryo. Once the looping process is complete, the atrial and ventricular portions of the heart can be distinguished. The chambers of the heart further develop and separate, generating the four chambers of the heart. The myocardium can be subdivided into sections such as the epicardium (outer region), the myocardium (middle region), and the endocardium (inner region). During this time, the cardiomyocytes migrate, proliferate,

Mimicking the Extracellular Matrix: The Intersection of Matrix Biology and Biomaterials
Edited by Gregory A. Hudalla and William L. Murphy
© The Royal Society of Chemistry 2020
Published by the Royal Society of Chemistry, www.rsc.org

differentiate, and organize to generate a highly structured myocardium capable of generating the physical and electrical impulses of the heart. Concurrently, cells from the proepicardial organ migrate and surround the heart to generate the embryonic epicardium. These cells then undergo an epithelial-to-mesenchymal transition, migrate into the myocardium, and differentiate into cardiac fibroblasts. In the heart, cardiac fibroblasts are the main source of extracellular matrix and contribute to the mechanical, biochemical, structural, and electrical activities of the heart.[1,2] Cells within the epicardium also differentiate into the vascular smooth muscle and endothelial cells of the heart, especially within the coronary vascular network.[3] Endothelial cells and smooth muscle cells may also contribute to the extracellular matrix (ECM) of the heart with development, either *via* production or remodeling of ECM, though this area is not well studied. ECM deposition and organization of ECM proteins by all cell types within the heart is crucial for development of the myocardium, the work force of the heart.

3.1.1.2 ECM and Myocardiogenesis. The myocardium experiences substantial mechanical stimulation with each beat of the heart. Pressure modestly increases as the ventricle is filled with blood during diastole, followed by a significant pressure increase (from ~10 mm Hg to 110 mm Hg in humans) during isovolumic contraction and ejection of blood from the ventricle during systole and finally a reduction in pressure back to baseline with isovolumic relaxation. Therefore, the myocardium must be strong enough to sustain tensile load and elastic enough to recoil following extension. These properties are endowed in part by the ECM within the myocardium which contains fibrillar collagens (*i.e.*, types I, II, III, V, XI) and proteins associated with elastic fibers. Collagen is the most abundant ECM within the body and is thought to play a major role in the tensile strength of tissues. The collagen family contains proteins composed of triple helices with high concentrations of glycine, proline, and hydroxyproline amino acids that self organize into fibrils and fibers. Structural integrity is conferred by the specific organization of the fibrillar collagens.[4] Collagen I is seen in the epicardium of the mouse embryonic heart and within the myocardium and endocardium at 2 days after birth. In the rat myocardium, collagen III is found in the interstitial space of the myocardium, in the media of medium-sized coronary arteries, and in the connective tissue surrounding blood vessels.[5] Interestingly, collagen III null mice have reduced numbers of collagen I fibrils within the epicardium and myocardium.[6] This observation indicates that collagen III not only individually provides mechanical properties to the heart, but may be vital in the organization of collagen I.

The network of elastic fibers in conjunction with cellular contraction machinery (*i.e.*, the sarcomeres) enable recoil of the heart after diastole. Elastin is the primary component of the elastic fiber and is rich in hydrophobic amino acids, which tend to aggregate until tension is applied. Elastin fibers are composed of crosslinked elastin molecules and microfibril containing glycoproteins such fibrillins and fibulins. Elastin expression peaks at embryonic day (ED) 16.5 of mouse cardiac development and aside from the lining of prominent blood vessels is seen predominantly in the epicardium 2 days after birth.[7] Fibrillin is a glycoprotein that is found closely associated with the elastic fibers and is thought to contribute to the formation of microfibrils. Fibrillin-1 is detected at ED 10.5 within the mouse myocardium, specifically associated with the endothelial cells and epithelial cells of the endocardium and epicardium, respectively. Fibrillin-2, however, is uniformly distributed throughout the myocardium.[8] Additionally, fibrillin-3 is observed within the myocardium and the endothelium of human heart tissues.[9] Perhaps the increase in elastin (and other proteins associated with elastic fiber assembly) at mid–late gestation is necessary for providing the elasticity for heart recoil as the heart is required to exert more ejection force with increasing size of the heart and fetal circulation. Soon after birth, the relative decrease in elastin may be the result of, or the stimulus for, complete maturation of cardiomyocytes and corresponding maturation of functional sarcomeres.[10]

Matricellular proteins (MCPs) are typically non-structural proteins of the extracellular environment that contain binding sites for structural ECM proteins, cell surface receptors, and growth factors. MCP, thrombospondin, contains three identical subunits linked through disulfide bonds. In the developing mouse myocardium, thrombospondin is initially observed in the epicardium. Expression then expands to the endocardium during later stages of development.[11] Fibronectin, another MCP, is a high molecular weight glycoprotein composed of two monomers linked by a pair of disulfide bonds. In murine cardiac development, fibronectin is observed in the external regions of the ventricles and throughout the developing trabeculae at ED 11 and 12.5. Expression persists throughout development and into the fetal stages wherein fibronectin is observed within all the regions of the ventricles.[7,12] Temporal expression of fibronectin mRNA with rat cardiac development has also been examined. Briefly, at ED 11 fibronectin was highly expressed within the heart tube but expression decreased thereafter with fibronectin mainly localized to the pericardium and the ventricular trabeculae.[13] With murine cardiac development, fibronectin is initially observed at the external regions of the heart, and then proceeds to populate the rest of the heart. This spatial distribution within development may indicate the role of fibronectin in cell migration into the heart since fibronectin gradients have been shown to influence cell migration.[14] Additionally, fibronectin has been shown to be important in collagen I fibril organization.[15] Thus fibronectin expression in the developing heart may also play a role in mechanical strengthening through its interaction with collagen I and III.

Proteoglycans are ECM proteins that contain a "core protein" with one or more glycosaminoglycans attached. They interact with other proteoglycans, glycosaminoglycans, ECM proteins, cations (*i.e.*, sodium and potassium), and water and generally serve to lubricate the extracellular space. Versican is a chondroitin sulfate proteoglycan belonging to the lectican protein family and has been implicated in trabeculae formation in the myocardium. Trabeculae are rounded structures that project from the inner surface of the ventricles and are thought to prevent suction that would occur between flat-surfaced walls of the ventricle with ejection. Versican mRNA expression was detected within the myocardium, particularly within the trabecular endocardium, at ED 9.5. The level of expression is high at ED 9.5, which subsequently declined through ED 16.5. Interestingly, the presence of versican coincides with the proliferation of cells of the myocardium, and the decline in versican expression signals a decrease in cellular proliferation in the myocardium within the trabeculae.[16] The location and expression level of fibulin-1 follows a similar trend. Fibulin-1, a MCP, was shown to be a co-factor for a disintegrin and metalloprotease with thrombospondin repeats (ADAMTS1)-mediated degradation of versican. Fibulin-1 null mice had decreased levels of versican degradation products and increased numbers of cardiomyocytes in the trabeculae.[16] ADAMTS1 is robustly expressed in the endocardium and is thought to be responsible for controlling trabeculation, since ADAMTS1 knockouts exhibit hypertrabeculae within the heart.[17] Additionally, a loss in ADAMTS9 results in defects in myocardial compaction thought to be associated with the build-up of versican.[18] From these observations, the levels of versican may affect cardiomyocyte proliferation in trabeculae, which can be modulated by fibulin-1, ADAMTS-1 and 9.

Glycosaminoglycans (GAGs) are polysaccharide chains, found in association with proteins or free floating; GAGs can affect cell behavior through interaction with growth factors, enzymes, cytokines, or other molecules present in the extracellular space. Hyaluronan is a non-sulfated glycosaminoglycan composed of D-glucuronic acid and D-*N*-acetylglucosamine chains that can bind with water and salt to expand the extracellular space. During mouse cardiac development hyaluronan has been detected within the interstitial space surrounding cardiomyocytes.[19] Furthermore, the expression of hyaluronan synthase-2 (Has-2) has also been examined. Has-2 mRNA is detected at ED 8.5 within the myocardium and endocardium of the heart. At ED 9.5 Has-2 expression is seen in the myocardium, particularly near the atrioventricular canal. Hyaluronan has been postulated to play a role in trabeculation, since Has-2 null mice exhibit a compact myocardium devoid of trabeculae.[20] Additionally, rat embryos cultured in hyaluronidase exhibit decreased ventricular function.[21] Thus hyaluronan expression and associated degradation are important in development of ventricular function *via* proper formation of trabeculae.

As described above, the development of the myocardium requires a number of ECM proteins, summarized in Table 3.1. Using knockout approaches to generate transgenic mice, the role of these proteins during development is emerging. And although there have been a number of studies examining changes of the ECM proteins during mammalian myocardial development, there have been a limited number of studies examining how these changes affect cell behavior. Knockout models utilized in these studies are expensive, and the rate of embryonic lethality sometimes limits the ability to study cell behavior in conjunction with ECM dynamics. Future approaches to study cell–matrix interactions with development could include the use of pluripotent cell types in embryoid bodies or other three-dimensional culture systems wherein a quasi-developmental state is replicated, often through the use of synthetic or natural biomaterials. Despite the heterogeneity of such systems, it is possible to conduct *in situ* analysis to include ECM content and distribution with cell behavior.[22]

3.1.2 Myocardial Disease

3.1.2.1 Overview. Heart disease is the leading cause of death in both men and women in the United States; a primary contributor is myocardial infarction (or heart attack), which occurs at a frequency of more than 715 000 annually (AHA Stats 2013). Myocardial infarction occurs due to the lack of oxygen supply to cardiomyocytes, mainly due to an occlusion within the coronary arteries, causing massive cardiomyocyte death. Cardiac healing post-infarct is generally described in

Table 3.1 Extracellular matrix proteins in the developing myocardium.[a]

ECM family	Stage of development			Ref.
	Embryonic	Neo-natal	Adult	
Collagen	Col IV constant in epi and endo; increased in myo E12.5 to E16.5[M]	Col IV in myo and endo; increased from E16.5 to d2; network formation d2[M]	Col IV in basement membrane spanning the entire myo[R]	5,7
	Col IV increased from E12.5 to E16.5[M]	*Col IV increased E16.5 to d2[M]*		
	Col I increased in epi, myo, and endo from E12.5 to E16.5; observed in myo and endo at E18.5[M]	Col I constant from E16.5 to d2 in myo and endo; decreased from E16.5 to d2 in epi[M]	Col I in epi and myo[R]	5,7
	Col I increased from E12.5 to E16.5[M]	*Col I increased from E16.5 to d2[M]*		
	Col III increased from E12.5 to E16.5[M]	*Col III increased from E16.5 to d2[M]*	Col III in epi and myo[R]	5,7
	Col III in myo[R]			
			Col VI in myo[R]	5
Elastic	Elastin increased at E16.5[M]	Elastin in fibrils at d2; decreased in myo and endo and increased in epi from E16.5 to d2[M]		7
	Elastin increased from E12.5 to E16.5[M]	*Elastin increased from E16.5 to d2[M]*		
Glycoprotein	*Fibrillin-1 in epi and endo at E10.5 and E13.5[M]*			8
	Fibrillin-2 throughout at E10.5 and E13.5[M]			8
	Fibrillin-3 in myo and endo[H]			9
	Laminin increased from E12.5 to E16.5[M]	*Laminin increased from E16.5 to d2[M]*	Laminin in myo basement membrane[R]	5,7

Matricellular	Thrombospondin in epi at E10; epi and endo at E13[M]			11
	Fibronectin in epi E12.5 and 14; endo at E12.5, E14.5, and E16.5; myo at E14.5 and E16.5 with increased organization[M]	Fibronectin in epi, myo, and endo with increased organization at d2[M]	Fibronectin in myo[R]	5,7
	Fibronectin increased from E12.5 to E14.5, then decreased at 16.5[M]	Fibronectin increased from E16.5 to d2[M]		
	Fibronectin increased from E12.5 to E16.5[M]	*Fibronectin increased from E16.5 to d2*[M]		
	Periostin in endo and epi at E11 to E13.5; diminished presence at E16.5[M]		Periostin absent from adult[M]	56
GAGs	Hyaluronan in myo[M]			19

[a]Epi, epicardium; endo, endocardium; myo, myocardium; E, embryonic day; d, day. GAGs, glycosaminoglycan. Italicized font indicates RNA expression. Regular font indicates protein expression. [M], mouse; [R], rat; [H], human.

four phases. First, large numbers of cardiomyocytes are lost. Second, inflammation ensues and includes ECM degradation and the clearing of cell and ECM debris. Third, granulation tissue forms and contains activated fibroblasts, macrophages, new blood vessels and ECM proteins. The fourth and final phase includes the generation of an acellular, collagen-rich scar. A number of ECM proteins, in addition to collagen I, have been shown to be present and important in this healing process.

3.1.2.2 ECM and Myocardial Disease. Proteoglycans have been observed during cardiac healing. Expression of syndecan-1, a heparan sulfate proteoglycan, increases in the event of myocardial infarction with maximal expression after 7 days. In syndecan-1 null mice, myocardial infarction is followed by an increase in the inflammatory response and levels of pro-matrix metalloproteinase 2 and 9 compared to wild-type mice. Matrix metalloproteinases (MMPs) are a family of zinc-dependent peptidases that specifically degrade ECM proteins. These null mice also have a higher rate of cardiac dilation and increased collagen production but with impaired organization. Furthermore, the overexpression of syndecan-1 reduces cardiac inflammation, improves collagen organization, and reduces cardiac dilation.[23,24] These findings suggest that syndecan-1 plays a role in regulating infarct inflammation, healing, and prevents cardiac dilation. Syndecan-4 expression is also upregulated within mouse models of myocardial infarction.[24] In particular, syndecan-4 null mice exhibit increased rates of mortality, mostly due to cardiac rupture, and a decrease in cardiac function compared to wild-type mice. The presence of fibroblasts, myofibroblasts, new capillaries, macrophages, leukocytes, and mature collagen I decreases in syndecan-4 null mice after myocardial infarction. Fibroblasts of syndecan-4 null mice exhibit reduced migration and limited ability to generate fibronectin-induced stress fibers. More specifically, syndecan-4 null mice exhibited impaired activation of focal adhesion kinase (FAK), Akt, and small G protein RhoA; signaling associated with FAK activation is crucial for differentiation of myofibroblasts. Syndecan-4 null endothelial cells also have a lower response to activation by bFGF, especially responses corresponding to cell viability, proliferation, and bFGF-induced tube formation.[25] These data emphasize the importance of syndecan-4 in the formation of granulation tissues during cardiac healing, in regulating cardiac fibroblast migration and differentiation behavior, and in mediating angiogenesis in cardiac healing.

A number of small leucine-rich proteoglycans such as biglycan and decorin have been observed within the post-infarction heart. Expression of decorin was observed within the rat after myocardial infarction. mRNA levels indicate that the presence of decorin was initially low and then slowly increased and reached a maximum level 14 days post-infarct. Additionally, decorin was typically found in the border zone and not within the infarct zone.[26,27] Mice deficient in decorin had a larger amount of ventricular hypertrophy and dilation as well as decline in ventricular function compared to wild-type mice. Decorin deficiency did not alter the collagen concentration and crosslinking post-infarct. However, decorin null mice did have impaired collagen organization post-infarct, particularly a looser packing and less uniform collagen fiber size compared to wild-type mice.[28] Biglycan mRNA expression appeared earlier and reached a maximum 14 days post-infarct in the rat and 7 days post-infarct in the mouse. Unlike decorin, biglycan expression was seen within the infarct zone.[26,29] Mice deficient in biglycan had a higher incidence of death due to ventricular rupture and mice that survived had impaired hemodynamic function compared to wild-type mice. Expression of MMPs 2, 4, 9, 13, and tissue inhibitor of MMPS (TIMP) 1 was different between wild-type and biglycan null mice. Although the collagen concentration and crosslinking did not vary between wild-type and biglycan null mice, collagen diameter and collagen packing was different in biglycan null mice.[29,30] From these observations it seems that biglycan and decorin play important roles in cardiac remodeling, particularly in the organization of collagen within the infarct zone.

In the remodeling phase, matricellular protein expression is increased, and includes expression of tenascin-C, osteopontin, and thrombospondin. After a myocardial infarction, tenascin-C expression is increased in the border zone of the infarct and then subsequently moves into the infarct zone. Cardiomyocytes cultured in laminin and tenascin had a higher number of attached cells than those on laminin alone. Cardiac fibroblasts depleted of tenascin-C had lower levels of migration and expression of α-smooth muscle actin.[31-33] These observations indicate that tenascin-C may facilitate the migration of myoblasts into the infarct zone and attachment of cardiomyocytes at the border zone. Osteopontin expression is also seen in myocardial infarction, typically around the inflammatory cells.[24,34] Expression was seen within the infarct zone 3 days post-infarct and peaked after 7 days. Myocardial infarction in mice deficient in osteopontin expression had higher levels of left ventricular cardiac dilation and lower levels of collagen production.[35] Additionally, the introduction of MMP inhibitors into osteopontin-null mice improved ventricular function after myocardial infarction.[36] These observations suggest that osteopontin is associated with collagen deposition and degradation in the infarct healing process. In the rat, thrombospondin-1 mRNA expression is detected after 6 hours and then decreases but is still detectable 28 days after infarction. Thrombospondin expression is seen at the border zone and associated with vimentin-positive (fibroblast) and CD68-positive cells. Thrombospondin-1 protein expression is initially observed but becomes undetectable 7 days post-infarct. Additionally, human peripheral blood mononuclear cells cultured with thrombospondin-1 induced higher expression of interleukin-6 and monocyte chemoattractant protein-1.[37] Therefore, thrombospondin-1 may play a role in cardiac healing by modulating the inflammatory process at early time points, particularly at the infarct zone. It is important here to note the decided difference in thrombospondin mRNA *vs.* protein expression. mRNA expression offers a view of current cellular activity with respect to ECM production. Protein expression offers a view of total accumulation of ECM protein in a particular space (*e.g.*, the infarct zone) and reflects the sum of production and degradation over time.

Another matricellular protein, periostin, has been shown to play a role in the post-infarct healing process. Periostin consists of four domains: a signal sequence, an N-terminal domain that contains cysteine residues, a carboxyl terminus that can be alternatively spliced, and four coiled fascilin domains. Periostin expression is upregulated during myocardial infarction and is often associated with cardiac fibroblasts. Periostin null mice have a lower rate of survival after infarction compared to wild-type mice. However, null mice that survive the myocardial infarction have better cardiac function than wild-type mice. In these null mice, there is a reduction in fibrosis, scar size, and recruitment of inflammatory cells. Gene analysis of the cells within periostin null mice show changes in expression of genes involved in fibrosis, cell adhesion, ECM, and those associated with fibroblasts.[38] Periostin null mice also exhibit a decrease in collagen expression, perhaps affecting the mechanical properties of the heart.[38,39] Additionally, Kühn *et al.* demonstrated that the delivery of human recombinant periostin to rats undergoing myocardial infarction improves cardiac function. In particular, hearts receiving periostin have a smaller scar volume, decreased cell hypertrophy, increased cardiomyocyte proliferation, and increased capillary and arteriolar density.[40] Taken together, these results show that periostin may regulate the cardiac healing process by modulating the recruitment of inflammatory cells and the deposition of ECM proteins, resulting in a decrease in fibrosis and scar tissue.

Not unlike development of cardiac tissue, degradation of ECM contributes to cardiac healing following infarction. In the inflammatory phase, increased MMP activity is seen resulting in massive ECM protein degradation.[41,42] Heymans *et al.* observed a reduction in death due to ventricular wall rupture post-myocardial infarction in mice deficient in MMP-9.[43] MMP-9 deficient mice also had limited left ventricular dilatation, reduced inflammatory responsiveness, and decreased collagen deposition and organization.[44] The presence of MMP inhibitors and other proteases can also modulate the response of the healing heart. In a mouse model, the overexpression of tissue inhibitor of metalloprotease-1 (TIMP-1) in mice after infarction resulted in diminished leukocyte infiltration, lower levels of angiogenesis, larger necrotic areas, and lower collagen content, which resulted in a lower incidence of cardiac rupture.[43] The reduction of MMP activity within the infarct zone resulted in a lower inflammatory response, reduced angiogenesis, and decreased levels of collagen deposition and organization. Plasmins are serine proteases within the blood that can degrade certain proteins, such as fibrin, or be used to activate certain other enzymes and proteins. Plasmin activity is dependent on the level of plasminogen, plasminogen activators (urokinase and tissue type), and plasminogen activator inhibitors. Mice deficient in plasminogen or urokinase-type plasminogen activator have a lower incidence of ventricular wall rupture, decreased neutrophil migration to the infarct zone, reduced levels of collagen clearance, persistence of necrotic cardiomyocytes, and lower levels of revascularization during cardiac healing. Furthermore, mice treated with a plasminogen activator inhibitor also had a reduced incidence of ventricular wall rupture.[43,45] Decreased plasmin activity yielded a lower inflammatory response, diminished collagen and necrotic cardiomyocyte clearance, and reduced vascularization. Thus, diminished activity of proteinases reduced the incidence of cardiac rupture and dilation, suggesting a potential role for MMP and plasmin inhibitors in clinical therapies. Of note, many animals did not survive following treatment with protease inhibitors, suggesting that certain insults may require MMP production and associated scar formation for survival.

In sum, proteoglycans, matricellular proteins, and proteinases have been shown to be important for tissue healing following myocardial infarction. These proteins can modulate the clearance of cell and matrix debris, the inflammatory response, and the deposition and organization of the final collagen-rich scar. The possibility of avoiding scar formation is the holy grail of myocardial healing following infarction and some of the work noted above related to protease inhibition has afforded steps toward this goal. In addition, some studies have shown regeneration of mammalian myocardium without scar formation in early postnatal periods, suggesting that it may be possible to modify the healing response in mammals. Of note, reproduction of these studies has not yet been accomplished.[46]

3.2 ECM DYNAMICS OF CARDIAC VALVES

3.2.1 Cardiac Valve Development

3.2.1.1 Overview. Accumulation of ECM proteins in specific parts of the heart tube occurs in the first stages of valvulogenesis, essentially forming the valve cushions. As the cushion develops there is an elongation and remodeling phase that extends past embryonic development, which organizes both the cells and the ECM proteins into the proper location. The development of the atrioventricular (AV) cushion results in the mitral and tricuspid valves while the outflow tract (OFT) cushion generates the aortic and pulmonary valves. The mature valve contains interstitial cells, ECM, and endothelial cells, with the ECM organized into distinct layers. For example, the aortic valve is composed of the fibrosa layer, containing mainly collagen fibers, the spongiosa layer, containing loosely arranged proteoglycans, and the ventricularis layer, containing elastin fibers.[47] This exact organization may differ between different types of valves. The reorganization and interaction between the extracellular matrix and surrounding mesenchyme is vital for proper valvulogenesis.

3.2.1.2 ECM and Valvulogenesis. Many collagen proteins are expressed during valvulogenesis at different stages and location within the developing valve. Collagen I is initially expressed throughout the leaflet but is then restricted to the fibrosa layer of the murine mitral valve after birth.[48] Collagen III is present in the aortic valve fibrosa layer throughout the developmental and juvenile stages of valvulogenesis.[47] Collagen V is present in the annulus fibrosus, a fibrous ring structure connecting the valve to the heart. Collagen V and VI are seen at the anchoring points of the developing mitral valve at ED 18.5 but not in the adult mitral valve. Collagen VI is present throughout the mitral leaflets starting from ED 18.5 to adulthood.[48] Collagen XVIII is expressed throughout the valve during early development but becomes restricted to the basement membrane of the endothelial cell layer surrounding the valves.[49] Like the myocardium, the presence and organization of collagen proteins provides mechanical stability to the valve. Furthermore, collagen proteins within the valve are found in locations that require strength, such as the fibrosa layer located on the ventricular side.

The development of an elastic network is also observed within valvulogenesis. Elastin expression increased slowly during the development of the mouse aorta.[50,51] Additionally, fibrillins, a family of glycoproteins associated with elastic fibers, are also observed. Fibrillin 1 and 2 are seen within the endocardial cushion tissue of the mouse heart during ED 10.5, and fibrillin-3 has been observed in the endocardial cushions in human heart tissues.[8,9] In addition to a structural role, fibrillin also directly impacts cell behavior. In particular, fibrillin deficient mice show an increase in cell proliferation, decrease in cell apoptosis, increase in BMP 2, 4, and 6 expression, and an increase in TGF-β activity. Furthermore, Marfan's syndrome, characterized by mutations in fibrillin-1, result in defects in the aorta and heart valves.

These results suggest that fibrillin may modulate cell proliferation and apoptosis through the activity of TGF-β and BMP, which could play a role in the degeneration of the mitral valve associated with Marfan's syndrome.[52] The elastin network, much like the collagen network, is also organized to maximize its mechanical properties, particularly to easily relax following leaflet opening to form a tight seal and avoid regurgitation.

Much like the development of the myocardium, proteoglycans also play a role in valvulogenesis. Aggrecan is a proteoglycan that is crucial in cartilage and joint structures. In aortic valve development, aggrecan is expressed during the elongation, remodeling, and juvenile stages.[47] The proteoglycan versican, isoforms V0 and V1, is present in the cardiac jelly, AV and OFT cushions, and the leaflets of the AV valves.[19,48,53,54] An examination of whole versican and its degradation product showed different spatial expression during valvulogenesis. During early development, versican is detected within the core of the valve and is associated with mesenchymal cushion cells. Versican degradation products, however, are seen in the outer regions of the cushion adjacent to the endocardium and are associated with the densely packed, round cells.[53] Additionally, mice deficient in ADAMTS9, a versican protease, have enlarged aortic and mitral valves.[18] These studies have led to the hypothesis that versican and its byproduct have two distinct functions: intact versican may be important for the differentiation of cushion cells and versican degradation product may be vital for the proliferation of these cells.

The GAG hyaluronan is found within the cardiac jelly and cardiac cushions.[20] It is expressed throughout the mitral valve from ED 15.5 to 18.5 and is restricted to the atrial side 6 days postnatal. In the septal tricuspid leaflet GAGs are strongly expressed throughout the leaflet from ED 18.5 to 3 weeks after birth. Accumulation of hyaluronan is largely restricted to the atrialis layer and nodular thickening at 8 weeks.[48] Mice without hyaluronan synthase-2 (Has-2) lack endocardial cushions and die during mid-gestation. Further analysis of these cells indicates defects with cell migration and endothelial–mesenchymal transition in Has-2 null mice.[20] Therefore, hyaluronan may play an important role within valvulogenesis by modulating the migration and transformation of endocardial cushion cells.

Expression of periostin, a matricellular protein, is seen in the AV and OFT cushions and continues to be expressed with mitral, tricuspid, pulmonary, and aortic valve development.[55,56] Periostin null mice have severe problems in valvular morphogenesis and maturation and display symptoms such as shorter and thicker valve leaflet, disorganized collagen bundles, lack of matrix stratification, extensive calcification, and ectopic aggrecan expression. These mice have a high number of unidentified, presumably undifferentiated, cells in the heart during the neonatal stage. Furthermore, periostin null mice display a subpopulation of MF20/myosin heavy chain positive myocytes and α-smooth muscle actin positive cells within the cushion mesenchyme; a population of cells not seen in normal cushion development. And *in vitro*, addition of purified periostin results in reduced expression of myocardial markers as well as an increase in fibroblast markers, indicating that periostin may play a role in encouraging the differentiation of cardiac fibroblasts while preventing the differentiation of valvular progenitor cells into cardiomyocytes and smooth muscle cells. By examining the activity of Smad6 and Foxc1, genes correlated with TGF-β activity, it is also hypothesized that the activity of TGF-β can regulate the activation of periostin and modulate valve maturation.[39,55–57] Interestingly, the introduction of human recombinant periostin to rat cardiomyocytes, both *in vitro* and *in vivo*, induced proliferation.[40] Thus periostin may play two different roles: (1) promoting the differentiation of cushion cells into fibroblast while discouraging differentiation of other cell types; and (2) promoting cardiomyocyte proliferation within the myocardium.

Other matricellular proteins, such as tenascin and fibronectin, have also been observed during valvulogenesis. Tenascin is localized to the annulus at ED 10 and within the fibrosa and the ventricular endothelium at later stages of valvulogenesis.[47] Fibronectin expression, particularly its EIIIA and EIIIB splice variants, was observed in mouse ED 10.5 valve cushions. EIIIA/EIIIB null mice presented defects in cardiac cushion formation, including lack of cushion cell differentiation, suggesting that FN containing splice variant EIIIA and EIIIB may modulate the differentiation of cushion cells.[58]

In sum, a number of ECM proteins are present and remodeled during the development of heart valves, summarized in Table 3.2. Collagen and elastin proteins are remodeled and organized within the fibrosa and ventricularis layers, respectively. This particular organization provides the valve with proper mechanical properties, particularly strength and elasticity, at precise locations. Proteins in the proteoglycan and matricellular family are vital for the migration, differentiation, and proliferation of the valvular interstitial and valvular endothelial cells. Of note, the literature relating to the modulation of cell behavior as a function of ECM deposition or degradation is more expansive than that related to myocardial development.

Table 3.2   Extracellular matrix protein in developing cardiac valves.[a]

ECM family	Stage of development			Ref.
	Embryonic	Neo-natal/juvenile	Adult	
Collagen	Col I throughout mitral valve at E15.5 and E18.5; throughout tricuspid valve at E18.5[M]	Col I restricted to ventricular side of mitral valve at d6.5; throughout the tricuspid valve at 3 weeks[M]	Col I restricted to ventricular side[M]	48,97
	Col V throughout mitral leaflet at E18.5[M]	Col V restricted to ventricular side[M]	Col V restricted to ventricular side[M]	48,97
	Col VI in AV endocardial cushion at E11 and E13; localized to ventricular side past E13[M]	Col VI in mitral valve[M]	Col VI in mitral valve[M]	48,98
	Col XI in the AV valves at E18.5[M]			97
	Col XVIII throughout in AV valves from E13.5 to E17.5; restricted to basement membrane at E18.5[M]	Col XVIII localized to basement membrane of endothelial cells in AV valves[M]	Col XVIII localized to basement membrane of endothelial cells in AV valves[M]	49
Glycoprotein	Fibrillin-1, 2 and 3 in endocardial cushion[M]			8,9
	Fibulin-2 in semilunar and AV valves[M]			78
Matricellular	Periostin in AV cushions; localized to AV and semilunar valves		Periostin mostly in ventricular side of the AV valves[M]	56
GAGs	Hyaluronan in AV valves at E13.5, and E17[M]			9
Proteoglycan	Versican in AV valves from E10.5 to 18.5,	Versican mostly restricted to atrial side of the mitral by d6.5; diffused throughout tricuspid valve until 3 weeks[M]	Versican mostly restricted to atrial side of the AV valves[M]	9,48,54

[a]AV, atrioventricular; E, embryonic day; d, day. GAGs, glycosaminoglycan. Italicized font indicates RNA expression. Regular font indicates protein expression. [M], mouse; [R], rat; [H], human.

3.2.2   Cardiac Valve Disease

3.2.2.1 Overview.   Calcific aortic valve disease (CAVD) is the most prevalent valvular disease within the United States (AHA Stats 2013). The cause of this disease is currently unknown, but factors such as age and mechanical properties of the heart valve have been implicated. CAVD is characterized by calcification, changes in ECM organization and composition, neoangiogenesis, inflammation, and cell infiltration. In this section we will examine some of the changes in the ECM environment of the valve during CAVD and how it can affect cell behavior.

3.2.2.2 ECM and Valvular Disease.   In the normal aortic valve, collagen I and III are located mainly within the fibrosa layers and organized circumferentially. In CAVD, accumulation of collagens within the aortic valve is increased and deposition is disorganized relative to the health valve.[59-61] The basement membrane containing collagen IV is also disrupted and can appear thin and frayed or sometimes completely absent.[62] In some *in-vitro* experiments the presence of collagen I has been shown to maintain a quiescence state in valvular interstitial cells perhaps due to the interaction between collagen I and TGF-β.[63,64] The change in the collagen network within CAVD may play a role in changing both the mechanical properties of the valve and the cellular response of the valvular cells.

Elastin in the normal valve is mainly found in the ventricularis layer and is organized in radially aligned fibers. In CAVD, there is a loss of elastin fiber content and organization.[47] Mice deficient in cathepsin S, an elastase, show reduced levels of calcification within the aortic valve compared to cathepsin S wild-type mice. Additionally, human vascular smooth muscle cells cultured with elastin peptide have an increased level of alkaline phosphatase, suggesting an increase in calcification.[65] Fibroblast cells cultured in elastin fragments have a higher expression level of α-smooth muscle actin, indicating differentiation of quiescent fibroblast to active myofibroblast. These cells express alkaline phosphatase and produce calcium deposits, indicative of the formation of calcium nodules.[66] These observations suggest that CAVD is associated with elastin network degradation; network degradation and subsequent cell signaling associated with degradation products play a role in the progression of valvular calcification.

The expression of proteoglycans and glycosaminoglycans has been shown to change in CAVD. In the normal valve, proteoglycans and glycosaminoglycans are predominantly located within the spongiosa layer of the valve. In CAVD, proteoglycans and hyaluronan accumulation is increased. Additionally, versican and hyaluronan are found within tissues immediately surrounding large calcific nodules.[47,67]

The increase in and reorganization of proteoglycan and glycosamino-glycans may disrupt the collagen and elastin network within the valve and thereby change the mechanical properties of the tissue.

Expression of matricellular proteins and bone-related proteins also increase in CAVD. Tenascin-C expression in the normal valve is low and generally restricted to the basement membrane.[60] However, in CAVD its expression increases and is seen within the interstitium, especially around calcified areas.[60,68] The exact role of tenascin-C in CAVD is not entirely known but some reports suggest a role in remodeling of the extracellular environment *via* induction of MMP-2 mRNA expression in valvular interstitial cells cultures and colocalization with MMP-2 in disease valves.[68] Chondromodulin-I, a glycoprotein predominantly found in avascular tissues, is expressed throughout the normal valve except in the outer endothelial cell layer. In disease states, however, chondromodulin-I expression is decreased in regions of neoangio-genesis. Endothelial cells cultured with chondromodulin-I had higher levels of apoptosis and decreased levels of migration and tube forma-tion, indicating that this glycoprotein may play a role in regulating angiogenesis within the diseased valve.[69] Periostin is observed within normal valves of the mouse, rat, and human.[55,70] In CAVD, periostin expression is greatly reduced.[55] Knockout experiments and CAVD-induced mouse models have indicated dual activity for periostin, namely in the promotion of angiogenesis and repression of calcifica-tion, perhaps through interaction with Notch1 signaling or changes in MMP activity.[57,70] Bone-related proteins, such as osteocalcin and osteopontin, are upregulated in CAVD.[71,72] Thus loss of matricellular proteins with CAVD leads to disregulated tissue function, stimulation of inflammation, and expression of bone-related proteins.

In sum, disruption in the ECM environment is prevalent in calcific aortic valve disease. Disruption has been attributed to changes in the amount of ECM degradation (MMPs and TIMPs), disorganized depo-sition of new ECM and physical disruption of the ECM due to calcified nodules. These changes are associated with the activation of valvu-lar interstitial cells leading to increase ECM deposition, infiltration of inflammatory cells, and disregulation of angiogenesis within the valve.[60,62,68,70,73–75]

3.3 ECM DYNAMICS OF THE VASCULATURE

3.3.1 Vasculature Development

3.3.1.1 Overview. Vasculogenesis requires the careful orchestra-tion of ECM degradation, deposition, and organization to generate the defined vascular network within the heart. Normal adult vessels contain endothelial cells, smooth muscle cells, pericytes, and extra-cellular matrix proteins organized into concentric layers. The intima is mainly composed of endothelial cells with some proteoglycans and hyaluronan. The media is separated from the intima by an elastic membrane and is composed of smooth muscle cells, elastic fibers, collagen, and proteoglycans. The adventitia is also separated from the media by an elastic membrane and is composed of fibroblasts and collagen.[76] The vasculature network within the heart is composed of the aorta and pulmonary arteries along with the coronary network. In the section below, we will focus on the ECM proteins during the development of the aorta.

3.3.1.2 ECM and Vasculogenesis. Like many other tissues, a number of collagen proteins are present in the development of the cardiac vasculature. mRNA expression analysis of the mouse aorta showed that collagen I, III, V and VI were the most abundant fibrillar collagens. Overall, the expression of these proteins dramatically increases beginning at ED 14, with high expression until 10 days post-birth. After this time, expression rapidly decreases and continues to fall during adult life.[50] In the media, collagen I is distributed around vascular smooth muscle cells but is absent in regions close to the elastic laminae. In the media, collagen III is localized near but not within the elastic laminae, while collagen V was seen within the entire medial layer.[77] The presence of fibrillar collagen within the mouse aorta is generally located in the medial layer of the artery and provides the mechanical strength required for this major artery.

The elasticity of the vasculature is vital to its function, such that it can recoil following distension. Therefore a number of proteins that endow elasticity are seen during vasculogenesis. Similar to collagen I, elastin mRNA expression within the mouse aorta increases from embryonic stages to 14 days post-birth, after which time expression decreases. mRNA expression of fibrillin-1 in the mouse developing aorta increases during embryonic development, with expression peaking at birth. This is followed by the decline of expression into adult life. Fibrillin-2 mRNA, however, is different with the highest expression at early embryonic stages and decreases throughout development.[50] In human cardiac development, fibrillin-3 has been observed throughout the vessels. Interestingly, fibrillin-3 expression was higher in locations without elastic fibers.[9] Fibulin is a calcium-binding ECM protein that can associate with fibronectin, proteoglycans, laminin, and elastic fibers. As the aorta develops, fibulin-2 expression is seen within mesenchymal cell components that eventually become vascular smooth muscle cells. It is also seen pericellular to proliferating smooth muscle progenitor cells of the media layer. At 13 days post-birth, fibulin-2 mRNA is present in the basement membrane, the elastic laminae, and the adventitia layer, and is associated with smooth muscle cells.[78] Elastin expression is maintained through the juvenile stages of mouse aorta development, while associated proteins, such as fibrillin 1 and 2, are generally seen during developmental stages. This may indicate that fibrillin proteins help generation of a native embryonic elastin network that matures during post-birth within the mouse aorta.

The development of the basement membrane has also been examined in the developing aorta. Laminin-1 appears early in development, while laminin-8 is typically seen at late embryonic stages and into fetal stages. In the juvenile period, laminin-8 and 9 contribute to the basement membrane.[50,51] Expression of collagen IV, a collagen subtype typically associated with laminin in the basement membrane, is constant during vasculogenesis in the mouse and is present in the medial layer of human tissues surrounding smooth muscle cells.[51,79] Entactin, or nidogen, is a glycoprotein that connects the laminin and collagen IV network within the basement membrane. In the developing mouse aorta, entactin-1 dramatically increases at ED 18 and begins to drop 7 days post-birth, with low expression levels throughout adulthood. Entactin-2 shows an increase at ED 14 and then decreases until day 21 post-birth, when it reaches the low expression level seen in the adult life.[51] Osteonectin, also known as secreted protein acidic and rich in cysteine (SPARC) or basement membrane protein-40, can bind to many ECM proteins such as collagen I–V and VIII, PDGF, and VEGF. In the developing mouse aorta its expression increases from embryogenesis to postnatal life, peaking at about 1 week after birth.[50] The main components of the basement membrane, laminin and collagen IV, are present throughout development and maturation of the aorta. Other proteins, such as entactin and osteonectin, are transiently expressed during embryonic development and early juvenile stages. This temporal distribution suggests that entactin and osteonectin aid the maturation of the basement membrane but do not contribute substantially to the mature aorta.

The presence of matricellular proteins tenascin, fibronectin, and thrombospondin has also been observed. In the mouse aorta, tenascin-X mRNA expression gradually increases from ED 16 until postnatal day 10. At this point it declines slowly and then increases again in adult life. Tenascin-C expression increases beginning at ED 14 and then gradually decreases starting at postnatal day 17. Fibronectin mRNA expression is relatively high and remains constant through development of the mouse aorta.[50] Fibronectin and its EIIIA and EIIIB variants are present in the dorsal aorta of the mouse at embryonic day 10.5. EIIIA/EIIIB null mice have a number of cardiovascular defects such as hemorrhage, thinned OFT, and anemia. Examination of vascular smooth muscle cells within EIIIA/EIIIB null mice indicated a delayed presence of vascular smooth muscle cells near the dorsal aortae. Additionally, vascular smooth muscle cells near the endothelial lining of the vessel of EIIIA/EIIIB null mice were rounded. These observations suggest that fibronectin, specifically the EIIIA and EIIIB splice variants, play a role in vascular smooth muscle cell migration and attachment.[58] Thrombospondins are typically localized to the vessel wall. *In situ* hybridization showed that only TSP-2 is present in the large vessels in the developing murine embryo, with expression in the dorsal aorta at ED 11[80]. Analysis of mRNA expression shows that TSP-2 levels rise through embryogenesis, remaining high during the first week, and then dropping afterwards. Gene array analysis also shows modest embryonic expression of TSP-3 in the mouse aorta with a temporary dip at birth and minor increase until 6 months post-birth.[50] Rat aorta cultured in three-dimensional fibrin/collagen gels with thrombospodin-1 demonstrated an increase in the number, length, and branching of microvessel outgrowth compared to control cultures. Interestingly, the angiogenic potential of thrombospondin is thought to act on fibroblasts and not endothelial cells.[81] Therefore, the expression of thrombospondin-1 during embryonic development may indicate periods of angiogenesis.

Immuno-stained tissues have shown the presence of versican in the medial layer of the fetal human aorta.[82] In the developing mouse aorta, versican mRNA expression decreases slightly from ED 12 to birth, rises sharply at birth, then falls dramatically by day 4 post-birth and is then maintained at low levels.[50] In mice with deficient levels of ADAMTS9, a versican protease, there is an increase in adventital thickness, disorganization of the aortic elastic lamellae, and an increase in cell infiltration.[18] Versican is also seen during the development of the OFT. In particular, intact versican protein is seen during mouse ED 9.5 and 10.5 correlating well with the migration of α-sarcomeric actin positive cells in the myocardium. During this period, only faint detection of the versican degradation product was seen in the distal OFT, aortic sac, and endocardium. Expression of whole versican decreased during ED 11.5 while expression of its degradation products increased and localized within areas of α-smooth muscle actin positive cells. At ED 12.5 there was no versican detected within the aorta while degradation products were associated with the medial layer. Concurrently, there was a decrease in α-sarcomeric actin positive cells and an increase in the α-smooth muscle actin positive cells within the aorta. This observation has led to the hypothesis that cleaved versican byproducts promote remodeling of the OFT by disaggregation of cardiomyocytes and thinning the OFT.[83]

Small leucine rich proteoglycan (SLRPs) have been observed in the cardiac vasculature. This family of proteins includes decorin, biglycan, osteoglycin, and lumican. mRNA expression profile of decorin in the mouse aorta shows increased levels of expression from ED 14 to birth, with decreased expression post-birth, but remains at a moderate level.[50] The mRNA expression profile of biglycan shows increased levels of expression during embryonic stages, with a peak 1 week after birth. Expression levels decrease over the first month but then increase as the animal enters adulthood.[50] Overexpression of biglycan in mice results in a higher number of proliferating cells within the aorta. Further *in-vitro* studies show an increase in proliferating vascular smooth muscle cells and a decrease in proliferating endothelial cells when biglycan is present. The migration of smooth muscle cells is also increased in the presence of biglycan.[84] These studies suggest that biglycan plays a role in modulating the proliferation and migration of vascular smooth muscle cells during vasculogenesis.

Another large component of the cardiac vasculature is the coronary system. Unfortunately, there are few data regarding the interaction between cells and the extracellular matrix in the development of this vital system and thus a potentially fruitful area of study given the high propensity for disease in these vessels, as described below.

Within vasculogenesis the temporal distribution of the ECM proteins seem to fall into two categories: (1) sustained expression and (2) dynamic expression. The distribution of these proteins is summarized in Table 3.3. ECM proteins such as fibronectin, laminin, and collagen IV are seen early in embryonic tissue and are present within the adult tissues. Perhaps this persistent expression is due to the role these proteins play in cell attachment and organization of the vasculature. The proteins of the collagen and elastin network are expressed dynamically, with collagen and elastin protein expression moving into the juvenile stage, while "helper" proteins are mainly expressed during embryonic stages. The presence of helper proteins may be vital in the initial organization of these two networks. The continued expression of collagen and elastin past the embryonic stage may indicate further maturation and organization of the network due to the change in mechanical requirements of the vasculature after birth.

Table 3.3 Extracellular matrix proteins in the developing cardiac vasculature.[a]

ECM family	Stage of development			Ref.
	Embryonic	Neo-natal/juvenile	Adult	
Collagen	Col I increased in aorta[M]	Col I decreased after d7/10 in aorta[M] Col I in adventitia of the aorta[H]	Col I low within aorta[M] Col I in adventitia of coronary arteries[R]; in media of aorta[H]	5,50,77,51,82
	Col III increased in aorta[M]	Col III increased through d7/10 then decreased within aorta[M]	Col III low within aorta[M] Col III in media of coronary arteries[R], in media of aorta[H]	5,50,77,51
	Col IV in aorta[M]	Col IV in aorta[M]	Col IV in aorta[M]	51
			Col V in media layer of aorta[H]	77
	Col VI increased in aorta[M]	Col VI increased through d7/10 then decreased into adulthood within aorta[M]	Col VI low within aorta[M] Col VI in coronary system[R]	5,50,51
Elastic	Elastin increased slowly in aorta[M]	Elastin increased until d14 in aorta[M] Elastin in vessels walls at d2[M]	Elastin at low levels in aorta[M]	7,50,51
Glycoprotein	Fibrillin-1 increased in aorta; in pulmonary artery at E13.5[M]	Fibrillin-1 decreased in aorta[M]	Fibrillin-1 low in aorta[M]	8,50,51
	Fibrillin-2 decreased in aorta; in pulmonary artery at E13.5[M]	Fibrillin-2 decreased in aorta[M]	Fibrillin-2 low in aorta[M]	8,51
	Fibulin-2 in aorta and pulmonary vessels[M]	Fibulin-2 in media and adventia of aorta[M]	Fibulin-2 in arteries and veins but not capillaries of coronary system[M]	78
	Fibulin-2 in coronary vessels[M]			9
	Fibrillin-3 in blood vessel walls[H]			
	Fibulin-5 in aorta and coronary arteries[M]		Fibulin-5 low in aorta[M]	99
	Laminin 8 in late stages in aorta[M]	Laminin 8, 9, and 10 in aorta[M]	Laminin 8, 9, and 10 in aorta[M]	51
	Entactin-1 increased at E18 in aorta[M]	Entactin-1 decreased after d7 in aorta[M]	Entactin-1 at lower levels in aorta[M]	51
	Entactin-2 decreased after E14 in aorta[M]	Entactin-2 decreased in aorta[M]	Entactin-2 at lower levels in aorta[M]	51
	Osteonectin increased in the aorta[M]	Osteonectin peaked at d14 in aorta[M]	Osteonectin at lower levels in aorta[M]	50

Matricellular	Thrombospondin -1 increased from E16 in aorta[M]	Thrombospondin-1 decreased to low levels in aorta[M]		50
	Thrombospondin-2 increased in aorta[M]	Thrombospondin-2 decreased after d7 in aorta[M]		50
	Thrombospondin-3 at moderate levels in aorta[M]	Thrombospondin-3 increased until 6 months in aorta[M]		50
	Tenascin-X at low levels until E16 in aorta[M]	Tenascin-X declined at E10.5 in aorta[M]	Tenascin-X increased in aorta[M]	50
	Tenascin-C increased at E14 in aorta[M]	Tenascin-C decreased at d7 in aorta[M]	Tenascin-C increased at 6m in aorta[M]	50
	Fibronectin at high levels aorta[M]	Fibronectin at high levels aorta[M]	Fibronectin at high levels aorta[M]	5,50,51
			Fibronectin in coronary capillaries[R]	
Proteoglycan	Versican decreased from E12 in aorta[M]	Versican decreased until d4 in aorta[M]	Versican at lower levels in aorta[M]	50,51,82,83
	Versican in outflow tract at E9.5–10.5[M]	Versican in media layer of aorta[H]		
	Decorin increased in aorta[M]	Decorin in aorta[M]	Decorin in aorta[M]	50,51
	Biglycan increased in aorta[M]	Biglycan peaked at d7 in aorta[M]	Biglycan increased in aorta[M]	50,51
	Lumican at low to moderate levels in aorta[M]	Lumican at low to moderate levels in aorta[M]	Lumican in adventia of coronary artery[H]	51,93
	Perlecan in aorta[M]	Perlecan in aorta[M]	Perlecan in the aorta[M]	51

[a]E, embryonic day; d, day. GAGs, glycosaminoglycan. Italicized font indicates RNA expression. Regular font indicates protein expression. [M], mouse; [R], rat; [H], human.

3.3.2   Vascular Disease

3.3.2.1   Overview.   The most pervasive disease state of the vessels of the heart and surrounding vasculature is atherosclerosis. In atherosclerosis, plaques are formed within the vessels, specifically in the intima layer, and can cause blockage of blood flow. Overall, plaque development begins with thickening of the intimal layer, which continues to build until a mature plaque develops. These plaques can be categorized as stable, typically asymptomatic, or unstable, which can rupture and lead to further complications such as stroke or myocardial infarctions. The ECM proteins within the plaque and its surrounding area play a role in the development and progression of the disease.

3.3.2.2   ECM and Vascular Disease.   The presence of the collagens has been observed in the progression of atherosclerosis. Diffuse intimal thickening contains low levels of collagen I and III and an elastin network. However, lipid accumulation deep within this thickening is associated with an increase of collagen I and III and elastic fibers. Immuno-histological analysis showed that plaques were typically devoid of cells and elastic tissue while being abundant in collagen proteins. Collagen I is seen closer to the intimal cells while collagen III is evenly distributed throughout the plaque. Additionally, collagen I expression was not found to localize with T lymphocytes. Collagen I and III have also been observed within the fibrous cap of certain plaque types. Collagen IV and V have also been observed within the plaque.[77,85–87] The lack of collagen I expression with T lymphocytes may be due to the expression of IFN-γ by T lymphocytes, which inhibits collagen synthesis by vascular smooth muscle cells.[88] The progression of atherosclerosis from diffuse intimal thickening to plaque coincides with the increase in expression of collagen proteins, especially collagens I and III, suggesting that collagen production, at least in part, enables the growth of atherosclerotic plaques.

The importance of elastin in atherosclerosis has also been examined. Mice deficient in cathepsin S, an elastase, have a lower level of calcification within the mouse aorta. Additionally, treatment of human vascular smooth muscles with elastin fragments induces an increase in alkaline phosphatase activity, indicating the activation of calcification processes.[65] Mice deficient in cathepsin S and low-density lipoprotein (LDL) receptor had a reduction in plaque size compared to LDL receptor deficient mice. Additionally, these double deficient mice had fewer smooth muscle cells within lesions, reduced collagen content in the aortic arches, a thinner fibrous cap, and a lower amount of monocyte/macrophages within the plaques. Peripheral blood cells from double null mice had a reduced level of infiltration through an endothelial cell layer.[89] Together, this indicates that an intact elastin network reduces cell infiltration, collagen deposition, and calcification of the vessels.

The presence of proteoglycans has been observed in the development of atherosclerosis. Interestingly, the level of proteoglycans within normal and coronary disease tissues is shown to be roughly the same. However, the properties of the proteoglycans with coronary disease are different. In particular, proteoglycans from disease tissues have higher affinity for LDLs, have a higher molecular size, and proteoglycan–LDL complexes from disease tissue induce a higher level of cholesteryl ester synthesis in macrophages.[90] This change in proteoglycan properties could result in the accumulation of extracellular lipids and the generation of foam cells within the plaque, indicating further progression of atherosclerosis.

Other ECM protein changes in atherosclerosis include changes in osteopontin, tenascin, thrombospondin, decorin, lumican, and fibronectin matricellular proteins. Within human atherosclerotic plaques, osteopontin was seen only in the intimal layer and associated with both cellular (smooth muscle cells and macrophages) and acellular regions of the plaque. Staining was seen surrounding the calcified deposits and the cholesterol clefts.[91] Thrombospondin-1 within atherosclerosis plaques was variable, with high intensity at hypocellular areas.[92] Decorin was seen in advanced plaques, associated with macrophages.[85] In the early stages of atherosclerosis, lumican is localized with the vascular smooth muscle cells in the media and intima layers. Staining of fibrolipid lesions showed lumican closely associated with foamy macrophages. In advanced stages of fibrous plaques, lumican is seen in thickened intimia near the calcifying plaques.[93] Interestingly, plaques containing collagen IV and V is associated with high fibronectin expression. Fibronectin has also been observed in regions of necrosis and where collagenous structures have been destroyed.[87] Tenascin-C was seen throughout lipid-rich plaques while fibrous plaques were devoid of tenascin-C. Within these lipid-rich plaques, tenascin-C expression correlated well with the presence of macrophages around the lipid core, at the plaque shoulders, and also in the fibrous cap. Plaques that had rupture also contained tenascin-C staining around the ruptured area. Additionally, macrophages cultured on tenascin-C had increased expression of pro-MMP9, indicating the importance of tenascin-C in the stability of the plaque.[94]

Atherosclerosis develops in numerous stages, from diffuse intimal thickening to fibrolipid lesions and advanced plaque, which can rupture causing further complications. The presence of ECM protein within the plaque varies spatially and perhaps also with the progression of the disease. Wanting in this area of research are studies specifically designed to delineate which cell types are responsible for ECM secretion and how cell behaviors are directly modified by local ECM.

3.4 CONCLUSION

ECM content, composition and distribution change dramatically with development and in disease states. Here we take the case of the tissues of the heart and describe the spatial and temporal dynamics of known ECM proteins, especially as they relate to cell behavior. Several ideas emerge from this compilation of years of study. In addition, common trends can be observed between these three tissues (*i.e.*, myocardium, valve, and vessel) and these can likely be extended to other tissue types. First, ECM proteins often work in concert to augment function. Take, for example, the case of type III collagen augmenting the function of type I collagen in the myocardium by ensuring proper fibril formation and thereby conferring mechanical strength to the tissue. Second, the role of a particular ECM protein can change dramatically depending on location in a tissue and/or whether the protein is intact, partially degraded, or completely degraded. Take, for example, versican expression in the valve. Versican protein is located within the valve cushion and is thought to play a role in cushion cell differentiation. Meanwhile, versican degradation products are localized to the outer region of the valve and are thought to play a role in the proliferation of the cushion cells. Third, and perhaps most importantly, there is a decided lack of information directly linking ECM content, composition, or distribution, with cell behavior, especially for studies of development. A summary of the cell behavior described in this section can be found in Table 3.4. In addition, to obtain a quantitative picture, we searched PubMed for articles associated with the following search terms: extracellular matrix and myocardium, extracellular matrix and heart valve, extracellular matrix and aorta. We placed each paper into categories according to their depth of study: structural or descriptive analysis, analysis of cell behavior associated with ECM and studies to uncover molecular mechanisms associated with cell–ECM interactions. Results according to year are shown in Figure 3.1. It is clear for all tissue types that structural analyses outweigh the others and in most cases reached peak values several years ago. Studies of cell behavior lag behind these, and only recently has there been a surge in studies of molecular mechanism. This may be due

Table 3.4 Summary of the impact of ECM proteins on behaviors of cardiac cell types.[a]

	Myocardium		Valve		Vasculature	
	Development	Disease	Development	Disease	Development	Disease
Migration		*Syndecan-4*: deficiency reduces migration of fibroblast *Tenascin-C*: deficiency reduces migration of fibroblast *Periostin*: deficiency reduces migration of inflammatory cells *TIMP*: overexpression reduces leukocyte infiltration *Plasminogen activator*: deficiency reduces neutrophil migration	*Hyaluronan*: deficiency inhibits cushion cell migration		*Fibronectin*: deficiency reduces migration of SMCs *Biglycan*: increases migration of SMCs	*Chondromodulin-1*: decreases migration of endothelial cells
Proliferation	*Versican*: increases in the number of cardiomyocytes	*Periostin*: increases cardiomyocyte proliferation	*Fibrillin*: deficiency increases proliferation *Versican degradation products*: increases proliferation of cushion cells		*Biglycan*: increases proliferation of SMC but decreases proliferation of ECs	

(continued)

Table 3.4 (*continued*)

	Myocardium		Valve		Vasculature	
	Development	Disease	Development	Disease	Development	Disease
Differentiation/activation		*Syndecan-4*: deficiency reduces activation of fibroblast and bFGF activation of ECs *Tenascin-C*: deficiency reduces activation of fibroblast	*Versican*: increases differentiation of cushion cells *Hyaluronan*: deficiency results in defects in EMT of cushion cells *Periostin*: deficiency reduces the number of differentiated cells and generates ectopic differentiation to myosin and α-smooth muscle actin (+) cells *Fibronectin*: deficiency reduces differentiation of cushion cells	*Elastin fragments*: increase activation of VICs *Collagen I*: maintains quiescent state of valve interstitial cells		*Elastin fragments*: induce differentiation of MSCs to osteoblasts

Other	Hyaluronan: deficiency results in reduced trabeculation	Syndecan-1: reduced inflammatory response	Chondromodulin-1: decrease apoptosis and tube formation in endothelial cells	Thrombospondin: increases the number, length, and branching of microvessels	Elastin fragments: deficiency reduces inflammation in the plaque
	Tenascin-C: increases cardiomyocyte attachment		Periostin: deficiency promotes angiogenesis	Versican degradation product: decreases CMs and increases SMC numbers	
	MMP-9: deficiency reduces inflammatory response				
	TIMP: overexpression reduces angiogenesis				

[a]Summary of the interactions between ECM proteins and cardiac cell types during development and disease states. A variety of ECM proteins can affect cell migration, proliferation, differentiation, and various other cell behaviors during both disease and developmental stages. The role of a particular ECM protein can change depending on location in a tissue and/or whether the protein is intact or degraded. It is important to note the lack of information directly linking how ECM proteins guide mammalian cell behavior, especially with development. bFGF, basic fibroblast growth factor; CM, cardiomyocyte; EC, endothelial cell; MMP, matrix metalloproteinase; MSC, mesenchymal stem cell; SMC, smooth muscle cell; TIMP, tissue inhibitor of metalloproteinase; VIC, valvular interstitial cell.

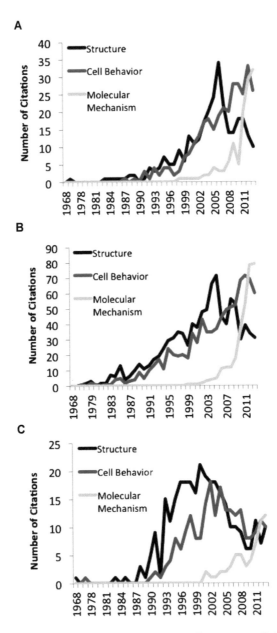

Figure 3.1 Trends in research toward understanding ECM–cell interactions. The literature was scanned for articles associated with: (A) ECM and myocardium; (B) ECM and heart valve; (C) ECM and aorta. These articles were subsequently divided into three categories: (1) articles examining the structure or description of the ECM environment; (2) articles relating the presence of ECM proteins to cell behavior; and (3) articles probing the molecular mechanisms guiding cell behavior. Overall, research of the ECM environment has been done for an extended period of time with cell behavior studies lagging slightly behind. Research examining the molecular mechanisms driving these cell behaviors, however, is a relatively new field of study, with a number of studies beginning in the early 2000s.

to technical limitations of such studies. Future advances will require advanced imaging and better *in-vitro* three-dimensional model systems to replicate development and disease.[95] This will require the ability to generate or replicate the spatial and temporal dynamics of the ECM during development and disease. One such technology is the use of multiphoton excited photochemistry to generate spatially defined three-dimensional environments composed of whole ECM proteins.[96] Additional technologies that can be used to replicate the temporal expression of the ECM under various biologic scenarios will be described in Chapter 9 by Kloxin.

REFERENCES

1. J. D. Lajiness and S. J. Conway, Origin, development, and differentiation of cardiac fibroblasts, *J. Mol. Cell. Cardiol.*, 2013.
2. E. M. Zeisberg and R. Kalluri, Origins of cardiac fibroblasts, *Circ. Res.*, 2010, **107**(11), 1304–1312.
3. T. Mikawa and R. G. Gourdie, Pericardial mesoderm generates a population of coronary smooth muscle cells migrating into the heart along with ingrowth of the epicardial organ, *Dev. Biol.*, 1996, **174**(2), 221–232.
4. B. A. Roeder, K. Kokini, J. E. Sturgis, J. P. Robinson and S. L. Voytik-Harbin, Tensile mechanical properties of three-dimensional type I collagen extracellular matrices with varied microstructure, *J. Biomech. Eng.*, 2002, **124**(2), 214–222.
5. R. I. Bashey, A. Martinez-Hernandez and S. A. Jimenez, Isolation, characterization, and localization of cardiac collagen type VI. Associations with other extracellular matrix components, *Circ. Res.*, 1992, **70**(5), 1006–1017.
6. X. Liu, H. Wu, M. Byrne, S. Krane and R. Jaenisch, Type III collagen is crucial for collagen I fibrillogenesis and for normal cardiovascular development, *Proc. Natl. Acad. Sci. U. S. A.*, 1997, **94**, 1852–1856.
7. K. Hanson, J. Jung, Q. Tran, *et al.*, Spatial and temporal analysis of extracellular matrix proteins in the developing murine heart: a blueprint for regeneration, *Tissue Eng., Part A*, 2013, **19**(9–10), 1132–1143.
8. H. Zhang, W. Hu and F. Ramirez, Developmental expression of fibrillin genes suggests heterogeneity of extracellular microfibrils, *J. Cell Biol.*, 1995, **129**(4), 1165–1176.
9. L. Sabatier, N. Miosge, D. Hubmacher, *et al.*, Fibrillin-3 expression in human development, *Matrix Biol.*, 2011, **30**(1), 43–52.

10. J. G. Jacot, A. D. McCulloch and J. H. Omens, Substrate stiffness affects the functional maturation of neonatal rat ventricular myocytes, *Biophys. J.*, 2008, **95**(7), 3479–3487.
11. K. S. O'Shea and V. M. Dixit, Unique distribution of the extracellular matrix component thrombospondin in the developing mouse embryo, *J. Cell Biol.*, 1988, **107**, 2737–2748.
12. J. Roman and J. A. McDonald, Expression of fibronectin, the integrin a-5, and a-smooth muscle actin in heart and lung development, *Am. J. Respir. Cell Mol. Biol.*, 1992, **6**(5), 472–480.
13. F. Farhadian, F. Contard, A. Corbier, *et al.*, Fibronectin expression during physiological and pathological cardiac growth, *J. Mol. Cell. Cardiol.*, 1995, **27**(4), 981–990.
14. J. T. Smith, J. K. Tomfohr, M. C. Wells, *et al.*, Measurement of cell migration on surface-bound fibronectin gradients, *Langmuir*, 2004, **20**(19), 8279–8286.
15. J. Sottile, F. Shi, I. Rublyevska, *et al.*, Fibronectin-dependent collagen I deposition modulates the cell response to fibronectin, *Am. J. Physiol.: Cell Physiol.*, 2007, **293**(6), 1934–1946.
16. M. A. Cooley, V. M. Fresco, M. E. Dorlon, *et al.*, Fibulin-1 is required during cardiac ventricular morphogenesis for versican cleavage, suppression of ErbB2 and Erk1/2 activation, and to attenuate trabecular cardiomyocyte proliferation, *Dev. Dyn.*, 2012, **241**(2), 303–314.
17. K. Stankunas, C. T. Hang, Z.-Y. Tsun, *et al.*, Endocardial Brg1 represses ADAMTS1 to maintain the microenvironment for myocardial morphogenesis, *Dev. Cell*, 2008, **14**(2), 298–311.
18. C. B. Kern, A. Wessels, J. Mcgarity, *et al.*, Reduced versican cleavage due to ADAMTS9 haploinsufficiency is associated with cardiac and aortic anomalies, *Matrix Biol.*, 2011, **29**(4), 304–316.
19. M. Lockhart, E. Wirrig, A. Phelps and A. Wessels, Extracellular matrix and heart development, *Birth Defects Res., Part A*, 2011, **91**(6), 535–550.
20. T. D. Camenisch, A. P. Spicer, T. Brehm-Gibson, *et al.*, Disruption of hyaluronan synthase-2 abrogates normal cardiac morphogenesis and hyaluronan-mediated transformation of epithelium to mesenchyme, *J. Clin. Invest.*, 2000, **106**(3), 349–360.
21. H. S. Baldwin, T. R. Lloyd and M. Solursh, Hyaluronate degradation affects ventricular function of the early postlooped embryonic rat heart in situ, *Circ. Res.*, 1994, **74**(2), 244–252.
22. R. Nair, A. V. Ngangan, M. L. Kemp and T. C. McDevitt, Gene expression signatures of extracellular matrix and growth factors during embryonic stem cell differentiation, *PLoS One*, 2012, 7(10), e42580.

23. D. Vanhoutte, M. W. M. Schellings, M. Götte, *et al.*, Increased expression of syndecan-1 protects against cardiac dilatation and dysfunction after myocardial infarction, *Circulation*, 2007, **115**(4), 475–482.
24. A. V. Finsen, P. R. Woldbaek, J. Li, *et al.*, Increased syndecan expression following myocardial infarction indicates a role in cardiac remodeling, *Physiol. Genomics*, 2004, **16**(3), 301–308.
25. Y. Matsui, M. Ikesue, K. Danzaki, *et al.*, Syndecan-4 prevents cardiac rupture and dysfunction after myocardial infarction, *Circ. Res.*, 2011, **108**(11), 1328–1339.
26. M. Doi, S. Kusachi, T. Murakami, *et al.*, Time-dependent changes of decorin in the infarct zone after experimentally induced myocardial infarction in rats: comparison with biglycan, *Pathol., Res. Pract.*, 2000, **196**(1), 23–33.
27. J. Hao, H. Ju, S. Zhao, *et al.*, Elevation of expression of Smads 2, 3, and 4, decorin and TGF-beta in the chronic phase of myocardial infarct scar healing, *J. Mol. Cell. Cardiol.*, 1999, **31**(3), 667–678.
28. S. M. Weis, S. D. Zimmerman, M. Shah, *et al.*, A role for decorin in the remodeling of myocardial infarction, *Matrix Biol.*, 2005, **24**(4), 313–324.
29. D. Westermann, J. Mersmann, A. Melchior, *et al.*, Biglycan is required for adaptive remodeling after myocardial infarction, *Circulation*, 2008, **117**(10), 1269–1276.
30. P. H. Campbell, D. L. Hunt, Y. Jones, *et al.*, Effects of biglycan deficiency on myocardial infarct structure and mechanis, *Mol. Cell. Biomech.*, 2008, **5**(1), 27–35.
31. I. E. M. G. Willems, J.-W. Arends and M. J. A. P. Daemen, Tenascin and fibronectin expression in healing human myocardial scars, *J. Pathol.*, 1996, **179**(3), 321–325.
32. M. Tamaoki, K. Imanaka-Yoshida, K. Yokoyama, *et al.*, Tenascin-C regulates recruitment of myofibroblasts during tissue repair after myocardial injury, *Am. J. Pathol.*, 2005, **167**(1), 71–80.
33. K. Imanaka-Yoshida, M. Hiroe, T. Nishikawa, *et al.*, Tenascin-C modulates adhesion of cardiomyocytes to extracellular matrix during tissue remodeling after myocardial infarction, *Lab. Invest.*, 2001, **81**(7), 1015–1024.
34. C. E. Murry, C. M. Giachelli, S. M. Schwartz and R. Vracko, Macrophages express osteopontin during repair of myocardial necrosis, *Am. J. Pathol.*, 1994, **145**(6), 1450–1462.
35. N. A. Trueblood, Z. Xie, C. Communal, *et al.*, Exaggerated left ventricular dilation and reduced collagen deposition after myocardial infarction in mice lacking osteopontin, *Circ. Res.*, 2001, **88**(10), 1080–1087.

36. P. Krishnamurthy, J. T. Peterson, V. Subramanian, M. Singh and K. Singh, Inhibition of matrix metalloproteinases improves left ventricular function in mice lacking osteopontin after myocardial infarction, *Mol. Cell. Biomech.*, 2009, **322**(1–2), 423–439.

37. S. Sezaki, S. Hirohata, A. Iwabu, *et al.*, Thrombospondin-1 is induced in rat myocardial infarction and its induction is accelerated by ischemia/reperfusion, *Exp. Biol. Med.*, 2005, **230**, 621–630.

38. T. Oka, J. Xu, R. A. Kaiser, *et al.*, Genetic manipulation of periostin expression reveals a role in cardiac hypertrophy and ventricular remodeling, *Circ. Res.*, 2007, **101**(3), 313–321.

39. R. A. Norris, T. K. Borg, J. T. Butcher, *et al.*, Neonatal and adult cardiovascular pathophysiological remodeling and repair: developmental role of periostin, *Ann. N. Y. Acad. Sci.*, 2008, **1123**, 30–40.

40. B. Kühn, F. del Monte, R. J. Hajjar, *et al.*, Periostin induces proliferation of differentiated cardiomyocytes and promotes cardiac repair, *Nat. Med.*, 2007, **13**(8), 962–969.

41. T. Etoh, C. Joffs, A. M. Deschamps, *et al.*, Myocardial and interstitial matrix metalloproteinase activity after acute myocardial infarction in pigs, *Am. J. Physiol.: Heart Circ. Physiol.*, 2001, **281**(3), H987–H994.

42. J. P. M. Cleutjens, J. C. Kandala, E. Guarda, R. V. Guntaka and K. T. Weber, Regulation of collagen degradation in the rat myocardium after infarction, *J. Mol. Cell. Cardiol.*, 1995, **27**(6), 1281–1292.

43. S. Heymans, A. Luttun, D. Nuyens, *et al.*, Inhibition of plasminogen activators or matrix metalloproteinases prevents cardiac rupture but impairs therapeutic angiogenesis and causes cardiac failure, *Nat. Med.*, 1999, **5**(10), 1135–1142.

44. A. Ducharme, S. Frantz, M. Aikawa, *et al.*, Targeted deletion of matrix metalloproteinase-9 attenuates left ventricular enlargement and collagen accumulation after experimental myocardial infarction, *J. Clin. Invest.*, 2000, **106**(1), 55–62.

45. E. Creemers, J. Cleutjens, J. Smits, *et al.*, Disruption of the plasminogen gene in mice abolishes wound healing after myocardial infarction, *Am. J. Pathol.*, 2000, **156**(6), 1865–1873.

46. D. C. Andersen, S. Ganesalingam, C. H. Jensen and S. P. Sheikh, Do neonatal mouse hearts regenerate following heart apex resection? *Stem Cell Rep.*, 2014, **2**(4), 406–413.

47. R. B. Hinton, J. Lincoln, G. H. Deutsch, *et al.*, Extracellular matrix remodeling and organization in developing and diseased aortic valves, *Circ. Res.*, 2006, **98**(11), 1431–1438.

48. B. P. Kruithof, S. A. Krawitz and V. Gaussin, Atrioventricular valve development during late embryonic and postnatal stages involves condensation and extracellular matrix remodeling, *Dev. Biol.*, 2007, **302**(1), 208–217.

49. L. S. Carvalhaes, O. L. Gervásio, C. Guatimosim, *et al.*, Collagen XVIII/endostatin is associated with the epithelial-mesenchymal transformation in the atrioventricular valves during cardiac development, *Dev. Dyn.*, 2006, **235**(1), 132–142.
50. C. M. Kelleher, S. E. Mclean and R. P. Mecham, *Vascular Extracellular Matrix and Aortic Development*, 2004.
51. S. E. Mclean, B. H. Mecham, C. M. Kelleher, T. J. Mariani and R. P. Mecham, Extracellular matrix gene expression in the developing mouse aorta, *Adv. Dev. Biol.*, 2005, **15**(05), 81–128.
52. C. M. Ng, A. Cheng, L. A. Myers, *et al.*, TGF- β–dependent pathogenesis of mitral valve prolapse in a mouse model of Marfan syndrome, *J. Clin. Invest.*, 2004, **114**(11), 1586–1592.
53. C. B. Kern, W. O. Twal, C. H. Mjaatvedt, *et al.*, Proteolytic cleavage of versican during cardiac cushion morphogenesis, *Dev. Dyn.*, 2006, **235**(8), 2238–2247.
54. E. E. Wirrig, B. S. Snarr, M. R. Chintalapudi, *et al.*, Cartilage link protein 1 (Crtl1), an extracellular matrix component playing an important role in heart development, *Dev. Biol.*, 2007, **310**(2), 291–303.
55. P. Snider, R. B. Hinton, R. A. Moreno-Rodriguez, *et al.*, Periostin is required for maturation and extracellular matrix stabilization of non-cardiomyocyte lineages of the heart, *Circ. Res.*, 2008, **102**(7), 752–760.
56. R. A. Norris, R. A. Moreno-Rodriguez, Y. Sugi, *et al.*, Periostin regulates atrioventricular valve maturation, *Dev. Biol.*, 2008, **316**(2), 200–213.
57. T. V. Tkatchenko, R. A. Moreno-Rodriguez, S. J. Conway, *et al.*, Lack of periostin leads to suppression of Notch1 signaling and calcific aortic valve disease, *Physiol. Genomics*, 2009, **39**(3), 160–168.
58. S. Astrof, D. Crowley and R. O. Hynes, Multiple cardiovascular defects caused by the absence of alternatively spliced segments of fibronectin, *Dev. Biol.*, 2007, **311**(1), 11–24.
59. O. Fondard, D. Detaint, B. Iung, *et al.*, Extracellular matrix remodelling in human aortic valve disease: the role of matrix metalloproteinases and their tissue inhibitors, *Eur. Heart J.*, 2005, **26**(13), 1333–1341.
60. J. Satta, J. Melkko, R. Pöllänen, *et al.*, Progression of human aortic valve stenosis is associated with tenascin-C expression, *J. Am. Coll. Cardiol.*, 2002, **39**(1), 96–101.
61. H. A. Eriksen, J. Satta, J. Risteli, *et al.*, Type I and type III collagen synthesis and composition in the valve matrix in aortic valve stenosis, *Atherosclerosis*, 2006, **189**(1), 91–98.
62. C. M. Otto, J. Kuusisto, D. D. Reichenbach, A. M. Gown and K. D. O'Brien, Characterization of the early lesion of "degenerative" valvular aortic stenosis. Histological and immunohistochemical studies, *Circulation*, 1994, **90**(2), 844–853.

63. M. C. Cushing, J.-T. Liao and K. S. Anseth, Activation of valvular interstitial cells is mediated by transforming growth factor-beta1 interactions with matrix molecules, *Matrix Biol.*, 2005, **24**(6), 428–437.

64. K. J. Rodriguez, L. M. Piechura, A. M. Porras and K. S. Masters, Manipulation of valve composition to elucidate the role of collagen in aortic valve calcification, *BMC Cardiovasc. Disord.*, 2014, **14**(1), 29.

65. E. Aikawa, M. Aikawa, P. Libby, *et al.*, Arterial and aortic valve calcification abolished by elastolytic cathepsin S deficiency in chronic renal disease, *Circulation*, 2009, **119**(13), 1785–1794.

66. A. Simionescu, D. T. Simionescu and N. R. Vyavahare, Osteogenic responses in fibroblasts activated by elastin degradation products and transforming growth factor-β1, *Am. J. Pathol.*, 2007, **171**(1), 116–123.

67. E. H. Stephens, J. G. Saltarrelli, L. S. Baggett, *et al.*, Differential Proteoglycan and hyaluronan distribution in calcified aortic valves, *Cardiovasc. Pathol.*, 2011, **20**(6), 334–342.

68. B. Jian, P. L. Jones, Q. Li, *et al.*, Matrix metalloproteinase-2 is associated with tenascin-C in calcific aortic stenosis, *Am. J. Pathol.*, 2001, **159**(1), 321–327.

69. M. Yoshioka, S. Yuasa, K. Matsumura, *et al.*, Chondromodulin-I maintains cardiac valvular function by preventing angiogenesis, *Nat. Med.*, 2006, **12**(10), 1151–1159.

70. D. Hakuno, N. Kimura, M. Yoshioka, *et al.*, Periostin advances atherosclerotic and rheumatic cardiac valve degeneration by inducing angiogenesis and MMP production in humans and rodents, *J. Clin. Invest.*, 2010, **120**(7), 2292–2306.

71. N. M. Rajamannan, M. Subramaniam, D. Rickard, *et al.*, Human aortic valve calcification is associated with an osteoblast phenotype, *Circulation*, 2003, **107**(17), 2181–2184.

72. V. Pohjolainen, P. Taskinen, Y. Soini, *et al.*, Noncollagenous bone matrix proteins as a part of calcific aortic valve disease regulation, *Hum. Pathol.*, 2008, **39**(11), 1695–1701.

73. O. Fondard, D. Detaint, B. Iung, *et al.*, Extracellular matrix remodelling in human aortic valve disease: the role of matrix metalloproteinases and their tissue inhibitors, *Eur. Heart J.*, 2005, **26**(13), 1333–1341.

74. J. J. Kaden, C.-E. Dempfle, R. Grobholz, *et al.*, Inflammatory regulation of extracellular matrix remodeling in calcific aortic valve stenosis, *Cardiovasc. Pathol.*, 2005, **14**(2), 80–87.

75. M. E. Edep, J. Shirani, P. Wolf and D. L. Brown, Matrix metalloproteinase expression in nonrheumatic aortic stenosis, *Cardiovasc. Pathol.*, 2000, **9**(5), 281–286.

76. E. W. Raines, The extracellular matrix can regulate vascular cell migration, proliferation, and survival: relationships to vascular disease, *Int. J. Exp. Pathol.*, 2000, **81**(3), 173–182.
77. K. G. McCullagh, V. C. Duance and K. A. Bishop, The distribution of collagen types I, III and V (AB) in normal and atherosclerotic human aorta, *J. Pathol.*, 1980, **130**(1), 45–55.
78. T. Tsuda, H. Wang, R. Timpl and M.-L. Chu, Fibulin-2 expression marks transformed mesenchymal cells in developing cardiac valves, aortic arch vessels, and coronary vessels, *Dev. Dyn.*, 2001, **222**(1), 89–100.
79. S. Katsuda, Y. Okada, T. Minamoto, *et al.*, Collagens in human atherosclerosis. Immunohistochemical analysis using collagen type-specific antibodies, *Arterioscler., Thromb., Vasc. Biol.*, 1992, **12**(4), 494–502.
80. M. L. Iruela-Arispe, D. J. Liska, E. H. Sage and P. Bornstein, Differential expression of thrombospondin 1, 2, and 3 during murine development, *Dev. Dyn.*, 1993, **197**(1), 40–56.
81. R. F. Nicosia and G. P. Tuszynski, Matrix-bound thrombospondin promotes angiogenesis in vitro, *J. Cell Biol.*, 1994, **124**(1–2), 183–193.
82. L. Y. Yao, C. Moody, E. Schönherr, T. N. Wight and L. J. Sandell, Identification of the proteoglycan versican in aorta and smooth muscle cells by DNA sequence analysis, in situ hybridization and immunohistochemistry, *Matrix Biol.*, 1994, **14**(3), 213–225.
83. C. B. Kern, R. a. Norris, R. P. Thompson, *et al.*, Versican proteolysis mediates myocardial regression during outflow tract development, *Dev. Dyn.*, 2007, **236**(3), 671–683.
84. R. Shimizu-Hirota, H. Sasamura, M. Kuroda, *et al.*, Extracellular matrix glycoprotein biglycan enhances vascular smooth muscle cell proliferation and migration, *Circ. Res.*, 2004, **94**(8), 1067–1074.
85. M. Fukuchi, J. Watanabe, K. Kumagai, *et al.*, Normal and oxidized low density lipoproteins accumulate deep in physiologically thickened intima of human coronary arteries, *Lab. Invest.*, 2002, **82**(10), 1437–1447.
86. M. D. Rekhter, K. Zhang, A. S. Narayanan, *et al.*, Type I collagen gene expression in human atherosclerosis. Localization to specific plaque regions, *Am. J. Pathol.*, 1993, **143**(6), 1634–1648.
87. B. V. Shekhonin, S. P. Domogatsky, G. L. Idelson, V. E. Koteliansky and V. S. Rukosuev, Relative distribution of fibronectin and type I, III, IV, V collagens in normal and atherosclerotic intima of human arteries, *Atherosclerosis*, 1987, **67**(1), 9–16.

88. E. P. Amento, N. Ehsani, H. Palmer and P. Libby, Cytokines and growth factors positively and negatively regulate interstitial collagen gene expression in human vascular smooth muscle cells, *Arterioscler., Thromb., Vasc. Biol.*, 1991, **11**(5), 1223–1230.

89. G. K. Sukhova, Y. Zhang, J.-H. Pan, *et al.*, Deficiency of cathepsin S reduces atherosclerosis in LDL receptor – deficient mice, *J. Clin. Invest.*, 2003, **111**(6), 897–906.

90. P. Vijayagopal, J. E. Figueroa, J. D. Fontenot and D. L. Glancy, Isolation and characterization of a proteoglycan variant from human aorta exhibiting a marked affinity for low density lipoprotein and demonstration of its enhanced expression in atherosclerotic plaques, *Atherosclerosis*, 1996, **127**(2), 195–203.

91. C. M. Giachelli, N. Bae, M. Almeida, *et al.*, Osteopontin is elevated during neointima formation in rat arteries and is a novel component of human atherosclerotic plaques, *J. Clin. Invest.*, 1993, **92**(4), 1686–1696.

92. R. Riessen, M. Kearney, J. Lawler and J. M. Isner, Immunolocalization of thrombospondin-1 in human atherosclerotic and restenotic arteries, *Am. Heart J.*, 1998, **135**(2), 357–364.

93. M. Onda, T. Ishiwata, K. Kawahara, *et al.*, Expression of lumican in thickened intima and smooth muscle cells in human coronary atherosclerosis, *Exp. Mol. Pathol.*, 2002, **72**(2), 142–149.

94. K. Wallner, C. Li, P. K. Shah, *et al.*, Tenascin-C is expressed in macrophage-rich human coronary atherosclerotic plaque, *Circulation*, 1999, **99**(10), 1284–1289.

95. J. P. Jung, J. M. Squirrell, G. E. Lyons, K. W. Eliceiri and B. M. Ogle, Imaging cardiac extracellular matrices: a blueprint for regeneration, *Trends Biotechnol.*, 2012, **30**(4), 233–240.

96. P.-J. Su, Q. A. Tran, J. J. Fong, *et al.*, Mesenchymal stem cell interactions with 3D ECM modules fabricated via multiphoton excited photochemistry, *Biomacromolecules*, 2012, **13**(9), 2917–2925.

97. J. Lincoln, J. B. Florer, G. H. Deutsch, R. J. Wenstrup and K. E. Yutzey, ColVa1 and ColXIa1 are required for myocardial morphogenesis and heart valve development, *Dev. Dyn.*, 2006, **235**(12), 3295–3305.

98. S. E. Klewer, S. L. Krob, S. J. Kolker and G. T. Kitten, Expression of Type VI Collagen in the Developing Mouse Heart, *Dev. Dyn.*, 1998, **211**, 248–255.

99. R. C. Kowal, J. A. Richardson, J. M. Miano and E. N. Olson, EVEC, a novel epidermal growth factor-like repeat-containing protein upregulated in embryonic and diseased adult vasculature, *Circ. Res.*, 1999, **84**(10), 1166–1176.

CHAPTER 4

Matrix Biology: Structure and Assembly of Laminin-Rich Matrices

KEVIN J. HAMILL[†b], SUSAN B. HOPKINSON[†a], NATALIE M. E. HILL[b], AND JONATHAN C. R. JONES*[a]

[a]School of Molecular Biosciences, Washington State University, Pullman, WA, USA; [b]Department of Eye and Vision Science, Institute of Ageing and Chronic Disease, University of Liverpool, Liverpool, UK
*E-mail: jcr.jones@vetmed.wsu.edu

4.1 INTRODUCTION

The extracellular matrix (ECM) has profound effects on cell behavior, with consequent influence on the regulation of development, organogenesis and tissue remodeling, maintenance of stem cells, tissue integrity and homeostasis, and angiogenesis. Moreover, matrix changes not only occur during disease, but reorganization and the laying down of new matrix is also a critical mechanism underlying pathogenesis of numerous syndromes.

The ECM is a complex structure and it would be impossible to provide a comprehensive review of matrix proteins in a single chapter. Hence, we will restrict our review to laminins. Laminins are heterotrimeric glycoproteins secreted by a variety of cell types and

---

[†]These authors contributed equally.

Mimicking the Extracellular Matrix: The Intersection of Matrix Biology and Biomaterials
Edited by Gregory A. Hudalla and William L. Murphy
© The Royal Society of Chemistry 2020
Published by the Royal Society of Chemistry, www.rsc.org

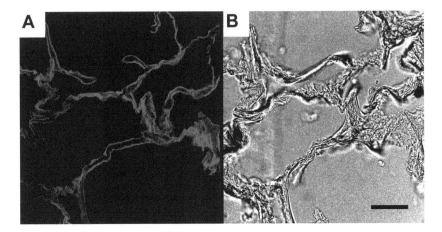

Figure 4.1   The localization of the α3 laminin subunit in frozen sections of human lung processed for immunofluorescence and viewed by confocal microscopy. Note that the α3 laminin subunit distributes in two clear lines within the walls surrounding the alveolar spaces (A). The image of the tissue in B is provided for orientation. Bar, 50 μm.

incorporated, along with other proteins, into specialized ECM structures, termed basement membranes (BMs) (Figure 4.1).[1-3] BMs are found beneath epithelia and endothelia and surround muscles and nerves. Laminin expression is dependent not only on cell type but also on context.[4] Moreover, domain composition may be modified as a result of alternative splicing or the structure may be changed following post-translational processing. In addition, laminin matrices may exhibit different assembly forms, depending on whether a tissue is quiescent or undergoing remodeling. The diversity in laminin matrix structure would imply that they exhibit a range of functions, evidence for which is discussed below. Most importantly, laminins provide the point of interaction between cells and their extracellular microenvironment. The laminins therefore are in an ideal location to influence directly cell behavior, which they do through two main mechanisms: (1) interaction with cell-surface receptors leading to modification of downstream signaling events; or (2) through physical modification of cellular substrates. We begin our review by detailing laminin nomenclature and the structure of laminin subunits.

4.2   NOMENCLATURE

Since their first description in 1979, there have been many variations on nomenclature used to describe the laminins. These were rationalized almost a decade ago and we will use the terminology suggested in 2005 throughout this review.[1,5]

The laminins are a multi-gene family generated by gene duplication from single ancestral copies.[6,7] In higher organisms 11 different laminin genes have been identified to date.[6] There are five α subunits (encoded by the genes LAMA1–5), three β subunits (encoded by the genes LAMB1–3), and three γ subunits (encoded for by the genes LAMC1–3) (Figure 4.2).[1] At the protein level, subunit names incorporate the gene from which they are derived and, if necessary, an Arabic letter describing the splice isoform. For example, the product of LAMA2 is termed α2 while the two major splice isoforms from LAMA3 are termed α3A and α3B.[1] Where studies have not distinguished between splice isoforms, or where the results are equally applicable to all, we will use the base term (for example, α3 to describe both α3A and α3B).

Laminins assemble into heterotrimeric structures, each consisting of one α, one β, and one γ subunit (Figure 4.3). Theoretically, 45

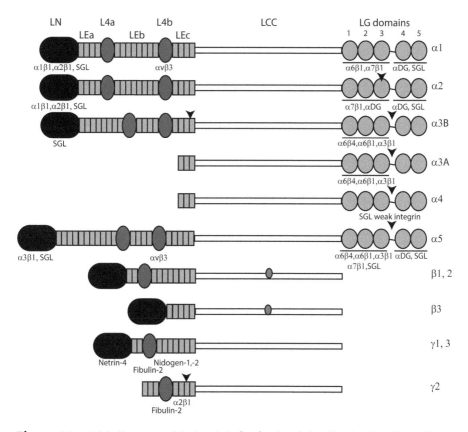

Figure 4.2 Stick diagrams of the laminin (LM) subunit family. The domain architecture and designations are indicated above the α1 subunit. Arrowheads indicate potential proteolytic cleavage sites. Receptor and ECM binding partners are indicated under each subunit. DG, dystroglycan; SGL, sulfated glycolipids. Adapted from ref. 3.

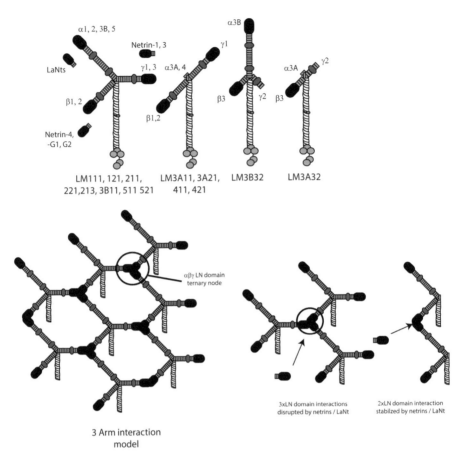

Figure 4.3   Stick diagrams of the different laminin heterotrimers are shown in the upper panel. In the lower panels, a putative model of laminin–laminin interactions is presented (left). The ability of netrins and LaNts either to disrupt or stabilize laminin–laminin interactions is depicted on the right. Adapted from ref. 1.

different laminin heterotrimers may be formed, although to date only 16 have been identified.[8–10] Each assembled trimer is named after its constituent subunits; thus laminin α2β1γ1 is termed LM211 while α3Bβ3γ2 is LM3B32.1. In terms of subunit mass, α subunits range from 200 to 440 kDa while β and γ subunits vary between 120 and 200 kDa. Assembled trimers therefore range from 400 to 800 kDa.[11–14]

4.3   THE STRUCTURE OF LAMININ SUBUNITS

4.3.1   Conserved Domain Architecture

The conserved ancestry of laminin genes means that each subunit shares regions of common protein architecture with other family members. The precise composition and number of these regions varies between different subunits and between different isoforms. However, working from N- to C-terminus, the archetypal subunit consists of a signal sequence, a laminin N-terminal domain (LN domain), a stretch of laminin-type epidermal growth factor-like repeats (LE domains) with interspersed globular domains (L4a and b in α subunits, LF and L4 in β and γ subunits, respectively), a laminin coiled-coil domain (LCC) and, in α subunits, a large globular C-terminal domain, which can be separated into five distinct regions (LG1–5) (Figure 4.2).[1]

Based on their domain composition, laminin subunits can be placed into one of two subgroups. "Full-length" subunits contain each of the domains described above, while "head-less" or "truncated" subunits lack some of the N-terminal domains (Figure 4.2). It should be noted, however, that to be classed as a laminin subunit, the subunit must contain a LCC domain and be capable of trimerization.[1,15] By rotary shadowing electron microscopy the archetypal laminin heterotrimers assemble into cross shaped molecules with a long arm of ~80 nm made from all three subunits and three short arms of 35–50 nm each from individual subunits.[16] Laminins containing "truncated" subunits lack one or more of these short arms and therefore assemble into Y-shaped or rod-like assemblies (Figures 4.2 and 4.3).[17]

4.3.1.1 LN Domains. LN domains are large globular regions of about 200 amino acids, found at the N-terminus of the α1, α2, α3B, α5, β1, β2, β3, γ1, and γ3 subunits (*i.e.*, not in the α3A, α4, and α2) (Figure 4.2). The crystal structures of the mouse α5, β1, and γ1 laminin subunit LN domains have recently been solved and are based around β sandwiches with elaborate loop regions, which differ between subunits.[18,19] LN domains are attached like the head of a flower onto the rod-like LE repeats and are tightly associated with the first LE repeat.[18,19] In contrast to the LCC domains, through which individual laminin subunits assemble into heterotrimers, the LN domains are the regions through which assembled heterotrimers directly interact with one another to form higher-order networks or polymers (Figure 4.3). These interactions are believed to be critical for BM assembly and function (Figure 4.3).[19–22] This will be described in greater detail below (Section 4.4.1).

The surfaces of the LN domain are not conserved across the α, β, and γ subfamilies. However, within each family there are distinct regions that are highly conserved. This likely indicates the functional importance of these conserved regions. Significantly, some pathogenic mutations in the LN domain of β2 laminin that lead to Pierson syndrome, a congenital disease affecting kidney and eyes, are in the conserved residues within these regions.[18,23,24] LN domains are also present in two other ECM protein families, the netrins and the LaNts (laminin N-terminus) (Figure 4.3). In the case of the former, the LN domains have been shown to mediate interaction between laminins and netrins and regulate BM formation.[15,25,26] This will be described in greater detail below (Section 4.4.2). The LN domains of some sub-units exhibit interaction with integrin receptors (Section 4.4.3.1) and sulfatides (Section 4.4.4.1) (Figure 4.1).[27–30]

4.3.1.2   LE Domains.   LE repeats are believed to play primarily the role of rigid spacers between the various globular domains of laminin subunits, effectively providing spatial separation between them. They exist as three to eight repeats, each of which resembles epidermal growth factor (EGF) folds but contain eight rather than six disulfide cross-linked cysteine residues between cysteines 1–3, 2–4, 5–6 and 7–8.[18,31] Interestingly, proteolytic processing of the N-terminus of γ2 subunit releases a protein fragment that demonstrates functionality in *in vitro* assays, by eliciting signaling through the epidermal growth factor receptor (EGFR) (Figure 4.2).[32] The γ2 subunit LE repeats clos-est to the LCC domain have also been shown to interact with α2β1 inte-grin and therefore could signal their integrin-binding partner (Figure 4.1).[33]

In addition to their potential signaling roles, the LE repeats of the γ1 and γ3 subunits contain a binding site for nidogens, and this interaction is important for BM formation (Section 4.4.2.1) (Figure 4.1).[34–36] In addi-tion, fibulin-2 binds LE repeats in γ2 and α1 subunits potentially mediat-ing a link between LM111 and LM332.[37] Fibulin-2 also interacts with type IV collagen and fibronectin, which raises the possibility that its interac-tions are important for the supramolecular organization of the BM.[38,39]

LE domains are also found in two other ECM families, the netrins and the LaNts, which are described in greater detail below (Section 4.4.2). Although no function has yet been attributed to LE domains in these protein families, it is interesting to note that addition of two or more LE domains increases the yield and stability of recombinantly expressed LN domains, suggesting that in these families, the LE repeats may play a stabilizing role for laminin-rich matrices (Figure 4.3).[40]

4.3.1.3   L4/LF Domains.   The LE repeats of "full-length" laminin sub-units are interrupted by large (~250 amino acid) globular domains termed L4 or LF. Little is currently known about the functional roles of L4 domains, and it is likely they are primarily important for breaking up the extended LE repeats, thereby permitting molecular flexibility. However, the L4 of α5 and α1 subunits have been demonstrated to interact with αvβ3 integrin (Figure 4.2).[41,42] The L4a (N-proximal) domain of the α5 subunit exhibits neurite outgrowth enhancement capabilities.[43] Moreover, a peptide sequence within L4a modulates the response of neuronal cells to the rest of the short arm.[43] Interestingly, the L4 domain of the α1 subunit masks a putative RGD site in the adjacent stretch of LE repeats. Thus, under certain conditions, such as tissue remodeling in wound healing or as a consequence of disease when matrix is modified by proteolysis, the masking effect may be reduced allowing integrin association and resultant receptor-mediated signalling.[42]

4.3.1.4 LCC Domains. The LCC domain mediates laminin trimer assembly. Sequences within the C-terminal half of the LCC domain determine which precise subunit combinations form heterotrimers.[44–46] Moreover, the LCC domain is also important in orienting the adjacent laminin G-like (LG) domains and N-terminal domains. Consistent with this, a glutamic acid at the third from last position of the LCC in γ subunits is required for integrin binding (see below) and the final 20 amino acids of the β subunit LCC determine LG domain integrin-binding affinity (see below).[44–46] In addition to these functions, exogenously added recombinant LCC domain of LM111 is capable of exerting anti-adhesive effects and induces up-regulation of genes involved in promoting cell migration and invasion.[47]

4.3.1.5 LG Domains. Five LG domains are located at the C-terminus of each α subunit. They represent the major site of cell-surface receptor interaction with high affinity binding sites for integrins, heparin, sulfatides and α-dystroglycan (Figure 4.2). Each LG domain consists of 180–200 amino acids, which assemble into a 14-stranded β sandwich with a calcium ion binding pocket.[48,49] Binding epitopes for heparin, sulfatides and α-dystroglycan are located in surface loops around the calcium site and show considerable overlap. Efficient ligand binding often requires LG modules to be present in tandem.[50,51] Proteolytic processing occurs within the LG domains of α2, α3, and α4 subunits (Section 4.5.1) (Figure 4.2). This cleavage releases the LG4–5 tandem, leading to organizational changes in the remaining LG1–3.[52] This alteration can impact receptor affinities, producing downstream consequences such as converting a particular laminin matrix from supporting cell migration to promoting adhesion. This is described in greater detail below (Section 4.5.1).

4.3.2   Splice Isoforms Enlarge the Laminin Family

The most dramatic example of alternative splicing within the laminin family occurs within the LAMA3 gene. Use of two distinct promoters gives rise to a "full-length" form, LAMA3B, consisting exons 1–38 + 40–72, while a "head-less" form is derived from a promoter in intron 38 giving rise to LAMA3A, which consists of exon 39 + 40–72. In other words, these two major forms share their C-terminal LCC and LG domains but differ in their N-terminal regions (Figure 4.2).[53,54] This has functional implications. Both the α3A and α3B subunits display the same heterotrimer assembly profile and C-terminal cell-surface interaction capabilities, but they differ in the presence of a LN domain and therefore ability to form networks (see below). The alternate promoter used in production of the two LAMA3 isoforms suggests they have distinct expression profiles, with α3B displaying more widespread, but generally a lower, level of expression than α3A.[55]

Two isoforms from the LAMC2 gene have been identified. The first is a full-length subunit derived from 23 exons, whereas the second (γ2*) is derived from splice site read-through from the exon 22 donor splice site into intron 22 that generates a shorter form, lacking the final 83 amino acids.[56,57] The two isoforms appear to have different expression profiles with the γ2* expressed at higher levels in cerebral cortex, lung, and distal tubules of the kidney.[56] As described above, the final amino acids of the LCC domain are required for trimer formation. Therefore it is unlikely that this splice isoform is capable of assembling into a heterotrimer.

Two isoforms of the α4 subunit (α4A and α4B) have been identified (Figure 4.2), with the B form containing seven extra amino acids toward the N-terminus of the LCC domain.[58] This heptapeptide insertion includes a cysteine residue located between two others cysteines thought to be important for intersubunit assembly. Therefore these two isoforms may differentially assemble.[58] *In vitro* studies using purified recombinant protein demonstrate both isoforms form trimers with β1 and γ1 subunits, with LM4B11 being more potent in promoting cell spreading than LM4A11.[58]

Alternate splicing at the 5′ end of the LAMB2 gene has also been described. The first intron on the human gene contains a nonconsensus 5′ splice site upstream of the translational start site suggesting that the upstream intron is inefficiently removed.[59] Interestingly, the different 5′ untranslated regions (UTRs) of these two isoforms might result in differences in translational efficiency of their respective mRNAs.[59]

4.3.3   Regulation of Trimer Assembly

Laminin heterotrimer assembly occurs intracellularly and proceeds from a βγ double stranded coiled-coil heterodimer, which is preferred over any homodimeric associations, to a thermodynamically stable αβγ triple coiled-coil heterotrimer.[9,60–62] In the parietal endoderm-like F9 cell line, the γ subunit is present at a lower level than its α and β binding partners.[62] In contrast, in the squamous cell cancer SCC25 line, the α3A subunit is the least abundant of the subunits comprising LM332, indicating cell-line specific differences in the trimerization rate limiting step.[61] Moreover, assembly is thought to involve a number of molecular chaperones including Bip, HSP70, protein-disulfide isomerase, GRP94, and cadnexin.[63] The process involved in assembly is rapidly completed, with pulse–chase experiments indicating that LM332 assembly occurs within 10 minutes and complete secretion by 24 hours.[61]

Assembly occurs prior to glycosylation, with assembled αβγ heterotrimers proceeding through intracellular trafficking for oligosaccharide side-chain processing and secretion.[62] Interestingly, blocking the processing of high-mannose type precursors on laminin subunits leads to reduced secretion and cytoplasmic accumulation in F9 cells.[62] However, blocking with tunicamycin, an inhibitor of N-linked glycosylation, does not.[61] Alpha subunits can be secreted independently (*i.e.*, without heterotrimerization) and provide the driving force for secretion of the trimer.[64] In contrast, in the absence of an α subunit, βγ dimers accumulate in the cytoplasm.[64]

4.3.4   Laminin Heterotrimer and Subunit Expression is Tightly Regulated

BMs are a feature of every tissue. Thus laminins, as major BM components, display widespread distribution profiles. However, since each BM contains a defined combination or ratio of different laminins, the properties of a BM reflect the nature of the cells which deposit and organize it, just as the characteristics of the BM impact the behavior of the cells which adhere to it.

The β1 and γ1 subunits are present in almost every tissue throughout development and in all adult tissues, and can form heterotrimers with every α subunit. The critical importance of these subunits during development is illustrated by the death of β1 and γ1 subunit knockout mice during embryogenesis (Table 4.1).[4,65,66]

The α1 subunit is the most prominent subunit during early development and its knockout is lethal in mouse (Table 4.1).[65,67,68] However,

Table 4.1 Diseases associated with laminin subunits. The human locus is
indicated.[a]

Laminin subunit	Human genetic locus	Phenotype
LAMA1	18p11.31	KO lethal in mouse
LAMA2	6p22–23	MDC1A-merosin-deficient congenital muscular dystrophy
LAMA3	18q11.2	JEB (LOC–LAMA3A specific)
LAMA4	6q21	Cardiomyopathy, glomerular sclerosis
LAMA5	20q13.2–13.3	KO lethal in mouse
LAMB1	7q31.1–31.3	KO lethal in mouse
LAMB2	3p21	Pierson syndrome
LAMB3	1q32	JEB
LAMC1	1q31	KO lethal in mouse
LAMC2	1q25–31	JEB
LAMC3	9q33–q34	Brain anomalies, no definite pathology in mouse

[a]JEB, junctional epidermolysis bullosa; KO, knockout; MDC1A, merosin-deficient congenital
muscular dystrophy.

in adult tissue, α1 is located only in the BM of liver, kidney, and repro-
ductive organs. The α2 subunit localizes to myofibers and Schwann
cells, as well as cerebral blood vessels.[69,70] α2-Null mice develop severe
muscular dystrophy which presents at 5 weeks, while in humans,
mutations in the LAMA2 gene give rise to merosin-deficient congenital muscular dystrophy (MDC1A) (Table 4.1).[71–73] The α3 subunits are
located in epithelial tissues including skin, cornea, gut, lung, and in
the oral mucosa, including those epithelial cells attached to the tooth
surface.[55,74] Mouse knockouts and human mutations in LAMA3 lead
to the skin fragility syndrome junctional epidermolysis bullosa (JEB)
(Table 4.1).[75,76] α3A isoform specific point mutations in the N-terminus
or an N-terminal truncation cause the chronic wound disorder laryngo-
onycho-cutaneous syndrome (LOCS), indicating the importance of
this domain in regulating the granulation tissue response.[53,77] Context is everything, however, since conditional α3 subunit knockout
in the mouse lung has an interesting phenotype. These animals are
protected from mechanically induced lung injury.[78] The α4 subunits
are expressed by cells of mesenchymal origin, most notably vascular
endothelial cells. LAMA4 mutations produce cardiomyopathy and glomerular sclerosis, though animals are viable and fertile if they survive
early anemia (Table 4.1).[79–81] Survival is thought to be dependent on
initiation of expression of α5 laminin and its assembly into the capillary ECM.[80] The α5 subunit is widely expressed and most adult BMs,
including the vasculature, contains at least some LM511. Knockout of

the α5 subunit is embryonic lethal in mouse.[82] α5-Null animals survive to mid-gestation, then arrest due to multiorgan failure, including exencephaly, syndactyly, placentophaly, and defective glomerulogenesis (Table 4.1).[82] Consistent with its role in development, LM511 expression appears to be associated with stem cell niches in various BMs. For example, in the hair follicle, loss of this subunit expression is associated with chemotherapy-induced alopecia, indicating an important role in hair morphogenesis.[83]

The β2 subunit is up-regulated during early organ maturation and is located in the glomerular BMs, neuromuscular junction, and smooth muscle cells of the aorta.[84,85] β2-Null mice die early due to kidney failure and human missense mutations are associated with Pierson syndrome, which presents with both kidney glomerularsclerosis and multiple ocular manifestations (Table 4.1).[86,87] The β3 and γ2 subunits are widely expressed in epithelial tissues. Mutations in either the β3 or γ2 subunit, similarly to mutations in α3, lead to JEB (Table 4.1).[88,89] The severity of presentation of the disease depends on the nature of the mutation, with loss of function mutations producing the severe, lethal form while missense mutations are associated with milder blistering.[89,90]

The γ3 subunit is the most recently identified laminin.[91] It is expressed in testis, kidney, brain, skin, and hair follicle.[36] Loss of function in humans is believed to result in brain anomalies (Table 4.1).[92] The γ3 subunit, along with β2 subunit-containing laminins, has recently been demonstrated to regulate astrocyte migration in the retina.[93] Interestingly, β2/γ3 double knockout mice display defective limiting membrane formation associated with retinal dysplasia.[94]

4.4 REGULATION OF LAMININ NETWORK/BM ASSEMBLY

4.4.1 Laminin–Laminin Interactions

When all three subunits of the laminin heterotrimer are full-length, and thus contain LN domains, laminin polymers assemble in the presence of calcium ions (Figure 4.3).[95–97] These polymers can be disrupted through addition of proteolytic or recombinantly produced laminin fragments containing LN domains.[27,96,98,99] Electron microscopical analyses of laminin heterotrimers have identified the globular LN domain at the end of the subunit short arms as the site of interaction among the laminin molecules, the "3 arm interaction model" (Figure 4.3).[96] The importance of the LN domains in laminin polymer assembly is also supported by the finding that missense mutations in the LN domains in humans are pathogenic.[87,100,101] Moreover, a destabilizing deletion of the α2 subunit LN domain results in muscular dystrophy in the dy2J mouse model.[102]

How laminins containing truncated or head-less subunits incorporate into BMs is less well known. However, one possibility is that they do so by interacting with laminin heterotrimers containing full-length subunits. Such may be the case where LM332 in the BM of the human amnion (and skin) binds to LM311 and LM321.[103] Under such circumstances, the stability of the BM would be presumed to be due to association of the nidogen-binding β1 and β2 subunits to a type IV collagen backbone (see below).

4.4.2   Role of Small Molecules and Laminin-Binding Proteins in Laminin Polymer and BM Assembly

4.4.2.1   Nidogen/Entactin.   The glycoproteins nidogen-1 and -2 are ubiquitous components of BMs.[104,105] Nidogens are capable of binding to laminins and type IV collagen, bridging these two major networks in the BM (Figure 4.2).[104–106] The two nidogen isoforms have low sequence homology. However, the absence of either is not lethal, as in most tissues there appears to be compensation by the other isoform.[107] In double knockout animals, the BM is lost or malformed in skin, heart, and lungs, which results in neonatal death.[107–110] Interestingly, though there is no change in BM composition in the adult kidney in the nidogen knockout, the double knockout BM is somewhat diffuse and thickened in appearance or may even be absent, indicating that the nidogens are likely not involved in the deposition of other BM components but may, in some way, regulate their integration into the BM.[108]

Nidogens have high affinity for the LE domain of γ1 and γ3 subunits, but extremely low affinity for the γ2 subunit (Figure 4.2).[34–36,111,112] This differential binding capability suggests BM context-specific effects where the local ratio of γ2 subunit-containing laminins (LM332) to γ1 or γ3 subunit-containing laminins may define the level of integration of the type IV collagen and laminin networks within the BM. Interestingly, treatment of 3D skin co-culture models with a fragment of laminin γ1 which competes for nidogen binding diminishes the accumulation of LM111 in the BM.[113] Moreover, it prevents the ultrastructural formation of hemidesmosomes and the lamina densa.[113] However, the non-nidogen binding LM332 still accumulates as normal in this model, as does type IV collagen.[113] Consistent with the importance of nidogen binding for skin BM organization, and an exception to the isoform compensation in mouse knockouts, targeted deletion of either nidogen isoform leads to disruption of the BM in skin, with a lack of nidogen-1 being associated with delayed wound re-epithelization.[114,115]

4.4.2.2 Netrins and LaNts.

Netrins are a family of secreted ECM proteins found both within and outside the central nervous system with defined roles in axonal guidance, interneuronal migration, tumor progression, angiogenesis, vasculogenesis, and branching morphogenesis.[116–122] Each netrin molecule is composed of a laminin-type LN domain followed by three LE repeats and a C-terminal domain unrelated to laminin (Figure 4.3).[118,123] In two family members, this C-terminal region also contains glycoslyphosphatidylinositol anchor sites.[123] In higher organisms there are five netrin genes; netrin-1 and -3 have highest sequence homology with laminin γ subunits, while netrin-4, -G1 and -G2 are more closely related to laminin β subunits.[124,125]

Although most netrin functions have been attributed to signaling cell-surface receptors, there is increasing evidence their interaction with laminin is an important mediator of this signalling.[25] For example, a functional complex of netrin-4 and laminin γ1 activates integrin α6β1 signaling to mitogen-activated protein kinase (MAPK) and controls proliferation and migration of adult neural stem cells destined for the mouse olfactory bulb.[25] Similarly, in endothelial cells, netrin-4 enhances α6 integrin–LM111 binding, resulting in increased cell migration, adhesion, and signaling to Src.[126]

In contrast to the netrins, the LaNts are derived by alternate splicing at the 5' end of laminin encoding genes.[127] Four LaNt mRNAs have been described so far, each encoding domains located in the N-terminal arms of a laminin subunit, *i.e.*, an LN domain followed by a stretch of LE repeats (Figure 4.3).[127] Knockdown of LaNt α31 expression results in a delay in the initial stages of attachment of cultured keratinocytes, as well as slower closure of scratch wounds.[127] In addition, cultured keratinocytes display up-regulation of LaNt α31 protein expression in a scratch wound assay.[127]

Both netrins and LaNts contain LN domains and thus their laminin interactions are predicted to occur in the regions laminins use to form networks. Hence, they are likely to modulate laminin network organization (Figure 4.3).[127,128] Recent studies support this idea. For example, netrin-4 interacts with γ1 and γ3 laminin, inhibiting LM111 self-assembly *in vitro* and its ability to integrate into the BM.[26] As netrins are closest to either β or γ LN domains, while LaNts are similar to α LN domains, the LaNts and the two netrin subsets are each likely to bind to different laminin trimers and elicit differential effects.

Interestingly, a major netrin receptor, the deleted in colorectal cancer protein, is enriched at sites of invadopodia and its recruitment there is required for and coincides with the physical separation of BMs during transmigration in worm development.[129] These findings suggest that a netrin/receptor complex may focally regulate laminin incorporation into the assembling BM.

4.4.3 Roles of Cell-Surface Receptors

BM formation and organization do not happen in isolation, as BMs are associated with cell layers. Thus laminin deposition, clustering, patterning, and integration into a coherent network, as well as the ability of cells to adhere to BMs and their behavior upon those matrices, are all dependent on the action of cell-surface receptors, which we will discuss next.

4.4.3.1 Integrins. Integrins are transmembrane, bi-directional signaling molecules.[130] As mentioned above, each α laminin subunit contains potential binding sites for different integrin heterodimers within its LG domains, primarily LG1–3 (Figure 4.2).[131–134] These interactions include α6β1 and α7β1 integrin with the α1 laminin subunit, α7β1 with the α2 subunit, α6β4, α3β1, and α6β1 integrin with the laminin α3A and α3B subunits, and αvβ3, α6β1, and α7β1 integrin with the α4 laminin subunit. The α5 laminin subunit possesses the most diverse binding repertoire and interacts with α6β1, α6β4, α3β1, and α7β1 integrins.[132,135–144] Interestingly, the strength of these interactions depends upon which β and γ subunit make up the heterotrimer, with β2-containing laminins displaying higher affinity for α3β1 and α7β1 integrin than β1-containing laminins.[46] Additionally, these laminin C-terminal–integrin interactions require a glutamic acid at the third from last position in the LCC domain of the γ subunits.[45] Since this residue is not present in the γ3 subunit, it has been suggested that γ3-containing laminins do not bind integrins.[44]

In addition to the C-terminal domain interactions described above, a number of integrin–laminin interactions have been described within the N-terminus of laminin subunits (Figure 4.2). Specifically, the LN domain of the α1 and α2 subunits interacts with α1β1 and α2β1 integrin, while the α5 laminin subunit LN and L4 domains bind to α3β1 and αvβ3 integrin, respectively.[27–29] Also, as mentioned above, the final stretch of LE repeats in the γ2 subunit interacts with α2β1 integrin.[33] In such a case, there is the potential for both "ends" of a laminin trimer to interact with integrin cell-surface receptors, allowing the laminin to play a bridging role between cells. This possibility, however, requires experimental validation.

The consequences of laminin–integrin interactions are multi-faceted, bi-directional, and cell-type and context specific. Ligation of integrin receptors by laminins modulates the activity of downstream signaling cascades leading ultimately to cytoskeletal rearrangement, as well as changes in proliferation rates and gene expression profiles.[145] Additionally there is increasing evidence that integrins are directly involved in defining either laminin deposition or in organizing its patterning. As an example of this phenomenon, we will next discuss how the deposition of LM332, the predominant laminin in many epithelial BMs, including skin, cornea, and gut, is potentially regulated by its integrin-binding partners, α6β4 and α3β1.

α6β4 integrin is a core component of a cell-matrix stable adhesive device termed the hemidesmosome, which not only tethers the keratin cytoskeleton to the cell surface in certain epithelial cells, but is also essential for maintaining epithelial cell/connective tissue interaction. There are numerous excellent reviews detailing hemidesmosome form and function.[89,90,146–152] In the context of this article, the β3 subunit of LM332 interacts with type VII collagen anchoring fibrils underlying each hemdesmosome.[153] This provides a vertical axis connecting the keratin cytoskeleton of an epithelial cell *via* the hemidesmosome to LM332 in the BM and anchoring fibrils. This connection may maintain the integrity of LM332 containing BMs, which, due to the sub-optimal LN domain composition of the LM332 heterotrimer, are unable to form stable laminin networks.[15,22,103] However, at least one missense mutation affecting the LN domain of LAMB3 has been demonstrated to be pathogenic in a skin blistering disease.[154] This raises the possibility that LN domain interactions are likely to be involved in LM332 BM formation, even if true laminin "polymerization" does not occur.

There is increasing evidence that α6β4 integrin, together with its binding partners type XVII collagen and the bullous pemphigoid antigen 1e (BPAG1e), play important roles in regulating keratinocyte migration, at least in part, by an impact on the deposition of LM332. Keratinocytes lacking expression of any of these components display an inability to establish or maintain polarity, reduced stability of lamellipodial protrusions, and assemble an aberrant LM332 matrix.[15,155–159] Specifically, whereas wild-type keratinocytes deposit LM332 in an almost linear track, cells deficient in expression of β4 integrin, type XVII collagen, or BPAG1e deposit LM332 in a circular pattern.[156,159,160] Interestingly, the latter cells also exhibit reduced activity of the small Rho family GTPase, Rac1, and its downstream effector protein cofilin.[156,159,160] Moreover, inhibition of Rac1 or cofilin activity through chemical means, expression of dominant negative forms, or inhibiting the slingshot phosphatases that activate cofilin, results in loss of front–rear polarity, with cells depositing LM332 in circles.[159,161] Together, these data suggest that α6β4 integrin and its associated proteins determine the pattern of LM332 deposition, although there remains the possibility that LM332 deposition is aberrant simply due to the lack of polarization of skin cells in which hemidesmosome protein levels are reduced.

Like α6β4 integrin, α3β1 integrin is a LM332 receptor.[162,163] However, it is found at focal contacts, cell-matrix adhesive complexes which link the actin cytoskeleton to the ECM, rather than at hemidesmosomes.[164] Indeed, LM332 is deposited aberrantly in spikes by keratinocytes lacking α3 integrin.[165] Interestingly, the impact of α3 integrin on LM332 organization may be dependent on its force transmission role, since inhibition of cytoskeletal contraction through targeting myosin II light chain or Rho-kinase leads to deposition of LM332 in dense patches at the cell periphery that co-localize with both α3β1 and α6β4 integrin.[166] These integrin/actin remodeling/contractile machinery effects are not restricted to the skin or to LM332, since it has been demonstrated that antibody based inhibition of β1 integrin or blocking of myosin light chain kinase prevents the organization of LM311 into fibers by type I alveolar epithelial cells.[167]

4.4.3.2 Dystroglycan. Dystroglycan provides a transmembrane linkage between the ECM and the cytoskeleton. The protein itself is generated from a single mRNA species and is subsequently post-translationally processed into α and β subunits.[168] The α subunit is associated with the extracellular portion of the plasma membrane and provides the binding sites for ECM molecules, including laminins.[169] The membrane spanning β dystroglycan, in contrast, interacts with a variety of cytoplasmic molecules and has signal transduction roles. α-Dystroglycan binds with high affinity, mainly through the laminin LG domains, to α1, α2, and α5 laminin subunits, but not to the α4 subunit.[170–172]

Laminins have been shown to preferentially polymerize when bound to cell-surface receptors including dystroglycan and α7β1 integrin.[173] Moreover, dystroglycan is required for the organization of LM111 on the cell surface.[169] Such receptor facilitated laminin network formation requires actin reorganization.[174] Dysfunctional glycosylation of dystroglycan results in defective laminin anchorage to the cell surface and correlates with aggressive breast and brain cancer presentation and poor outcomes.[175,176] *In vitro* studies demonstrate that restoration of the defective glycosyltransferase expression rescues laminin anchorage, reducing cell proliferation and tumor growth.[175]

Interaction of dystroglycan with the laminin network can also occur indirectly, for example perlecan binding (see below). These interactions are important for mechanosignal transduction independent of integrin activity in alveolar epithelial cells.[177]

4.4.3.3 Tetraspanins. The tetraspanins are a large family of proteins characterized by four membrane-spanning domains with cytoplasmic N- and C-ends and two extracellular loops. The larger loop contains four to six cysteine residues which form disulfide bonds necessary for proper folding.[178] The membrane-proximal region also has conserved cysteines that undergo palmitoylation, which is thought to be important for tetraspanin–tetraspanin interactions.[178] In addition, tetraspanins serve to recruit various membrane proteins, as well as signaling molecules and adaptor proteins, to form distinct membrane microdomains termed tetraspanin-enriched microdomains (TEMs).[179] Thus tetraspanins are involved in many aspects of cell functions, including signaling, cell adhesion, molecular trafficking, and tissue organization.

The laminin-binding integrins, including α3β1, α6β1, α7β1, and α6β4, interact with the tetraspanins. These integrins are targeted to the TEM by direct interaction with CD151.[180] Indeed, it is thought that CD151 and integrins interact during synthesis and traffic to the cell surface together, thus allowing the integrins to directly link with other tetraspanins and indirectly associate with other molecules, forming the so-called "tetraspanin web".[180,181]

The role of tetraspanins in modulating integrin–laminin interactions is still unresolved. One study comparing LM511/521 binding to liposomes containing α3β1 integrin alone or paired with CD151 demonstrated increased ligand binding in the presence of CD151.[182] In addition, loss of laminin binding ability is rescued by addition of CD151, providing evidence that CD151 may stabilize the active conformation of α3β1 integrin to increase ligand interactions.[182] In addition, adhesion and/or migration on laminin substrates is negatively impacted in carcinoma cell lines deficient in expression of CD151, as well as in NIH3T3 fibroblasts.[183–187] In addition, CD151 null mice demonstrate delayed wound healing, re-epithelialization, abnormal expression of α6β4 integrin, and disorganized laminin deposition.[188] Keratinocytes isolated from CD151 null mice also have defects in cell adhesion and migration on laminin, but not type 1 collagen.[189] Taken together, these results imply that CD151 is a mediator of integrin/laminin interactions, thereby promoting cell spreading, integrin clustering, and integrin activation.[180]

4.4.3.4 Syndecans. The four members of the syndecan family are transmembrane proteoglycans. They associate intracellularly with G protein-coupled receptors and extracellularly, *via* association with heparan sulfate and/or chondroitin sulfate chains, with a wide range of ECM proteins, including the G domains of all the laminin α subunits. These interactions lead to regulation of a wide repertoire of signaling pathways.[190,191]

As described below (Section 4.5), proteolytic processing of the α2, α3, and α4 subunits occurs at the C-terminus, removing the syndecan-binding LG4–5 module. The loss in their ability to bind syndecans provides an explanation for some observed functional differences between processed and unprocessed laminin heterotrimers. For example, the finding that the unprocessed form of the α3 subunit in LM332 preferentially supports migration over stable attachment may be due to its ability to interact with syndecan-1.[155,192]

Interestingly, in addition, or as a result of their signaling roles, there is an apparent feedback mechanism in which syndecans influence laminin matrix organization. Specifically, Chinese hamster ovary cells transfected with full-length syndecan-2 deposit fibrillar laminin and fibronectin matrix, while cells transfected with the C-terminally truncated form of syndecan-2 do not.[193] Moreover, syndecan-1 null keratinocytes deposit arrowhead-like arrays of LM332 in their matrix compared with the cloud-like trails observed in wild-type keratinocytes, indicating a potential role for syndecan-1 in LM332 matrix assembly.[194]

4.4.3.5 Lectins.   Lectins, from the latin *legere* 'to select', are proteins that bind to specific carbohydrates $(C_nH_{2n}O_n)$ and have been associated with regulating glycoprotein production, controlling cell division and death, and modulating cell–cell, cell–molecule, and cell–BM interactions.[195] Lectins are classified into multiple groups, depending upon the type of carbohydrate recognition domain present. Laminin subunits contain binding sites for a subgroup of lectins, the galectins, and galectin-3 preferentially binds LM111, LM511, and LM332.[196]

Galectin-3 binding to LM111 mediates signaling the PI-3 kinase pathway, stimulating changes in cell polarization, increasing migration, and inhibiting apoptosis.[197] Galectin-3 also exhibits affinity for LM332, and interaction between galectin-3 and LM332 leads to a reduction in focal adhesion kinase, paxillin phosphorylation, and increased cell motility.[197] Moreover, galectin-3 modulates the formation of complexes of LM332, integrins, and EGFR, as well as influencing integrin clustering, focal contact formation, and EGFR phosphorylation.[198,199] Together, these findings indicate that galectin–laminin interactions add an extra layer of regulation to laminin-mediated signaling events and may, in part, explain some of the context dependent cellular responses to superficially similar ECM assemblies.

In addition to laminins, galectins interact with a wide range of intra-cellular and extracellular proteins and, through binding, modulate the activity of pathways important for cell attachment, migration, prolif-eration, and the regulation of apoptosis.[200,201] Consistent with these findings, up-regulation in galectin-3 expression is associated with aggressive metastatic tumor formation.[196] Interestingly, galectin-3 over-expression in breast cancer is associated with increased surface expression of α4 and β7 integrins, suggesting that galectin–laminin binding may stabilize integrin complexes on cell surfaces, reducing their turnover.[202] Galectin-binding ligands, such as laminins, induce a positive feedback loop which promotes secretion of galectins through a non-standard secretory pathway.[197] These data indicate that, in addi-tion to the direct impact on laminin function, galectin–laminin inter-actions may indirectly impact signaling pathways.

4.4.4 Other Laminin-Binding Proteins

4.4.4.1 Heparan Sulfate Proteoglycans – Perlecan and Agrin. Perlecan is a major heparan sulfate proteoglycan present in almost all BMs, and interacts with a wide range of ECM molecules including nidogen, fibronectin, type IV collagen, laminins, and nidogen 1.[203–205] Although not required for matrix assembly, perlecan is essential for maintaining BM integrity, particularly in highly stressed tissues such as the heart.[203,206] Perlecan also co-localizes with LM311 fibers in alveolar epithelial cells and work from our lab indicates that the perlecan–LM311 complex may be involved in mechanosignal transduction.[177]

A second heparan sulfate proteoglycan, agrin, also interacts with laminins, specifically LM111, 211, and 221, and is thought to be important for formation of neuromuscular junctions.[207–209] These interactions occur *via* the LCC domain and, based on the widespread expression profile of agrin, suggest that laminin–agrin interactions are a feature of most, if not all, BMs.[210,211]

Interestingly, heparin- and heparan-sulfate-binding capabilities have been described for the LN domains of α1 and α2 subunits, with weaker interactions also described for α3B.[27,28,212] Whether these interactions influence the ability of laminins to form higher order net-works or whether they affect their mechanical properties has not yet been determined. However, it is possible that these interactions are a means *via* which context dependent network organizational changes are mediated.[98]

4.4.4.2   Type VII Collagen.   Type VII collagen is a critical component of skin integrity as it is a major component of anchoring fibrils, which interact with the β3 subunit of LM332.[153,213] In corneal keratinocytes *in vitro* anchoring fibrils appear involved in determining where hemidesmosomes form.[214] This raises the possibility that one function of type VII collagen is to target LM332 or organize LM332. In support of this, immunoelectron microscopy reveals that LM332 is distributed in a patchy pattern along the BM, with patches of LM332 located directly beneath hemidesmosomal plaques.[215] However, laminin deposition does not depend on the action of type VII collagen since LM332 is deposited normally from type VII collagen null-keratinocytes in skin grafts.[216]

4.5   LAMININ FUNCTIONAL MODIFICATIONS

4.5.1   Proteolytic Processing

Proteolytic processing is a major form of post-translational functional modification in laminins. These processing events may remove specific functionally active sites, may cause conformational changes in other binding sites that modify laminin receptor affinities, and may release active fragments. In addition to removal of the secretion signal sequence, major processing events have been described for the C-terminus of the α2, α3, and α4 subunits, and the N-terminus of α3A, α3B, and γ2 (Figure 4.2).[11,52,98,144,170,217–221]

Processing of the α2 subunit occurs within the LG3 domain by a furin-like protease (Figure 4.2).[222] LM211 binds to cell surfaces, mediates BM assembly, and induces acetylcholine receptor (AChR) clustering. Although all three of these activities depend on the LG1–3 module, only AChR clustering requires its processing.[223]

The C-terminal cleavage event for α3A and α3B occurs in the linker region between LG3 and LG4, leading to release of LG4 and 5 (Figure 4.2).[52,98,221,224] This domain appears to be particularly amenable to processing and is a target of several enzymes, including plasmin and the bone morphogenetic protein-1 isoenzymes (BMP1, mammalian tolloid and tolloid-like 1 and 2).[52,225] Indeed, plasminogen and tissue plasminogen activator (tPA) exhibit high-affinity, specific binding to the LM332 α3 LG1. Binding to this region leads to a 32-fold increase in the catalytic efficiency of tPA-catalyzed plasminogen activation, generating increased matrix associated plasmin.[219] Thus the LM332–tPA interaction effectively acts as a targeting mechanism to ensure efficient LM332 processing. With regard to LM332, this C-terminal processing has functional implications, since the unprocessed form preferentially supports keratinocyte migration, whereas the processed form promotes stable attachment.[52,226] Consistent with this, while normal undamaged skin predominantly contains the processed form of LM332, in wounded or squamous cell carcinoma the unprocessed form is up-regulated.[227,228] Antibodies targeting the LG4–5 module have potential therapeutic utility, since they inhibit tumorigenesis in mouse models.[228]

Cleavage of the α4 subunit also occurs at its C-terminus in the linker region between LG1–3 and LG4–5, with loss of the released fragment (Figure 4.2).[170] The functional significance of this cleavage is unclear. However, the syndecan-binding site resides within the LG4–5 region, and therefore this cleavage event may change receptor interactions.[229] Interestingly, peptides derived from the α4 and α5 LG domains display antimicrobial activities, suggesting that cleavage and breakdown of the LG4–5 domain may have important immune response functions.[230]

In addition to C-terminal processing events, both the α3A and α3B subunits are processed at their N-termini (Figure 4.2).[220] In the case of α3B, this leads to release of a 190 kDa fragment capable of promoting cell adhesion, migration, and proliferation through interaction with α3β1 integrin.[220] N-terminal processing of α3A removes two LE domains totaling approximately 20 kDa.[217,224] As described above, the N-terminus of α3A is mutated in the chronic granulation disorder laryngo-onycho-cutaneous syndrome, raising the possibility that this released fragment plays an important role as a soluble signaling molecule and may terminate the granulation tissue response.[53]

Cleavage of the N-terminal 50 kDa of γ2 is necessary for deposition of LM332 by cultivated keratinocytes (Figure 4.2).[218] Cleavage is also thought to permit interaction with type VII collagen.[213] Indeed, the released γ2 subunit short arm promotes adhesion to processed LM332 and suppresses migration stimulated by epidermal growth factor *via* its LE domain in *in vitro* recombinant protein studies.[231]

4.5.2 Solid Phase Agonists

There is increasing evidence that information transmission from the ECM to cells is dependent not only ligand density but also on the physical (mechanical) characteristics of the polymer itself, with substrate stiffness stimulating integrin expression and activation.[232,233] Indeed, substrate stiffness impacts the differentiation path down which cells progress.[234–238] Stiffness is also influenced by the composition of the

BM itself, and analyses of embryonal carcinoma cell BM reveals a viscosity increase as a function of LM111 or type IV collagen concentration.[239-241] Interestingly, substrate stiffness also influences matrix deposition. For example, plating type II alveolar epithelial cells of the lung onto stiffer substrates promotes expression of laminin α3 and LM311 fiber assembly, which unusually follows the pattern of fibronectin fiber organization (Figure 4.4).[242]

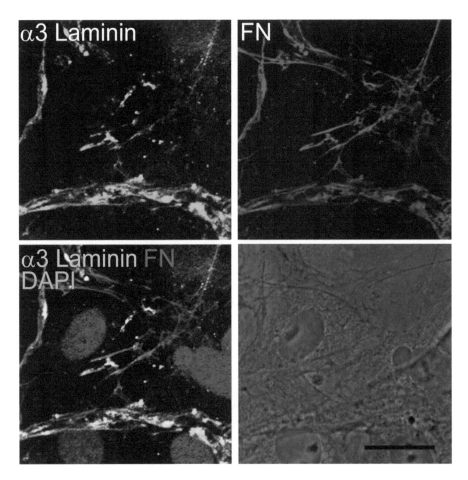

Figure 4.4   Primary rat type II alveolar cells at 4 days after culturing on a stiff substrate (glass coverslip) were prepared for double label immunofluorescence using antibodies against the α3 laminin subunit (top panel, left) and fibronectin (FN) (top panel, right). An overlay of the stained cells stained with DAPI to mark nuclei reveals the overlapping localization of laminin and fibronectin in a fibrous matrix (bottom panel, left). A phase contrast image of the cells is shown in the bottom panel on the right. Bar, 20 μm.

4.5.3   Glycosylation

Laminins are heavily glycosylated molecules, with a reported 13 to 30% of the apparent molecular weight of laminins due to N-linked oligosaccharide chains.[243] Not surprisingly, glycosylation impacts the entire gamut of laminin function in cell-type-, laminin-, and context-specific ways. For example, LM111 purified from tunicamycin-treated mouse embryonal carcinoma-derived M1536 B3 cells is impaired in its ability to support cell spreading and neurite outgrowth, but not in its ability to support cell adhesion when compared with normally glycosylated LM111.[244] Interestingly, although glycosylation fails to impact heterotrimer assembly, it inhibits laminin secretion.[245]

In the case of LM332, the length and branching of its oligosaccharides affects the behavior of cells with which it interacts. Specifically, Kariya and co-workers have demonstrated that LM332 purified from cells over-expressing the bisecting GlcNAc N-acetylglucosaminyltransferase III (GnT-III) is an inferior substrate for supporting cell attachment, cell scattering, and migration when compared with LM332 from wild-type cells or those over-expressing the β1,6 GlcNAc GnT-V.[246] This is likely a consequence of impaired α3β1 integrin clustering.[246]

4.6   SOURCES OF PURIFIED LAMININ

One approach for analyzing the properties of laminin heterotrimers involves the use of purified laminin molecules. Below we discuss their use in *in vitro* and *in vivo* studies.

4.6.1   EHS Laminin/Matrigel

The Engelbreth-Holm-Swarm (EHS) tumor produces large amounts of ECM with BM properties and was initially identified in mouse as a spontaneously arising, poorly differentiated benign tumor.[247,248] Discovery of this property of the tumor allowed workers to isolate relatively large amounts of BM components from it, enabling the characterization of the proteins both biochemically and functionally. Laminin was isolated from the EHS tumor matrix, in addition to type IV collagen.[5,249] The EHS-purified laminin (later identified as LM111) promotes adhesion of a variety of cell types, including endothelial and epithelial cells, as well as tumor cells.[250]

The crude BM preparation isolated from the EHS tumor and commercialized as Matrigel is composed primarily of laminin, type IV collagen, and entactin, as well as a variety of growth factors, proteases, and other proteins.[251] Matrigel has been used extensively as a matrix for the culture of primary cells, immortalized cell lines, embryonic stem cells, induced pluripotent stem cells from a variety of tissues, and tissue explants, with the advantage of using Matrigel as a culture substrate being that there is no need for a fibroblast feeder layer.[252–254] Cells can be plated on or within the gel depending upon the assay. Remarkably, Matrigel supports organization of cells into structures resembling the tissues from which they were originally isolated. For example, endothelial cells assemble into structures resembling capillaries, complete with lumen, while mammary epithelial cells assemble duct-like or alveolar-like structures with lumens into which they secrete casein.[255,256] Matrigel is also routinely used in invasion assays using Boyden chambers.[257] In addition, Matrigel can be used as a platform when introducing tumor cells into mice, as it enhances the ability and speed of these cells to form tumors.[258] Matrigel has utility in enhancing cell differentiation in *in vivo* models. For example, following subcutaneous implantation with Matrigel into dogs, mesenchymal stem cells derived from adipose cells grown in 2D and 3D culture show enhanced osteogenic potential.[259]

Though Matrigel has been a "work horse" of cell culture systems, it is not without its problems. It is a crude preparation that may vary from lot to lot. For example, varying the diet or age of the mice from which the Matrigel is generated can impact on its characteristics and functions.[260] Moreover, a recent in-depth analysis of its composition has identified 1976 unique proteins.[261] Many of these proteins are growth factors, which may explain why Matrigel supports such robust growth. Surprisingly, there is also variation in the growth and behavior of cells plated on Matrigel depending upon the underlying substrate upon which the Matrigel is coated.[262] Finally, the major laminin present in Matrigel is LM111, which is not present in most adult tissues. Hence, its effects on cells and their differentiation may reflect processes that occur during development rather than in adult tissues.

4.6.2 Placentally Derived Laminin

Human placental tissue-derived laminin preparations exhibit similar properties to Matrigel. Such preparations are now commercially available and have been used in the culture of immortalized cell lines, primary cells, and stem cells. The addition of human placental-derived laminins to the media enhances the proliferation and survival of neural stem cells.[263] Moreover, injection of human placental-derived laminins into the lesional areas of injured spinal cord of rats enhances spinal cord regeneration.[264] Despite its successful use in many functional studies *in vitro* and *in vivo*, human placental-derived laminins share the same drawback as Matrigel in that they are mixtures of many proteins and vary in composition from one batch to another. For example, analyses of these preparations have identified multiple laminin isoforms, including LM211, 411, and 511, as well as contaminating proteins such as fibronectin.[265]

4.6.3 LM111

LM111 has been isolated from EHS tumors.[266] This mouse LM111 has been extensively studied at the biophysical and biochemical levels since procedures for its isolation were first described.[266] Moreover, this laminin is now commercially available. However, these preparations are not 100% pure and may contain other ECM proteins such as nidogen.[266]

4.6.4 Laminin Isolated from Cells in Culture

Conditioned medium of a number of different cell types has been used as a source of laminin isoforms. For example, the human lung adenocarcinoma cell line A549 expresses the α5 laminin subunit, but no other α subunit, thus facilitating purification of LM511 and LM521 from its media using immunoaffinity chromatography.[267,268] Similarly, the glioblastoma cell line T98G expresses α4 laminin, making its conditioned media a source from which to purify LM411.[269] These cultured-cell-derived laminins have been used to characterize receptor interactions and their influence on cell behavior.[140,163,268,269] Using column chromatography, LM332 has also been purified from conditioned media of human keratinocytes and the rat-bladder-derived cell line 804G by several groups.[11,22,268,270] Addition of soluble LM332 purified from 804G cells rapidly induces cell attachment and spreading in a variety of cell lines, as well as inducing formation of hemidesmosomes in corneal explants.[270] Moreover, LM332 purified by column chromatography from conditioned media of 184A1 mammary epithelial cells induces increased migration and invasion of the breast carcinoma cell line MCF7.[271] Generally speaking, laminins purified from conditioned media are not commercially available and are only obtained from research laboratories.

4.6.5 Recombinant Laminin

In an effort to develop conditions under which high levels of a specific laminin trimer can be produced without other contaminants, many groups have developed systems for expression and purification of recombinant laminins. This approach involves sequential transformation of a mammalian cell line, frequently HEK293, with an expression plasmid encoding each laminin subunit and selectable antibiotic marker. One subunit of the trimer can be tagged with sequence encoding a small epitope, such as Flag or His residues, for ease of purification. Large volumes of cells can be cultured under triple antibiotic selection and media containing the secreted laminin collected for easy purification *via* column chromatography using an antibody-containing matrix, nickel resin, or heparin-sepharose. Such approaches have been successfully used to isolate preparations of LM121, LM411, LM511, and LM332.[16,17,143,272–274] Examples of the utility of recombinant laminin include its ability to support the adhesion and migration of endothelial cells.[16] Recombinant LM511 has been used successfully as a substrate for growing human embryonic stem cells, as well as induced pluripotent stem cells.[274] Moreover, recombinant LM511 can rescue hair growth in mice deficient in endogenous LM511 expression.[275] Recombinant LM332, containing an unprocessed γ2 subunit, induces directional hypermotility accompanied by expression of p16 and growth arrest in cultured keratinocytes.[273]

Until recently, recombinant laminins were mostly available in the labs of individual investigators. However, purified recombinant laminin isoforms are now available commercially. For example, BioLamina sells several recombinant, purified laminin isoforms. The ability to purify large quantities of recombinant laminins using a standardized protocol will make these extremely valuable reagents. However, even these recombinantly expressed laminins may not exhibit the entire repertoire of functions of their native counterparts in tissues. Specifically, the cell line used in the preparation of recombinant laminin may not glycosylate the recombinant protein in the same way as the cell which produces the native molecule.[276] Since glycosylation modulates laminin function (Section 4.5.3), it is important that the glycosylation patterns mimic those *in vivo*.

4.7   FUTURE DIRECTIONS

4.7.1   Expression of Tagged Laminin Subunits

An emerging approach for the study and modeling of laminin trimer assembly is through the use of tagged laminin subunits exogenously expressed by cells in culture or in intact tissue.[277] For example, we have evaluated LM332 matrix assembly and functions in motility using tagged laminin subunits.[159,277] In an exciting *in vivo* approach, Currie and co-workers have exogenously expressed mCherry-tagged laminin α2 or α4 to investigate myotome formation in laminin α2-deficient zebrafish, a model for congenital muscular dystrophy.[278] In addition, laminin molecules labeled with a fluorophore have been used to analyze the relationship between defective laminin anchorage at the cell surface of mammary cells and tumor progression, as well as for examining laminin–membrane or laminin–pathogen interactions.[175,279–281] More work in this area will likely provide new insight into how laminins become incorporated into BMs.

4.7.2 Laminin Peptides as a Therapeutic in Wound Healing

There are numerous reports in which investigators have attempted to mimic laminin function using laminin peptides, following the pioneering work of Yamada and his co-workers.[282–284] Indeed, a number of laminin peptides are commercially available and have been used for coating substrates for the culture of cells *in vitro*. There are also a number of reports that laminin-derived peptides may have therapeutic utility.[285] In a rat full-thickness wound model, the addition of a specific 12-mer peptide derived from either the laminin α1 or γ1 subunit enhances wound healing.[285,286] In addition, Min and co-workers coated α3 laminin peptides onto chitin microfibers and used the resultant material to coat full-thickness cutaneous wounds in rats and rabbits.[286] Peptide-treated wounds heal faster than the untreated wounds.[286] More recently, an α3 laminin subunit peptide incorporated into a collagen scaffold has been used to treat rat full-thickness wounds.[287] Wounds treated with the peptide show significant improvement in wound healing, as well as decreased inflammation and increased fibroblast proliferation.[287] The disadvantage of laminin-based peptides is that their function is unlikely to mimic that of a complex three-subunit molecule. On the other hand, the utility of laminin-based peptides with specific receptor binding affinities has advantages when receptor-mediated activation of pathways is needed.

4.7.3 Laminin Molecules to Enhance Cell Repopulation of Tissue Scaffolds

The demand to replace tissues and organs has far outstripped the present availability of donor organs and has driven an explosion in research to find biological substitutes. ECM scaffolds repopulated with donor cells have recently been described for a number of organs, including the lung, and some of these have been used therapeutically in humans (Figure 4.5).[288–295] With regard to laminin, an ECM-based protein scaffold has been developed as a vehicle to deliver neural stem

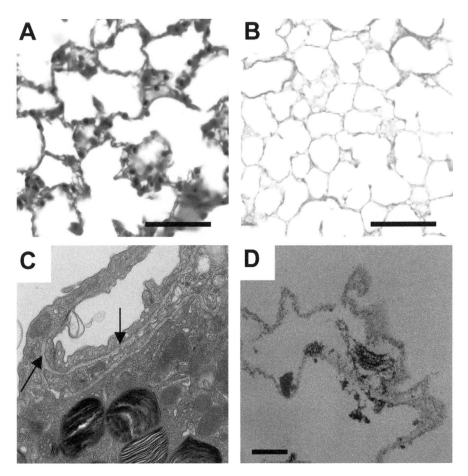

Figure 4.5 Light and electron microscopical analyses of an untreated (A and C) and the decellularized scaffold of a mouse lung (B and D). The material in A and B was fixed and stained with hematoxylin/eosin following paraffin embedding while the samples in C and D were prepared for ultrastructural examination. Arrows in C indicate BM. Bars in A and B, 200 μm. Bar in D, 500 nm.

cells directly into a mouse brain.[296] Either laminin or fibronectin is added to a collagen-based scaffold, which is then injected into mice who had received traumatic brain injuries. Survivability of mice who receive the laminin-coated scaffold is enhanced over mice treated with the fibronectin-coated scaffolds, and they also perform better on cognitive tests than non-treated animals.[296]

An alternative to an ECM-based scaffold is the use of hydrogel materials. This technology requires that the proper ECM milieu exists within the gel.[297] *In vitro* studies indicate that the inclusion of laminin in collagen-based hydrogels enhances neurite outgrowth from dorsal root ganglia and improves the function of Schwann cells.[298,299] Furthermore, the formation of nascent blood vessels is promoted both *in vitro* and *in vivo* in polyethylene glycol (PEG) diacrylate hydrogels in which laminin peptides have been incorporated.[300] Another use of laminin–PEG hydrogels has recently been described for the treatment of invertebral disc degeneration.[301] In this system, the laminin hydrogel can be used to used deliver cells to the disc to enhance its regeneration.[301] Taken together, these studies emphasize the utility of laminins in both bioengineering and therapeutic fields. The following chapters will discuss artificial ECMs fabricated from hydrogels and other biomaterials in greater detail.

ACKNOWLEDGEMENTS

KJH, SBH and JCRJ were supported by the National Institute of Arthritis and Musculoskeletal and Skin Diseases and the National Heart, Lung, and Blood Institute of the National Institutes of Health under award numbers K99AR060242 (KJH), RO1 AR054184 (JCRJ and SBH) and RO1 HL092963 (JCRJ). The content is solely the responsibility of the authors and does not necessarily represent the official views of the National Institutes of Health. We thank Jessica Eisenberg, Alex Woychek and Sho Hiroyasu for help in preparing the figures.

REFERENCES

1. M. Aumailley, L. Bruckner-Tuderman, W. G. Carter, R. Deutzmann, D. Edgar, P. Ekblom, J. Engel, E. Engvall, E. Hohenester, J. C. R. Jones, H. K. Kleinman, M. P. Marinkovich, G. R. Martin, U. Mayer, G. Meneguzzi, J. H. Miner, K. Miyazaki, M. Patarroyo, M. Paulsson, V. Quaranta, J. R. Sanes, T. Sasaki, K. Sekiguchi, L. M. Sorokin, J. F. Talts, K. Tryggvason, J. Uitto, I. Virtanen, K. von der Mark, U. M. Wewer, Y. Yamada and P. D. Yurchenco, A simplified laminin nomenclature, *Matrix Biol.*, 2005, **24**, 326–332.
2. P. D. Yurchenco, P. S. Amenta and B. L. Patton, Basement membrane assembly, stability and activities through the developmental lens, *Matrix Biol.*, 2004, **22**, 521–538.
3. A. Domogatskaya, S. Rodin and K. Tryggvason, Functional diversity of laminins, *Annu. Rev. Cell Dev. Biol.*, 2012, **28**, 523–553.
4. J. H. Miner and P. D. Yurchenco, Laminin functions in tissue morphogenesis, *Annu. Rev. Cell Dev. Biol.*, 2004, **20**, 255–284.

5. R. Timpl, H. Rohde, P. G. Robey, S. I. Rennard, J. M. Foidart and G. R. Martin, Laminin–a glycoprotein from basement membranes, *J. Biol. Chem.*, 1979, **254**, 9933–9937.

6. B. Fahey and B. M. Degnan, Origin and evolution of laminin gene family diversity, *Mol. Biol. Evol.*, 2012, **29**, 1823–1836.

7. J. D. Lopes, G. F. Da-Mota, C. R. Carneiro, L. Gomes, F. Costa-e-Silva-Filho and R. R. Brentani, Evolutionary conservation of laminin-binding proteins, *Braz. J. Med. Biol. Res.*, 1988, **21**, 1269–1273.

8. I. Hunter, T. Schulthess, M. Bruch, K. Beck and J. Engel, Evidence for a specific mechanism of laminin assembly, *Eur. J. Biochem.*, 1990, **188**, 205–211.

9. I. Hunter, T. Schulthess and J. Engel, Laminin chain assembly by triple and double stranded coiled-coil structures, *J. Biol. Chem.*, 1992, **267**, 6006–6011.

10. P. R. Macdonald, A. Lustig, M. O. Steinmetz and R. A. Kammerer, Laminin chain assembly is regulated by specific coiled-coil interactions, *J. Struct. Biol.*, 2010, **170**, 398–405.

11. M. P. Marinkovich, G. P. Lunstrum and R. E. Burgeson, The anchoring filament protein kalinin is synthesized and secreted as a high molecular weight precursor, *J. Biol. Chem.*, 1992, **267**, 17900–17906.

12. M. Sasaki, S. Kato, K. Kohno, G. R. Martin and Y. Yamada, Sequence of the cDNA encoding the laminin B1 chain reveals a multidomain protein containing cysteine-rich repeats, *Proc. Natl. Acad. Sci. U. S. A.*, 1987, **84**, 935–939.

13. H. Rohde, G. Wick and R. Timpl, Immunochemical characterization of the basement membrane glycoprotein laminin, *Eur. J. Biochem.*, 1979, **102**, 195–201.

14. L. Risteli and R. Timpl, Isolation and characterization of pepsin fragments of laminin from human placental and renal basement membranes, *Biochem. J.*, 1981, **193**, 749–755.

15. K. J. Hamill, K. Kligys, S. B. Hopkinson and J. C. Jones, Laminin deposition in the extracellular matrix: a complex picture emerges, *J. Cell Sci.*, 2009, **122**, 4409–4417.

16. M. Doi, J. Thyboll, J. Kortesmaa, K. Jansson, A. Iivanainen, M. Parvardeh, R. Timpl, U. Hedin, J. Swedenborg and K. Tryggvason, Recombinant human laminin-10 (alpha5beta1gamma1). Production, purification, and migration-promoting activity on vascular endothelial cells, *J. Biol. Chem.*, 2002, **277**, 12741–12748.

17. J. Kortesmaa, P. Yurchenco and K. Tryggvason, Recombinant laminin-8 (α4β1γ1). Production, purification, and interactions with integrins, *J. Biol. Chem.*, 2000, **275**, 14853–14859.

18. F. Carafoli, S. A. Hussain and E. Hohenester, Crystal structures of the network-forming short-arm tips of the laminin beta1 and gamma1 chains, *PLoS One*, 2012, 7, e42473.
19. S. A. Hussain, F. Carafoli and E. Hohenester, Determinants of laminin polymerization revealed by the structure of the alpha5 chain amino-terminal region, *EMBO Rep.*, 2011, **12**, 276–282.
20. A. Purvis and E. Hohenester, Laminin network formation studied by reconstitution of ternary nodes in solution, *J. Biol. Chem.*, 2012, **287**, 44270–44277.
21. K. K. McKee, D. Harrison, S. Capizzi and P. D. Yurchenco, Role of laminin terminal globular domains in basement membrane assembly, *J. Biol. Chem.*, 2007, **282**, 21437–21447.
22. Y. S. Cheng, M. F. Champliaud, R. E. Burgeson, M. P. Marinkovich and P. D. Yurchenco, Self-assembly of laminin isoforms, *J. Biol. Chem.*, 1997, **272**, 31525–31532.
23. H. J. Choi, B. H. Lee, J. H. Kang, H. J. Jeong, K. C. Moon, I. S. Ha, Y. S. Yu, V. Matejas, M. Zenker, Y. Choi and H. I. Cheong, Variable phenotype of Pierson syndrome, *Pediatr. Nephrol.*, 2008, **23**, 995–1000.
24. M. Kagan, A. H. Cohen, V. Matejas, C. Vlangos and M. Zenker, A milder variant of Pierson syndrome, *Pediatr. Nephrol.*, 2008, **23**, 323–327.
25. F. I. Staquicini, E. Dias-Neto, J. Li, E. Y. Snyder, R. L. Sidman, R. Pasqualini and W. Arap, Discovery of a functional protein complex of netrin-4, laminin gamma1 chain, and integrin alpha6beta1 in mouse neural stem cells, *Proc. Natl. Acad. Sci. U. S. A.*, 2009, **106**, 2903–2908.
26. F. I. Schneiders, B. Maertens, K. Bose, Y. Li, W. J. Brunken, M. Paulsson, N. Smyth and M. Koch, Binding of netrin-4 to laminin short arms regulates basement membrane assembly, *J. Biol. Chem.*, 2007, **282**, 23750–23758.
27. H. Colognato-Pyke, J. J. O'Rear, Y. Yamada, S. Carbonetto, Y. S. Cheng and P. D. Yurchenco, Mapping of network-forming, heparin-binding, and alpha 1 beta 1 integrin-recognition sites within the alpha-chain short arm of laminin-1, *J. Biol. Chem.*, 1995, **270**, 9398–9406.
28. H. Colognato, M. MacCarrick, J. J. O'Rear and P. D. Yurchenco, The laminin alpha2-chain short arm mediates cell adhesion through both the alpha1beta1 and alpha2beta1 integrins, *J. Biol. Chem.*, 1997, **272**, 29330–29336.
29. P. K. Nielsen and Y. Yamada, Identification of cell-binding sites on the Laminin alpha 5 N-terminal domain by site-directed mutagenesis, *J. Biol. Chem.*, 2001, **276**, 10906–10912.

30. K. K. McKee, S. Capizzi and P. D. Yurchenco, Scaffold-forming and adhesive contributions of synthetic laminin-binding proteins to basement membrane assembly, *J. Biol. Chem.*, 2009, **284**, 8984–8994.

31. G. Panayotou, P. End, M. Aumailley, R. Timpl and J. Engel, Domains of laminin with growth-factor activity, *Cell*, 1989, **56**, 93–101.

32. S. Schenk, E. Hintermann, M. Bilban, N. Koshikawa, C. Hojilla, R. Khokha and V. Quaranta, Binding to EGF receptor of a laminin-5 EGF-like fragment liberated during MMP-dependent mammary gland involution, *J. Cell Biol.*, 2003, **161**, 197–209.

33. F. Decline and P. Rousselle, Keratinocyte migration requires alpha2beta1 integrin-mediated interaction with the laminin 5 gamma2 chain, *J. Cell Sci.*, 2001, **114**, 811–823.

34. M. Gerl, K. Mann, M. Aumailley and R. Timpl, Localization of a major nidogen-binding site to domain III of laminin B2 chain, *Eur. J. Biochem.*, 1991, **202**, 167–174.

35. U. Mayer, R. Nischt, E. Poschl, K. Mann, K. Fukuda, M. Gerl, Y. Yamada and R. Timpl, A single EGF-like motif of laminin is responsible for high affinity nidogen binding, *EMBO J.*, 1993, **12**, 1879–1885.

36. N. Gersdorff, E. Kohfeldt, T. Sasaki, R. Timpl and N. Miosge, Laminin gamma3 chain binds to nidogen and is located in murine basement membranes, *J. Biol. Chem.*, 2005, **280**, 22146–22153.

37. A. Utani, M. Nomizu and Y. Yamada, Fibulin-2 binds to the short arms of laminin-5 and laminin-1 via conserved amino acid sequences, *J. Biol. Chem.*, 1997, **272**, 2814–2820.

38. K. Balbona, H. Tran, S. Godyna, K. C. Ingham, D. K. Strickland and W. S. Argraves, Fibulin binds to itself and to the carboxyl-terminal heparin-binding region of fibronectin, *J. Biol. Chem.*, 1992, **267**, 20120–20125.

39. T. Sasaki, H. Larsson, D. Tisi, L. Claesson-Welsh, E. Hohenester and R. Timpl, Endostatins derived from collagens XV and XVIII differ in structural and binding properties, tissue distribution and anti-angiogenic activity, *J. Mol. Biol.*, 2000, **301**, 1179–1190.

40. U. Odenthal, S. Haehn, P. Tunggal, B. Merkl, D. Schomburg, C. Frie, M. Paulsson and N. Smyth, Molecular analysis of laminin N-terminal domains mediating self-interactions, *J. Biol. Chem.*, 2004, **279**, 44504–44512.

41. T. Sasaki and R. Timpl, Domain IVa of laminin alpha5 chain is cell-adhesive and binds beta1 and alphaVbeta3 integrins through Arg-Gly-Asp, *FEBS Lett.*, 2001, **509**, 181–185.

42. B. Schulze, K. Mann, E. Poschl, Y. Yamada and R. Timpl, Structural and functional analysis of the globular domain IVa of the laminin alpha 1 chain and its impact on an adjacent RGD site, *Biochem. J.*, 1996, **314**, 847–851.

43. F. Katagiri, M. Sudo, T. Hamakubo, K. Hozumi, M. Nomizu and Y. Kikkawa, Identification of active sequences in the L4a domain of laminin alpha5 promoting neurite elongation, *Biochemistry*, 2012, **51**, 4950–4958.

44. H. Ido, S. Ito, Y. Taniguchi, M. Hayashi, R. Sato-Nishiuchi, N. Sanzen, Y. Hayashi, S. Futaki and K. Sekiguchi, Laminin isoforms containing the gamma3 chain are unable to bind to integrins due to the absence of the glutamic acid residue conserved in the C-terminal regions of the gamma1 and gamma2 chains, *J. Biol. Chem.*, 2008, **283**, 28149–28157.

45. H. Ido, A. Nakamura, R. Kobayashi, S. Ito, S. Li, S. Futaki and K. Sekiguchi, The requirement of the glutamic acid residue at the third position from the carboxyl termini of the laminin gamma chains in integrin binding by laminins, *J. Biol. Chem.*, 2007, **282**, 11144–11154.

46. Y. Taniguchi, H. Ido, N. Sanzen, M. Hayashi, R. Sato-Nishiuchi, S. Futaki and K. Sekiguchi, The C-terminal region of laminin beta chains modulates the integrin binding affinities of laminins, *J. Biol. Chem.*, 2009, **284**, 7820–7831.

47. P. Santos-Valle, I. Guijarro-Munoz, A. M. Cuesta, V. Alonso-Camino, M. Villate, A. Alvarez-Cienfuegos, F. J. Blanco, L. Sanz and L. Alvarez-Vallina, The heterotrimeric laminin coiled-coil domain exerts anti-adhesive effects and induces a pro-invasive phenotype, *PLoS One*, 2012, **7**, e39097.

48. E. Hohenester, D. Tisi, J. F. Talts and R. Timpl, The crystal structure of a laminin G-like module reveals the molecular basis of alpha-dystroglycan binding to laminins, perlecan, and agrin, *Mol. Cell*, 1999, **4**, 783–792.

49. D. Tisi, J. F. Talts, R. Timpl and E. Hohenester, Structure of the C-terminal laminin G-like domain pair of the laminin alpha2 chain harbouring binding sites for alpha-dystroglycan and heparin, *EMBO J.*, 2000, **19**, 1432–1440.

50. H. Ido, K. Harada, S. Futaki, Y. Hayashi, R. Nishiuchi, Y. Natsuka, S. Li, Y. Wada, A. C. Combs, J. M. Ervasti and K. Sekiguchi, Molecular dissection of the alpha-dystroglycan- and integrin-binding sites within the globular domain of human laminin-10, *J. Biol. Chem.*, 2004, **279**, 10946–10954.

51. F. Yokoyama, N. Suzuki, Y. Kadoya, A. Utani, H. Nakatsuka, N. Nishi, M. Haruki, H. K. Kleinman and M. Nomizu, Bifunctional peptides derived from homologous loop regions in the laminin alpha chain LG4 modules interact with both alpha 2 beta 1 integrin and syndecan-2, *Biochemistry*, 2005, **44**, 9581–9589.

52. L. E. Goldfinger, M. S. Stack and J. C. R. Jones, Processing of laminin-5 and its functional consequences: role of plasmin and tissue-type plasminogen activator, *J. Cell Biol.*, 1998, **141**, 255–265.

53. W. H. I. McLean, A. D. Irvine, K. J. Hamill, N. V. Whittock, C. M. Coleman-Campbell, J. E. Mellerio, G. S. Ashton, P. J. H. Dopping-Hepenstal, R. A. J. Eady, T. Jamil, R. J. Phillips, S. G. Shabbir, T. S. Haroon, K. Khurshid, J. E. Moore, B. Page, J. Darling, D. J. Atherton, M. A. M. van Steensel, C. S. Munro, F. J. D. Smith and J. A. McGrath, An unusual N-terminal deletion of the laminin 3a isoform leads to the chronic granulation tissue disorder laryngo-onycho-cutaneous syndrome, *Hum. Mol. Genet.*, 2003, **12**, 2395–2409.

54. M. C. Ryan, R. Tizard, D. R. VanDevanter and W. G. Carter, Cloning of the LamA3 gene encoding the alpha 3 chain of the adhesive ligand epiligrin. Expression in wound repair, *J. Biol. Chem.*, 1994, **269**, 22779–22787.

55. R. Doliana, I. Bellina, F. Bucciotti, M. Mongiat, R. Perris and A. Colombatti, The human α3b is a 'full-sized' laminin chain variant with a more widespread tissue expression than the truncated α3a, *FEBS Lett.*, 1997, **417**, 65–70.

56. T. Airenne, H. Haakana, K. Sainio, T. Kallunki, P. Kallunki, H. Sariola and K. Tryggvason, Structure of the human laminin gamma 2 chain gene LAMC2: alternative splicing with different tissue distribution of two transcripts, *Genomics*, 1996, **32**, 54–64.

57. T. Airenne, Y. Lin, M. Olsson, P. Ekblom, S. Vainio and K. Tryggvason, Differential expression of mouse laminin gamma2 and gamma2* chain transcripts, *Cell Tissue Res.*, 2000, **300**, 129–137.

58. Y. Hayashi, K. H. Kim, H. Fujiwara, C. Shimono, M. Yamashita, N. Sanzen, S. Futaki and K. Sekiguchi, Identification and recombinant production of human laminin alpha4 subunit splice variants, *Biochem. Biophys. Res. Commun.*, 2002, **299**, 498–504.

59. M. E. Durkin, M. Gautam, F. Loechel, J. R. Sanes, J. P. Merlie, R. Albrechtsen and U. M. Wewer, Structural organization of the human and mouse laminin beta2 chain genes, and alternative splicing at the 5' end of the human transcript, *J. Biol. Chem.*, 1996, **271**, 13407–13416.

60. R. A. Kammerer, P. Antonsson, T. Schulthess, C. Fauser and J. Engel, Selective chain recognition in the C-terminal alpha-helical coiled-coil region of laminin, *J. Mol. Biol.*, 1995, **250**, 64–73.
61. C. Matsui, C. K. Wang, C. F. Nelson, E. A. Bauer and W. K. Hoeffler, The assembly of laminin-5 subunits, *J. Biol. Chem.*, 1995, **270**, 23496–23503.
62. A. Morita, E. Sugimoto and Y. Kitagawa, Post-translational assembly and glycosylation of laminin subunits in parietal endoderm-like F9 cells, *Biochem. J.*, 1985, **229**, 259–264.
63. C. Kumagai and Y. Kitagawa, Potential molecular chaperones involved in laminin chain assembly, *Cytotechnology*, 1997, **25**, 173–182.
64. P. D. Yurchenco, Y. Quan, H. Colognato, T. Mathus, D. Harrison, Y. Yamada and J. J. O'Rear, The alpha chain of laminin-1 is independently secreted and drives secretion of its beta- and gamma-chain partners, *Proc. Natl. Acad. Sci. U. S. A.*, 1997, **94**, 10189–10194.
65. J. H. Miner, C. Li, J. L. Mudd, G. Go and A. E. Sutherland, Compositional and structural requirements for laminin and basement membranes during mouse embryo implantation and gastrulation, *Development*, 2004, **131**, 2247–2256.
66. P. Murray and D. Edgar, Regulation of programmed cell death by basement membranes in embryonic development, *J. Cell Biol.*, 2000, **150**, 1215–1221.
67. C. Heng, O. Lefebvre, A. Klein, M. M. Edwards, P. Simon-Assmann, G. Orend and D. Bagnard, Functional role of laminin alpha1 chain during cerebellum development, *Cell Adhes. Migr.*, 2011, **5**, 480–489.
68. F. Alpy, I. Jivkov, L. Sorokin, A. Klein, C. Arnold, Y. Huss, M. Kedinger, P. Simon-Assmann and O. Lefebvre, Generation of a conditionally null allele of the laminin α1 gene, *Genesis*, 2005, **43**, 59–70.
69. A. Malandrini, M. Villanova, P. Sabatelli, S. Squarzoni, J. Six, P. Toti, G. Guazzi and N. M. Maraldi, Localization of the laminin alpha 2 chain in normal human skeletal muscle and peripheral nerve: an ultrastructural immunolabeling study, *Acta Neuropathol.*, 1997, **93**, 166–172.
70. M. Villanova, A. Malandrini, P. Sabatelli, C. A. Sewry, P. Toti, S. Torelli, J. Six, G. Scarfó, L. Palma, F. Muntoni, S. Squarzoni, P. Tosi, N. M. Maraldi and G. C. Guazzi, Localization of laminin alpha 2 chain in normal human central nervous system: an immunofluorescence and ultrastructural study, *Acta Neuropathol.*, 1997, **94**, 567–571.

71. L. T. Guo, X. U. Zhang, W. Kuang, H. Xu, L. A. Liu, J. T. Vilquin, Y. Miyagoe-Suzuki, S. Takeda, M. A. Ruegg, U. M. Wewer and E. Engvall, Laminin α2 deficiency and muscular dystrophy; genotype-phenotype correlation in mutant mice, *Neuromuscular Disord.*, 2003, **13**, 207–215.

72. A. Helbling-Leclerc, X. Zhang, H. Topaloglu, C. Corinne, F. Tesson, W. Jean, F. M. S. Tomé, K. Schwartz, M. Fardeau, K. Tryggvason and P. Guicheney, Mutations in the laminin alpha2–chain gene LAMA2 cause merosin-deficient congenital muscular dystrophy, *Nat. Genet.*, 1995, **11**, 216–218.

73. Y. Miyagoe-Suzuki, M. Nakagawa and S. Takeda, Merosin and congenital muscular dystrophy, *Microsc. Res. Tech.*, 2000, **48**, 181–191.

74. J. H. Miner, B. L. Patton, S. I. Lentz, D. J. Gilbert, W. D. Snider, N. A. Jenkins, N. G. Copelnad and J. R. Sanes, The laminin alpha chains: expression, developmental transitions, and chromsomal locations of alpha1-5, identification of heterotrimeric laminins 8-11, and cloning of a novel alpha3 isoform, *J. Cell Biol.*, 1997, **137**, 685–701.

75. J. Kuster, M. Guarnieri, J. Ault, L. Flaherty and P. Swiatek, IAP insertion in the murine LamB3 gene results in junctional epidermolysis bullosa, *Mamm. Genome*, 1997, **8**, 673–681.

76. M. C. Ryan, K. Lee, Y. Miyashita and W. G. Carter, Targeted disruption of the LAMA3 gene in mice reveals abnormalities in survival and late stage differentiation of epithelial cells, *J. Cell Biol.*, 1999, **145**, 1309–1323.

77. M. Barzegar, N. Mozafari, A. Kariminejad, Z. Asadikani, L. Ozoemena and J. A. McGrath, A new homozygous nonsense mutation in LAMA3A underlying laryngo-onycho-cutaneous syndrome, *Br. J. Dermatol.*, 2013, **169**, 1353–1356.

78. D. Urich, J. L. Eisenberg, K. J. Hamill, D. Takawira, S. E. Chiarella, S. Soberanes, A. Gonzalez, F. Koentgen, T. Manghi, S. B. Hopkinson, A. V. Misharin, H. Perlman, G. M. Mutlu, G. R. Budinger and J. C. Jones, Lung-specific loss of the laminin alpha3 subunit confers resistance to mechanical injury, *J. Cell Sci.*, 2011, **124**, 2927–2937.

79. C. K. Abrass, K. M. Hansen and B. L. Patton, Laminin α4-null mutant mice develop chronic kidney disease with persistent overexpression of platelet-derived growth factor, *Am. J. Pathol.*, 2010, **176**, 839–849.

80. J. Thyboll, J. Kortesmaa, R. Cao, R. Soininen, L. Wang, A. Iivanainen, L. Sorokin, M. Risling, Y. Cao and K. Tryggvason,

Deletion of the laminin α4 chain leads to impaired microvessel maturation, *Mol. Cell. Biol.*, 2002, **22**, 1194–1202.

81. J. Wang, M. Hoshijima, J. Lam, Z. Zhou, A. Jokiel, N. D. Dalton, K. Hultenby, P. Ruiz-Lozano, J. Ross, K. Tryggvason and K. R. Chien, Cardiomyopathy associated with microcirculation dysfunction in laminin α4 chain-deficient mice, *J. Biol. Chem.*, 2006, **281**, 213–220.

82. J. H. Miner, J. Cunningham and J. R. Sanes, Roles for laminin in embryogenesis: exencephaly, syndactyly, and placentopathy in mice lacking the laminin α5 chain, *J. Cell Biol.*, 1998, **143**, 1713–1723.

83. H. Imanishi, D. Tsuruta, C. Tateishi, K. Sugawara, R. Paus, T. Tsuji, M. Ishii, K. Ikeda, H. Kunimoto, K. Nakajima, J. C. R. Jones and H. Kobayashi, Laminin-511, inducer of hair growth, is down-regulated and its suppressor in hair growth, laminin-332, up-regulated in chemotherapy-induced alopecia, *J. Dermatol. Sci.*, 2010, **58**, 578–583.

84. M. Glukhova, V. Koteliansky, C. Fondacci, F. Marotte and L. Rappaport, Laminin variants and integrin laminin receptors in developing and adult human smooth muscle, *Dev. Biol.*, 1993, **157**, 437–447.

85. B. L. Patton, Laminins of the neuromuscular system, *Microsc. Res. Tech.*, 2000, **51**, 247–261.

86. P. G. Noakes, J. H. Miner, M. Gautam, J. M. Cunningham, J. R. Sanes and J. P. Merlie, The renal glomerulus of mice lacking s-laminin/laminin [beta]2: nephrosis despite molecular compensation by laminin [beta]1, *Nat. Genet.*, 1995, **10**, 400–406.

87. M. Zenker, T. Aigner, O. Wendler, T. Tralau, H. Muntefering, R. Fenski, S. Pitz, V. Schumacher, B. Royer-Pokora, E. Wuhl, P. Cochat, R. Bouvier, C. Kraus, K. Mark, H. Madlon, J. Dotsch, W. Rascher, I. Maruniak-Chudek, T. Lennert, L. M. Neumann and A. Reis, Human laminin beta2 deficiency causes congenital nephrosis with mesangial sclerosis and distinct eye abnormalities, *Hum. Mol. Genet.*, 2004, **13**, 2625–2632.

88. C. Muhle, A. Neuner, J. Park, F. Pacho, Q. Jiang, S. N. Waddington and H. Schneider, Evaluation of prenatal intra-amniotic LAMB3 gene delivery in a mouse model of Herlitz disease, *Gene Ther.*, 2006, **13**, 1665–1676.

89. J. D. Fine, Inherited epidermolysis bullosa, *Orphanet J. Rare Dis.*, 2010, **5**, 12.

90. J. D. Fine, Inherited epidermolysis bullosa: past, present, and future, *Ann. N. Y. Acad. Sci.*, 2010, **1194**, 213–222.

91. M. Koch, P. F. Olson, A. Albus, W. Jin, D. D. Hunter, W. J. Brunken, R. E. Burgeson and M.-F. Champliaud, Characterization and expression of the laminin γ3 chain: a novel, non-Basement membrane–associated, laminin chain, *J. Cell Biol.*, 1999, **145**, 605–618.

92. T. Barak, K. Y. Kwan, A. Louvi, V. Demirbilek, S. Saygi, B. Tuysuz, M. Choi, H. Boyaci, K. Doerschner, Y. Zhu, H. Kaymakcalan, S. Yilmaz, M. Bakircioglu, A. O. Caglayan, A. K. Ozturk, K. Yasuno, W. J. Brunken, E. Atalar, C. Yalcinkaya, A. Dincer, R. A. Bronen, S. Mane, T. Ozcelik, R. P. Lifton, N. Sestan, K. Bilguvar and M. Gunel, Recessive LAMC3 mutations cause malformations of occipital cortical development, *Nat. Genet.*, 2011, **43**, 590–594.

93. G. Gnanaguru, G. Bachay, S. Biswas, G. Pinzón-Duarte, D. D. Hunter and W. J. Brunken, Laminins containing the β2 and γ3 chains regulate astrocyte migration and angiogenesis in the retina, *Development*, 2013, **140**, 2050–2060.

94. G. Pinzón-Duarte, G. Daly, Y. N. Li, M. Koch and W. J. Brunken, Defective formation of the inner limiting membrane in laminin β2- and γ3-null Mice produces retinal dysplasia, *Invest. Ophthalmol. Visual Sci.*, 2010, **51**, 1773–1782.

95. M. Paulsson, The role of Ca2+ binding in the self-aggregation of laminin-nidogen complexes, *J. Biol. Chem.*, 1988, **263**, 5425–5430.

96. P. D. Yurchenco and Y. S. Cheng, Self-assembly and calcium-binding sites in laminin. A three-arm interaction model, *J. Biol. Chem.*, 1993, **268**, 17286–17299.

97. P. D. Yurchenco, E. C. Tsilibary, A. S. Charonis and H. Furthmayr, Laminin polymerization in vitro. Evidence for a two-step assembly with domain specificity, *J. Biol. Chem.*, 1985, **260**, 7636–7644.

98. J. H. O. Garbe, W. Göhring, K. Mann, R. Timpl and T. Sasaki, Complete sequence, recombinant analysis and binding to laminins and sulphated ligands of the N-terminal domains of laminin alpha3B and alpha5 chains, *Biochem. J.*, 2002, **362**, 213–221.

99. J. C. Schittny and P. D. Yurchenco, Terminal short arm domains of basement membrane laminin are critical for its self-assembly, *J. Cell Biol.*, 1990, **110**, 825–832.

100. K. Hasselbacher, R. C. Wiggins, V. Matejas, B. G. Hinkes, B. Mucha, B. E. Hoskins, F. Ozaltin, G. Nurnberg, C. Becker, D. Hangan, M. Pohl, E. Kuwertz-Broking, M. Griebel, V. Schumacher, B. Royer-Pokora, A. Bakkaloglu, P. Nurnberg, M. Zenker and F. Hildebrandt, Recessive missense mutations in LAMB2 expand the clinical spectrum of LAMB2-associated disorders, *Kidney Int.*, 2006, **70**, 1008–1012.

101. V. Matejas, B. Hinkes, F. Alkandari, L. Al-Gazali, E. Annexstad, M. B. Aytac, M. Barrow, K. Bláhová, D. Bockenhauer, H. I. Cheong, I. Maruniak-Chudek, P. Cochat, J. Dötsch, P. Gajjar, R. C. Hennekam, F. Janssen, M. Kagan, A. Kariminejad, M. J. Kemper, J. Koenig, J. Kogan, H. Y. Kroes, E. Kuwertz-Bröking, A. F. Lewanda, A. Medeira, J. Muscheites, P. Niaudet, M. Pierson, A. Saggar, L. Seaver, M. Suri, A. Tsygin, E. Wühl, A. Zurowska, S. Uebe, F. Hildebrandt, C. Antignac and M. Zenker, Mutations in the human laminin β2 LAMB2 gene and the associated phenotypic spectruma, *Hum. Mutat.*, 2010, **31**, 992–1002.
102. B. L. Patton, B. Wang, Y. S. Tarumi, K. L. Seburn and R. W. Burgess, A single point mutation in the LN domain of LAMA2 causes muscular dystrophy and peripheral amyelination, *J. Cell Sci.*, 2008, **121**, 1593–1604.
103. M. F. Champliaud, G. P. Lunstrum, P. Rousselle, T. Nishiyama, D. R. Keene and R. E. Burgeson, Human amnion contains a novel laminin variant, laminin-7, which like laminin-6, covalently associates with laminin-5 to promote stable epithelial-stromal attachment, *J. Cell Biol.*, 1996, **132**, 1189–1198.
104. K. Mann, R. Deutzmann, M. Aumailley, R. Timpl, L. Raimondi, Y. Yamada, T. C. Pan, D. Conway and M. L. Chu, Amino acid sequence of mouse nidogen, a multidomain basement membrane protein with binding activity for laminin, collagen IV and cells, *EMBO J.*, 1989, **8**, 65–72.
105. R. Timpl, S. Fujiwara, M. Dziadek, M. Aumailley, S. Weber and J. Engel, Laminin, proteoglycan, nidogen and collagen IV: structural models and molecular interactions, *Ciba Found. Symp.*, 1984, **108**, 25–43.
106. M. Aumailley, H. Wiedemann, K. Mann and R. Timpl, Binding of nidogen and the laminin-nidogen complex to basement membrane collagen type IV, *Eur. J. Biochem.*, 1989, **184**, 241–248.
107. N. Miosge, T. Sasaki and R. Timpl, Evidence of nidogen-2 compensation for nidogen-1 deficiency in transgenic mice, *Matrix Biol.*, 2002, **21**, 611–621.
108. N. Gersdorff, S. Otto, M. Roediger, J. Kruegel and N. Miosge, The absence of one or both nidogens does not alter basement membrane composition in adult murine kidney, *Histol. Histopathol.*, 2007, **22**, 1077–1084.
109. S. Mokkapati, A. Baranowsky, N. Mirancea, N. Smyth, D. Breitkreutz and R. Nischt, Basement membranes in skin are differently affected by lack of nidogen 1 and 2, *J. Invest. Dermatol.*, 2008, **128**, 2259–2267.

110. J. Schymeinsky, S. Nedbal, N. Miosge, E. Poschl, C. Rao, D. R. Beier, W. C. Skarnes, R. Timpl and B. L. Bader, Gene structure and functional analysis of the mouse nidogen-2 gene: nidogen-2 is not essential for basement membrane formation in mice, *Mol. Cell. Biol.*, 2002, **22**, 6820–6830.

111. U. Mayer, E. Kohfeldt and R. Timpl, Structural and genetic analysis of laminin-nidogen interaction, *Ann. N. Y. Acad. Sci.*, 1998, **857**, 130–142.

112. U. Mayer, E. Poschl, D. R. Gerecke, D. W. Wagman, R. E. Burgeson and R. Timpl, Low nidogen affinity of laminin-5 can be attributed to two serine residues in EGF-like motif gamma 2III4, *FEBS Lett.*, 1995, **365**, 129–132.

113. D. Breitkreutz, N. Mirancea, C. Schmidt, R. Beck, U. Werner, H. J. Stark, M. Gerl and N. E. Fusenig, Inhibition of basement membrane formation by a nidogen-binding laminin gamma1-chain fragment in human skin-organotypic cocultures, *J. Cell Sci.*, 2004, **117**, 2611–2622.

114. A. Baranowsky, S. Mokkapati, M. Bechtel, J. Krugel, N. Miosge, C. Wickenhauser, N. Smyth and R. Nischt, Impaired wound healing in mice lacking the basement membrane protein nidogen 1, *Matrix Biol.*, 2010, **29**, 15–21.

115. C. C. Yates, D. Whaley, S. Hooda, P. A. Hebda, R. J. Bodnar and A. Wells, Delayed reepithelialization and basement membrane regeneration after wounding in mice lacking CXCR3, *Wound Repair Regener.*, 2009, **17**, 34–41.

116. S. Alcantara, M. Ruiz, F. De Castro, E. Soriano and C. Sotelo, Netrin 1 acts as an attractive or as a repulsive cue for distinct migrating neurons during the development of the cerebellar system, *Development*, 2000, **127**, 1359–1372.

117. A. Bernet and J. Fitamant, Netrin-1 and its receptors in tumour growth promotion, *Expert Opin. Ther. Targets*, 2008, **12**, 995–1007.

118. J. G. Culotti and D. C. Merz, DCC and netrins, *Curr. Opin. Cell Biol.*, 1998, **10**, 609–613.

119. M. Dakouane-Giudicelli, C. Duboucher, J. Fortemps, H. Missey-Kolb, D. Brule, Y. Giudicelli and P. de Mazancourt, Characterization and expression of netrin-1 and its receptors UNC5B and DCC in human placenta, *J. Histochem. Cytochem.*, 2010, **58**, 73–82.

120. J. A. Meyerhardt, K. Caca, B. C. Eckstrand, G. Hu, C. Lengauer, S. Banavali, A. T. Look and E. R. Fearon, Netrin-1: interaction with deleted in colorectal cancer DCC and alterations in brain tumors and neuroblastomas, *Cell Growth Differ.*, 1999, **10**, 35–42.

121. M. Nacht, T. B. St Martin, A. Byrne, K. W. Klinger, B. A. Teicher, S. L. Madden and Y. Jiang, Netrin-4 regulates angiogenic responses and tumor cell growth, *Exp. Cell Res.*, 2009, **315**, 784–794.

122. H. Saueressig, J. Burrill and M. Goulding, Engrailed-1 and netrin-1 regulate axon pathfinding by association interneurons that project to motor neurons, *Development*, 1999, **126**, 4201–4212.

123. T. Nakashiba, T. Ikeda, S. Nishimura, K. Tashiro, T. Honjo, J. G. Culotti and S. Itohara, Netrin-G1: a novel glycosyl phosphatidylinositol-linked mammalian netrin that is functionally divergent from classical netrins, *J. Neurosci.*, 2000, **20**, 6540–6550.

124. M. Koch, J. R. Murrell, D. D. Hunter, P. F. Olson, W. Jin, D. R. Keene, W. J. Brunken and E. Burgeson, A novel member of the netrin family, beta-netrin, shares homology with the beta chain of laminin: identification, expression, and functional characterization, *J. Cell Biol.*, 2000, **151**, 221–234.

125. Y. Yin, J. H. Miner and J. R. Sanes, Laminets: laminin- and netrin-related genes expressed in distinct neuronal subsets, *Mol. Cell. Neurosci.*, 2002, **19**, 344–358.

126. F. Larrieu-Lahargue, A. L. Welm, K. R. Thomas and D. Y. Li, Netrin-4 activates endothelial integrin α6β1, *Circ. Res.*, 2011, **109**, 770–774.

127. K. J. Hamill, L. Langbein, J. C. Jones and W. H. McLean, Identification of a novel family of laminin N-terminal alternate splice isoforms: structural and functional characterization, *J. Biol. Chem.*, 2009, **284**, 35588–35596.

128. K. J. Hamill and W. H. McLean, The alpha-3 polypeptide chain of laminin 5: insight into wound healing responses from the study of genodermatoses, *Clin. Exp. Dermatol.*, 2005, **30**, 398–404.

129. E. J. Hagedorn, J. W. Ziel, M. A. Morrissey, L. M. Linden, Z. Wang, Q. Chi, S. A. Johnson and D. R. Sherwood, The netrin receptor DCC focuses invadopodia-driven basement membrane transmigration in vivo, *J. Cell Biol.*, 2013, **201**, 903–913.

130. R. O. Hynes, Integrins: versitility, modulation, and signaling in cell adhesion, *Cell*, 1992, **69**, 11–29.

131. M. Aumailley, R. Timpl and A. Sonnenberg, Antibody to integrin alpha 6 subunit specifically inhibits cell-binding to laminin fragment 8, *Exp. Cell Res.*, 1990, **188**, 55–60.

132. M. Shang, N. Koshikawa, S. Schenk and V. Quaranta, The LG3 module of laminin-5 harbors a binding site for integrin alpha-3beta1 that promotes cell adhesion, spreading, and migration, *J. Biol. Chem.*, 2001, **276**, 33045–33053.

133. A. P. Skubitz, P. C. Letourneau, E. Wayner and L. T. Furcht, Synthetic peptides from the carboxy-terminal globular domain of the A chain of laminin: their ability to promote cell adhesion and neurite outgrowth, and interact with heparin and the beta 1 integrin subunit, *J. Cell Biol.*, 1991, **115**, 1137–1148.

134. A. Sonnenberg, K. R. Gehlsen, M. Aumailley and R. Timpl, Isolation of alpha 6 beta 1 integrins from platelets and adherent cells by affinity chromatography on mouse laminin fragment E8 and human laminin pepsin fragment, *Exp. Cell Res.*, 1991, **197**, 234–244.

135. M. A. Chernousov, S. J. Kaufman, R. C. Stahl, K. Rothblum and D. J. Carey, Alpha7beta1 integrin is a receptor for laminin-2 on Schwann cells, *Glia*, 2007, **55**, 1134–1144.

136. I. de Curtis and G. Gatti, Identification of a large complex containing the integrin alpha 6 beta 1 laminin receptor in neural retinal cells, *J. Cell Sci.*, 1994, **107**, 3165–3172.

137. K. R. Gehlsen, P. Sriramarao, L. T. Furcht and A. P. Skubitz, A synthetic peptide derived from the carboxy terminus of the laminin A chain represents a binding site for the alpha 3 beta 1 integrin, *J. Cell Biol.*, 1992, **117**, 449–459.

138. K. Hozumi, N. Suzuki, P. K. Nielsen, M. Nomizu and Y. Yamada, Laminin alpha1 chain LG4 module promotes cell attachment through syndecans and cell spreading through integrin alpha-2beta1, *J. Biol. Chem.*, 2006, **281**, 32929–32940.

139. M. G. Jasiulionis, R. Chammas, A. M. Ventura, L. R. Travassos and R. R. Brentani, alpha6beta1-Integrin, a major cell surface carrier of beta1-6-branched oligosaccharides, mediates migration of EJ-ras-transformed fibroblasts on laminin-1 independently of its glycosylation state, *Cancer Res.*, 1996, **56**, 1682–1689.

140. Y. Kikkawa, H. Yu, E. Genersch, N. Sanzen, K. Sekiguchi, R. Fässler, K. P. Campbell, J. F. Talts and P. Ekblom, Laminin isoforms differentially regulate adhesion, spreading, proliferation, and ERK activation of β1 integrin-null cells, *Exp. Cell Res.*, 2004, **300**, 94–108.

141. J. Lian, X. Dai, X. Li and F. He, Identification of an active site on the laminin alpha4 chain globular domain that binds to alphavbeta3 integrin and promotes angiogenesis, *Biochem. Biophys. Res. Commun.*, 2006, **347**, 248–253.

142. J. E. Rooney, P. B. Gurpur, Z. Yablonka-Reuveni and D. J. Burkin, Laminin-111 restores regenerative capacity in a mouse model for alpha7 integrin congenital myopathy, *Am. J. Pathol.*, 2009, **174**, 256–264.

143. T. Sasaki, J. Takagi, C. Giudici, Y. Yamada, E. Arikawa-Hirasawa, R. Deutzmann, R. Timpl, A. Sonnenberg, H. P. Bachinger and D. Tonge, Laminin-121–recombinant expression and interactions with integrins, *Matrix Biol.*, 2010, **29**, 484–493.
144. J. F. Talts, Z. Andac, W. Göhring, A. Brancaccio and R. Timpl, Binding of the G domains of laminin alpha1 and alpha2 chains and perlecan to heparin, sulfatides, alpha-dystroglycan and several extracellular matrix proteins, *EMBO J.*, 1999, **18**, 863–870.
145. R. O. Hynes, Integrins: bidirectional, allosteric signaling machines, *Cell*, 2002, **110**, 673–687.
146. J. D. Fine, Epidermolysis bullosa: a genetic disease of altered cell adhesion and wound healing, and the possible clinical utility of topically applied thymosin beta4, *Ann. N. Y. Acad. Sci.*, 2007, **1112**, 396–406.
147. J. D. Fine, Inherited epidermolysis bullosa: recent basic and clinical advances, *Curr. Opin. Pediatr.*, 2010, **22**, 453–458.
148. J. C. R. Jones, S. B. Hopkinson and L. E. Goldfinger, Structure and assembly of hemidesmosomes, *BioEssays*, 1998, **20**, 488–494.
149. M. S. Lin, J. M. Mascaro, Jr., Z. Liu, A. Espana and L. A. Diaz, The desmosome and hemidesmosome in cutaneous autoimmunity, *Clin. Exp. Immunol.*, 1997, **107**(suppl. 1), 9–15.
150. A. M. Mercurio, I. Rabinovitz and L. M. Shaw, The alpha 6 beta 4 integrin and epithelial cell migration, *Curr. Opin. Cell Biol.*, 2001, **13**, 541–545.
151. D. Tsuruta, T. Hashimoto, K. J. Hamill and J. C. R. Jones, Hemidesmosomes and focal contact proteins: functions and cross-talk in keratinocytes, bullous diseases and wound healing, *J. Dermatol. Sci.*, 2011, **62**, 1–7.
152. H. Zhang and M. Labouesse, The making of hemidesmosome structures in vivo, *Dev. Dyn.*, 2010, **239**, 1465–1476.
153. M. Chen, M. P. Marinkovich, J. C. Jones, E. A. O'Toole, Y. Y. Li and D. T. Woodley, NC1 domain of type VII collagen binds to the beta3 chain of laminin 5 via a unique subdomain within the fibronectin-like repeats, *J. Invest. Dermatol.*, 1999, **112**, 177–183.
154. J. E. Mellerio, R. A. Eady, D. J. Atherton, B. D. Lake and J. A. McGrath, E210K mutation in the gene encoding the beta3 chain of laminin-5 LAMB3 is predictive of a phenotype of generalized atrophic benign epidermolysis bullosa, *Br. J. Dermatol.*, 1998, **139**, 325–331.
155. L. E. Goldfinger, S. B. Hopkinson, G. W. deHart, S. Collawn, J. R. Couchman and J. C. Jones, The alpha3 laminin subunit, alpha-6beta4 and alpha3beta1 integrin coordinately regulate wound healing in cultured epithelial cells and in the skin, *J. Cell Sci.*, 1999, **112**, 2615–2629.

156. K. J. Hamill, S. B. Hopkinson, M. F. Jonkman and J. C. Jones, Type XVII collagen regulates lamellipod stability, cell motility, and signaling to Rac1 by targeting bullous pemphigoid antigen 1e to alpha6beta4 integrin, *J. Biol. Chem.*, 2011, **286**, 26768–26780.

157. C. E. Pullar, B. S. Baier, Y. Kariya, A. J. Russell, B. A. Horst, M. P. Marinkovich and R. R. Isseroff, beta4 integrin and epidermal growth factor coordinately regulate electric field-mediated directional migration via Rac1, *Mol. Biol. Cell*, 2006, **17**, 4925–4935.

158. I. Rabinovitz and A. M. Mercurio, The integrin alpha6beta4 functions in carcinoma cell migration on laminin-1 by mediating the formation and stabilization of actin-containing motility structures, *J. Cell Biol.*, 1997, **139**, 1873–1884.

159. B. U. Sehgal, P. J. DeBiase, S. Matzno, T. L. Chew, J. N. Claiborne, S. B. Hopkinson, A. Russell, M. P. Marinkovich and J. C. Jones, Integrin beta4 regulates migratory behavior of keratinocytes by determining laminin-332 organization, *J. Biol. Chem.*, 2006, **281**, 35487–35498.

160. K. J. Hamill, S. B. Hopkinson, P. DeBiase and J. C. Jones, BPAG1e maintains keratinocyte polarity through beta4 integrin-mediated modulation of Rac1 and cofilin activities, *Mol. Biol. Cell*, 2009, **20**, 2954–2962.

161. K. Kligys, J. N. Claiborne, P. J. DeBiase, S. B. Hopkinson, Y. Wu, K. Mizuno and J. C. Jones, The slingshot family of phosphatases mediates Rac1 regulation of cofilin phosphorylation, laminin-332 organization, and motility behavior of keratinocytes, *J. Biol. Chem.*, 2007, **282**, 32520–32528.

162. R. Nishiuchi, O. Murayama, H. Fujiwara, J. Gu, T. Kawakami, S. Aimoto, Y. Wada and K. Sekiguchi, Characterization of the ligand-binding specificities of integrin α3β1 and α6β1 using a panel of purified laminin isoforms containing distinct α chains, *J. Biochem.*, 2003, **134**, 497–504.

163. R. Nishiuchi, J. Takagi, M. Hayashi, H. Ido, Y. Yagi, N. Sanzen, T. Tsuji, M. Yamada and K. Sekiguchi, Ligand-binding specificities of laminin-binding integrins: a comprehensive survey of laminin–integrin interactions using recombinant α3β1, α6β1, α7β1 and α6β4 integrins, *Matrix Biol.*, 2006, **25**, 189–197.

164. C. M. DiPersio, S. Shah and R. O. Hynes, alpha 3A beta 1 integrin localizes to focal contacts in response to diverse extracellular matrix proteins, *J. Cell Sci.*, 1995, **108**, 2321–2336.

165. G. W. deHart, K. E. Healy and J. C. Jones, The role of alpha3beta1 integrin in determining the supramolecular organization of laminin-5 in the extracellular matrix of keratinocytes, *Exp. Cell Res.*, 2003, **283**, 67–79.

166. G. W. deHart and J. C. Jones, Myosin-mediated cytoskeleton contraction and Rho GTPases regulate laminin-5 matrix assembly, *Cell Motil. Cytoskeleton*, 2004, **57**, 107–117.

167. P. J. DeBiase, K. Lane, S. Budinger, K. Ridge, M. Wilson and J. C. Jones, Laminin-311 Laminin-6 fiber assembly by type I-like alveolar cells, *J. Histochem. Cytochem.*, 2006, **54**, 665–672.

168. K. H. Holt, R. H. Crosbie, D. P. Venzke and K. P. Campbell, Biosynthesis of dystroglycan: processing of a precursor propeptide, *FEBS Lett.*, 2000, **468**, 79–83.

169. M. D. Henry and K. P. Campbell, A role for dystroglycan in basement membrane assembly, *Cell*, 1998, **95**, 859–870.

170. J. F. Talts, T. Sasaki, N. Miosge, W. Gohring, K. Mann, R. Mayne and R. Timpl, Structural and functional analysis of the recombinant G domain of the laminin alpha4 chain and its proteolytic processing in tissues, *J. Biol. Chem.*, 2001, **275**, 35192–35199.

171. R. Timpl, D. Tisi, J. F. Talts, Z. Andac, T. Sasaki and E. Hohenester, Structure and function of laminin LG modules, *Matrix Biol.*, 2000, **19**, 309–317.

172. H. Shimizu, H. Hosokawa, H. Ninomiya, J. H. Miner and T. Masaki, Adhesion of cultured bovine aortic endothelial cells to laminin-1 mediated by dystroglycan, *J. Biol. Chem.*, 1999, **274**, 11995–12000.

173. H. Colognato and P. D. Yurchenco, The laminin alpha2 expressed by dystrophic dy 2J mice is defective in its ability to form polymers, *Curr. Biol.*, 1999, **18**, 1327–1330.

174. H. Colognato, D. A. Winkelmann and P. D. Yurchenco, Laminin polymerization induces a receptor-cytoskeleton network, *J. Cell Biol.*, 1999, **145**, 619–631.

175. A. Akhavan, O. L. Griffith, L. Soroceanu, D. Leonoudakis, M. G. Luciani-Torres, A. Daemen, J. W. Gray and J. L. Muschler, Loss of cell-surface laminin anchoring promotes tumor growth and is associated with poor clinical outcomes, *Cancer Res.*, 2012, **72**, 2578–2588.

176. X. Bao, M. Kobayashi, S. Hatakeyama, K. Angata, D. Gullberg, J. Nakayama, M. N. Fukuda and M. Fukuda, Tumor suppressor function of laminin-binding α- dystroglycan requires a distinct β3-N-acetylglucosaminyltransferase, *Proc. Natl. Acad. Sci. U. S. A.*, 2009, **106**, 12109–12114.

177. J. C. R. Jones, K. Lane, S. B. Hopkinson, E. Lecuona, R. C. Geiger, D. A. Dean, E. Correa-Meyer, M. Gonzales, K. Campbell, J. I. Sznajder and S. Budinger, Laminin-6 assembles into multimolecular fibrillar complexes with perlecan and participates in mechanosignal transduction via a dystroglycan-dependent, integrin-independent, mechanism, *J. Cell Sci.*, 2005, **118**, 2557–2565.

178. C. S. Stipp, T. V. Kolesnikova and M. E. Hemler, Functional domains in tetraspanin proteins, *Trends Biochem. Sci.*, 2003, **28**, 106–112.

179. C. Boucheix and E. Rubinstein, Tetraspanins, *Cell. Mol. Life Sci.*, 2001, **58**, 1189–1205.

180. C. S. Stipp, Laminin-binding integrins and their tetraspanin partners as potential antimetastatic targets, *Expert Rev. Mol. Med.*, 2010, **12**, e3.

181. M. E. Hemler, Tetraspanin functions and associated micro-domains, *Nat. Rev. Mol. Cell Biol.*, 2005, **6**, 801–811.

182. R. Nishiuchi, N. Sanzen, S. Nada, Y. Sumida, Y. Wada, M. Okada, J. Takagi, H. Hasegawa and K. Sekiguchi, Potentiation of the ligand-binding activity of integrin $\alpha 3\beta 1$ via association with tetraspanin CD151, *Proc. Natl. Acad. Sci. U. S. A.*, 2005, **102**, 1939–1944.

183. M. Hasegawa, M. Furuya, Y. Kasuya, M. Nishiyama, T. Sugiura, T. Nikaido, Y. Momota, M. Ichinose and S. Kimura, CD151 dynamics in carcinoma-stroma interaction: integrin expression, adhesion strength and proteolytic activity, *Lab. Invest.*, 2007, **87**, 882–892.

184. N. E. Winterwood, A. Varzavand, M. N. Meland, L. K. Ashman and C. S. Stipp, A critical role for tetraspanin CD151 in $\alpha 3\beta 1$ and $\alpha 6\beta 4$ integrin–dependent tumor cell functions on laminin-5, *Mol. Biol. Cell*, 2006, **17**, 2707–2721.

185. M. Yamada, Y. Sumida, A. Fujibayashi, K. Fukaguchi, N. Sanzen, R. Nishiuchi and K. Sekiguchi, The tetraspanin CD151 regulates cell morphology and intracellular signaling on laminin-511, *FEBS J.*, 2008, **275**, 3335–3351.

186. S. Zevian, N. E. Winterwood and C. S. Stipp, Structure-function analysis of tetraspanin CD151 reveals distinct requirements for tumor cell behaviors mediated by $\alpha 3\beta 1$ versus $\alpha 6\beta 4$ integrin, *J. Biol. Chem.*, 2011, **286**, 7496–7506.

187. J. Lammerding, A. R. Kazarov, H. Huang, R. T. Lee and M. E. Hemler, Tetraspanin CD151 regulates $\alpha 6\beta 1$ integrin adhesion strengthening, *Proc. Natl. Acad. Sci. U. S. A.*, 2003, **100**, 7616–7621.

188. A. J. Cowin, D. Adams, S. M. Geary, M. D. Wright, J. C. R. Jones and L. K. Ashman, Wound healing is defective in mice lacking tetraspanin CD151, *J. Invest. Dermatol.*, 2006, **126**, 680–689.

189. S. M. Geary, A. J. Cowin, B. Copeland, R. M. Baleato, K. Miyazaki and L. K. Ashman, The role of the tetraspanin CD151 in primary keratinocyte and fibroblast functions: implications for wound healing, *Exp. Cell Res.*, 2008, **314**, 2165–2175.

190. J. R. Couchman, Syndecans: proteoglycan regulators of cell-surface microdomains? *Nat. Rev. Mol. Cell Biol.*, 2003, **4**, 926–938.

191. N. Suzuki, N. Ichikawa, S. Kasai, M. Yamada, N. Nishi, H. Morioka, H. Yamashita, Y. Kitagawa, A. Utani, M. P. Hoffman and M. Nomizu, Syndecan binding sites in the laminin α1 chain G domain, *Biochemistry*, 2003, **42**, 12625–12633.

192. S. Bachy, F. Letourneur and P. Rousselle, Syndecan-1 interaction with the LG4/5 domain in laminin-332 is essential for keratinocyte migration, *J. Cell. Physiol.*, 2008, **214**, 238–249.

193. C. M. Klass, J. R. Couchman and A. Woods, Control of extracellular matrix assembly by syndecan-2 proteoglycan, *J. Cell Sci.*, 2000, **113**, 493–506.

194. M. A. Stepp, Y. Liu, S. Pal-Ghosh, R. A. Jurjus, G. Tadvalkar, A. Sekaran, K. LoSicco, L. Jiang, M. Larsen, L. Li and S. H. Yuspa, Reduced migration, altered matrix and enhanced TGFβ1 signaling are signatures of mouse keratinocytes lacking Sdc1, *J. Cell Sci.*, 2007, **120**, 2851–2863.

195. N. Sharon and H. Lis, History of lectins: from hemagglutinins to biological recognition molecules, *Glycobiology*, 2004, **14**, 53R–62R.

196. J. Dumic, S. Dabelic and M. Flögel, Galectin-3: an open-ended story, *Biochim. Biophys. Acta*, 2006, **1760**, 616–635.

197. F. Melo, D. Butera, M. Junqueira, D. Hsu, A. Moura da Silva, F. Liu, M. Santos and R. Chammas, The promigratory activity of the matricellular protein galectin-3 depends on the activation of PI-3 kinase, *PLoS One*, 2011, **6**, e29313.

198. Y. Kariya, C. Kawamura, T. Tabei and J. Gu, Bisecting GlcNAc residues on raminin-332 down-regulate galectin-3-dependent keratinocyte motility, *J. Biol. Chem.*, 2010, **285**, 3330–3340.

199. T. Weissenbacher, T. Vrekoussis, D. Roeder, A. Makrigiannakis, D. Mayr, N. Ditsch, K. Friese, U. Jeschke and D. Dian, Analysis of Epithelial Growth Factor-Receptor (EGFR) phosphorylation in uterine smooth muscle tumors: correlation to mucin-1 and galectin-3 expression, *Int. J. Mol. Sci.*, 2013, **14**, 4783–4792.

200. F.-T. Liu, R. J. Patterson and J. L. Wang, Intracellular functions of galectins, *Biochim. Biophys. Acta*, 2002, **1572**, 263–273.

201. H. Leffler, S. Carlsson, M. Hedlund, Y. Qian and F. Poirier, Introduction to galectins, *Glycoconjugate J.*, 2002, **19**, 433–440.

202. P. Matarrese, O. Fusco, N. Tinari, C. Natoli, F.-T. Liu, M. L. Semeraro, W. Malorni and S. Iacobelli, Galectin-3 overexpression protects from apoptosis by improving cell adhesion properties, *Int. J. Cancer*, 2000, **85**, 545–554.

203. M. Costell, E. Gustafsson, A. Aszódi, M. Mörgelin, W. Bloch, E. Hunziker, K. Addicks, R. Timpl and R. Fässler, Perlecan maintains the integrity of cartilage and some basement membranes, *J. Cell Biol.*, 1999, **147**, 1109–1122.

204. D. M. Noonan, A. Fulle, P. Valente, S. Cai, E. Horigan, M. Sasaki, Y. Yamada and J. R. Hassell, The complete sequence of perlecan, a basement membrane heparan sulfate proteoglycan, reveals extensive similarity with laminin A chain, low density lipoprotein-receptor, and the neural cell adhesion molecule, *J. Biol. Chem.*, 1991, **266**, 22939–22947.

205. M. Hopf, W. Göhring, E. Kohfeldt, Y. Yamada and R. Timpl, Recombinant domain IV of perlecan binds to nidogens, laminin–nidogen complex, fibronectin, fibulin-2 and heparin, *Eur. J. Biochem.*, 1999, **259**, 917–926.

206. I. Sher, S. Zisman-Rozen, L. Eliahu, J. M. Whitelock, N. Maas-Szabowski, Y. Yamada, D. Breitkreutz, N. E. Fusenig, E. Arikawa-Hirasawa, R. V. Iozzo, R. Bergman and D. Ron, Targeting perlecan in human keratinocytes reveals novel roles for perlecan in epidermal formation, *J. Biol. Chem.*, 2006, **281**, 5178–5187.

207. A. J. Denzer, M. Gesemann, B. Schumacher and M. A. Ruegg, An amino-terminal extension is required for the secretion of chick agrin and its binding to extracellular matrix, *J. Cell Biol.*, 1995, **131**, 1547–1560.

208. M. A. Ruegg and J. L. Bixby, Agrin orchestrates synaptic differentiation at the vertebrate neuromuscular junction, *Trends Neurosci.*, 1998, **21**, 22–27.

209. A. J. Denzer, T. Schulthess, C. Fauser, B. Schumacher, R. A. Kammerer, J. Engel and M. A. Ruegg, Electron microscopic structure of agrin and mapping of its binding site in laminin-1, *EMBO J.*, 1998, **17**, 335–343.

210. J. B. Mascarenhas, M. A. Rüegg, U. Winzen, W. Halfter, J. Engel and J. Stetefeld, Mapping of the laminin-binding site of the N-terminal agrin domain (NtA), *EMBO J.*, 1993, **22**, 529–536.

211. J. Stetefeld, M. Jenny, T. Schulthess, R. Landwehr, B. Schumacher, S. Frank, M. A. Rüegg, J. Engel and R. A. Kammerer, The laminin-binding domain of agrin is structurally related to N-TIMP-1, *Nat. Struct. Biol.*, 2001, **8**, 705–709.

212. N. Ettner, W. Göhring, T. Sasaki, K. Mann and R. Timpl, The N-terminal globular domain of the laminin α1 chain binds to α1β1 and α2β1 integrins and to the heparan sulfate-containing domains of perlecan, *FEBS Lett.*, 1998, **430**, 217–221.

213. P. Rousselle, D. R. Keene, F. Ruggiero, M. F. Champliaud, M. Rest and R. E. Burgeson, Laminin 5 binds the NC-1 domain of type VII collagen, *J. Cell Biol.*, 1997, **138**, 719–728.

214. I. K. Gipson, S. M. Grill, S. J. Spurr and S. J. Brennan, Hemides- mosome formation in vitro, *J. Cell Biol.*, 1983, **97**, 849–857.

215. J. C. R. Jones, J. Asmuth, S. E. Baker, M. Langhofer, S. I. Roth and S. B. Hopkinson, Hemidesmosomes: extracellular matrix/inter- mediate filament connectors, *Exp. Cell Res.*, 1994, **213**, 1–11.

216. M. Chen, N. Kasahara, D. R. Keene, L. Chan, W. K. Hoeffler, D. Finlay, M. Barcova, P. M. Cannon, C. Mazurek and D. T. Wood- ley, Restoration of type VII collagen expression and function in dystrophic epidermolysis bullosa, *Nat. Genet.*, 2002, **32**, 670–675.

217. S. Amano, I. A. Scott, K. Takahara, M. Koch, D. R. Gerecke, D. R. Keene, D. L. Hudson, T. Nishiyama, S. Lee, D. S. Greenspan and R. E. Burgeson, Bone morphogenetic protein-1 BMP-1 is an extracellular processing enzyme of the laminin 5 g2 chain, *J. Biol. Chem.*, 2000, **275**, 22728–22735.

218. L. Gagnoux-Palacios, M. Allegra, F. Spirito, O. Pommeret, C. Romero, J. P. Ortonne and G. Meneguzzi, The short arm of the laminin gamma2 chain plays a pivotal role in the incorporation of laminin 5 into the extracellular matrix and in cell adhesion, *J. Cell Biol.*, 2001, **153**, 835–850.

219. L. E. Goldfinger, L. Jiang, S. B. Hopkinson, M. S. Stack and C. Jones, Spatial regulation and activity modulation of plasmin by high affinity binding to the G domain of the alpha 3 subunit of laminin-5, *J. Biol. Chem.*, 2000, **275**, 34887–34893.

220. Y. Kariya, C. Yasuda, Y. Nakashima, K. Ishida, Y. Tsubota and K. Miyazaki, Characterization of laminin 5B and NH2-terminal proteolytic fragment of its α3B chain: promotion of cellular adhesion, migration, and proliferation, *J. Biol. Chem.*, 2004, **279**, 24774–24784.

221. Y. Tsubota, H. Mizushima, T. Hirosaki, S. Higashi, H. Yasumitsu and K. Miyazaki, Isolation and activity of proteolytic fragment of laminin-5 a3 chain, *Biochem. Biophys. Res. Commun.*, 2000, **278**, 614–620.

222. J. F. Talts and R. Timpl, Mutation of a basic sequence in the laminin alpha2LG3 module leads to a lack of proteolytic pro- cessing and has different effects on beta1 integrin-mediated cell adhesion and alpha-dystroglycan binding, *FEBS Lett.*, 1999, **458**, 319–323.

223. S. P. Smirnov, E. L. McDearmon, S. Li, J. M. Ervasti, K. Tryggvason and P. D. Yurchenco, Contributions of the LG modules and furin processing to laminin-2 functions, *J. Biol. Chem.*, 2002, **277**, 18928–18937.

224. M. P. Marinkovich, G. P. Lunstrum, D. R. Keene and R. E. Burgeson, The dermal-epidermal junction of human skin contains a novel laminin variant, *J. Cell Biol.*, 1992, **119**, 695–703.

225. D. P. Veitch, P. Nokelainen, K. A. McGowan, T.-T. Nguyen, N. E. Nguyen, R. Stephenson, W. N. Pappano, D. R. Keene, S. M. Spong, D. S. Greenspan, P. R. Findell and M. P. Marinkovich, Mammalian tolloid metalloproteinase, and not matrix metalloprotease 2 or membrane type 1 metalloprotease, processes laminin-5 in keratinocytes and skin, *J. Biol. Chem.*, 2003, **278**, 15661–15668.

226. D. E. Frank and W. G. Carter, Laminin 5 deposition regulates keratinocyte polarization and persistent migration, *J. Cell Sci.*, 2004, **117**, 1351–1363.

227. Y. Baba, K.-i. Iyama, K. Hirashima, Y. Nagai, N. Yoshida, N. Hayashi, N. Miyanari and H. Baba, Laminin-332 promotes the invasion of oesophageal squamous cell carcinoma via PI3K activation, *Br. J. Cancer*, 2008, **98**, 974–980.

228. M. Tran, P. Rousselle, P. Nokelainen, S. Tallapragada, N. T. Nguyen, E. F. Fincher and M. P. Marinkovich, Targeting a tumor-specific laminin domain citical for human carcinogenesis, *Cancer Res.*, 2008, **68**, 2885–2894.

229. A. Utani, M. Nomizu, H. Matsuura, K. Kato, T. Kobayashi, U. Takeda, S. Aota, P. K. Nielsen and H. Shinkai, A unique sequence of the laminin alpha 3 G domain binds to heparin and promotes cell adhesion through syndecan-2 and -4, *J. Biol. Chem.*, 2001, **276**, 28779–28788.

230. I. Senyürek, G. Klein, H. Kalbacher, M. Deeg and B. Schittek, Peptides derived from the human laminin α4 and α5 chains exhibit antimicrobial activity, *Peptides*, 2010, **31**, 1468–1472.

231. T. Ogawa, Y. Tsubota, J. Hashimoto, Y. Kariya and K. Miyazaki, The short arm of laminin γ2 chain of laminin-5 (laminin-332) binds syndecan-1 and regulates cellular adhesion and migration by suppressing phosphorylation of integrin β4 chain, *Mol. Biol. Cell*, 2007, **18**, 1621–1633.

232. A. Katsumi, T. Naoe, T. Matsushita, K. Kaibuchi and M. A. Schwartz, Integrin activation and matrix binding mediate cellular responses to mechanical stretch, *J. Biol. Chem.*, 2005, **280**, 16546–16549.

233. A. Katsumi, A. W. Orr, E. Tzima and M. A. Schwartz, Integrins in mechanotransduction, *J. Biol. Chem.*, 2004, **279**, 12001–12004.

234. J. C. Chen and C. R. Jacobs, Mechanically induced osteogenic lineage commitment of stem cells, *Stem Cell Res. Ther.*, 2013, **4**, 107.

235. J. G. Jacot, J. C. Martin and D. L. Hunt, Mechanobiology of cardiomyocyte development, *J. Biomech.*, 2010, **43**, 93–98.

236. N. D. Leipzig and M. S. Shoichet, The effect of substrate stiffness on adult neural stem cell behavior, *Biomaterials*, 2009, **30**, 6867–6878.

237. A. S. Rowlands, P. A. George and J. Cooper-White, Directing osteogenic and myogenic differentiation of MSCs: interplay of stiffness and adhesive ligand presentation, *Am. J. Physiol.: Cell Physiol.*, 2008, **295**, C1037–C1044.

238. M. Witkowska-Zimny, K. Walenko, E. Wrobel, P. Mrowka, A. Mikulska and J. Przybylski, Effect of substrate stiffness on the osteogenic differentiation of bone marrow stem cells and bone-derived cells, *Cell Biol. Int.*, 2013, **37**, 608–616.

239. P. D. Yurchenco, Y. S. Cheng and H. Colognato, Laminin forms an independent network in basement membranes, *J. Cell Biol.*, 1992, **117**, 1119–1133.

240. P. D. Yurchenco, Y. S. Cheng and J. C. Schittny, Heparin modulation of laminin polymerization, *J. Biol. Chem.*, 1990, **265**, 3981–3991.

241. P. D. Yurchenco and H. Furthmayr, Self-assembly of basement membrane collagen, *Biochemistry*, 1984, **23**, 1839–1850.

242. J. L. Eisenberg, A. Safi, X. Wei, H. D. Espinosa, G. S. Budinger, D. Takawira, S. B. Hopkinson and J. C. Jones, Substrate stiffness regulates extracellular matrix deposition by alveolar epithelial cells, *Res. Rep. Biol.*, 2011, **2011**, 1–12.

243. S. Fujiwara, H. Shinkai, R. Deutzmann, M. Paulsson and R. Timpl, Structure and distribution of N-linked oligosaccharide chains on various domains of mouse tumour laminin, *Biochem. J.*, 1988, **252**, 453–461.

244. J. W. Dean, S. Chandrasekaran and M. L. Tanzer, A biological role of the carbohydrate moieties of laminin, *J. Biol. Chem.*, 1990, **265**, 12553–12562.

245. C. C. Howe, Functional role of laminin carbohydrate, *Mol. Cell. Biol.*, 1984, **4**, 1–7.

246. Y. Kariya, R. Kato, S. Itoh, T. Fukuda, Y. Shibukawa, N. Sanzen, K. Sekiguchi, Y. Wada, N. Kawasaki and J. Gu, N-glycosylation of laminin-332 regulates its biological functions: a novel function of the bisecting GlcNAc, *J. Biol. Chem.*, 2008, **283**, 33036–33045.

247. R. W. Orkin, P. Gehron, E. B. McGoodwin, G. R. Martin, T. Valentine and R. Swarm, A murine tumor producing a matrix of basement membrane, *J. Exp. Med.*, 1977, **145**, 204–220.

248. R. L. Swarm, Transplantation of a murine chondrosarcoma in mice of differential inbred strains, *J. Natl. Cancer Inst.*, 1963, **31**, 953–975.

249. R. Timpl, G. R. Martin, P. Bruckner, G. Wick and H. Wiedemann, Nature of the collagenous protein in a tumor basement membrane, *Eur. J. Biochem.*, 1978, **84**, 43–52.

250. V. P. Terranova, M. Aumailley, L. H. Sultan, G. R. Martin and H. K. Kleinman, Regulation of cell attachment and cell number by fibronectin and laminin, *J. Cell. Physiol.*, 1986, **127**, 473–479.

251. H. K. Kleinman and G. R. Martin, Matrigel: basement membrane matrix with biological activity, *Semin. Cancer Biol.*, 2005, **15**, 378–386.

252. E. S. Rosler, G. J. Fisk, X. Ares, J. Irving, T. Miura, M. S. Rao and M. K. Carpenter, Long-term culture of human embryonic stem cells in feeder-free conditions, *Dev. Dyn.*, 2004, **229**, 259–274.

253. T. J. Rowland, L. M. Miller, A. J. Blaschke, E. L. Doss, A. J. Bonham, S. T. Hikita, L. V. Johnson and D. O. Clegg, Roles of integrins in human induced pluripotent stem cell growth on matrigel and vitronectin, *Stem Cells Dev.*, 2010, **18**, 1231–1240.

254. C. Xu, M. S. Inokuma, J. Denham, K. Golds, P. Kundu, J. D. Gold and M. K. Carpenter, Feeder-free growth of undifferentiated human embryonic stem cells, *Nat. Biotechnol.*, 2001, **19**, 971–974.

255. Y. Kubota, H. K. Kleinman, G. R. Martin and T. J. Lawley, Role of laminin and basement membrane in the morphological differentiation of human endothelial cells into capillary-like structures, *J. Cell Biol.*, 1988, **107**, 1589–1598.

256. M. L. Li, J. Aggeler, D. A. Farson, C. Hatier, J. Hassell and M. J. Bissell, Influence of a reconstituted basement membrane and its components on casein gene expression and secretion in mouse mammary epithelial cells, *Proc. Natl. Acad. Sci. U. S. A.*, 1987, **84**, 136–140.

257. A. Albini, Y. Iwamoto, H. K. Kleinman, G. R. Martin, S. A. Aaronson, J. M. Kozlowski and R. N. McEwan, A rapid in vitro assay for quantitating the invasive potential of tumor cells, *Cancer Res.*, 1987, **47**, 3239–3245.

258. R. Fridman, G. Benton, I. Aranoutova, H. K. Kleinman and R. D. Bonfil, Increased initiation and growth of tumor cell lines, cancer stem cells and biopsy material in mice using basement membrane matrix protein (Cultrex or Matrigel) co-injection, *Nat. Protoc.*, 2012, **7**, 1138–1144.

259. B.-J. Kang, H.-H. Ryu, S.-S. Park, Y. Kim, H.-M. Woo, W. H. Kim and O.-K. Kweon, Effect of matrigel on the osteogenic potential

of canine adipose tissue-derived mesenchymal stem cells, *J. Vet. Med. Sci.*, 2012, **74**, 827–836.

260. R. Pili, Y. Guo, J. Chang, H. Nakanishi, G. R. Martin and A. Passaniti, Altered angiogenesis underlying age-dependent changes in tumor growth, *J. Natl. Cancer Inst.*, 1994, **86**, 1303–1314.

261. C. S. Hughes, L. M. Postovit and G. A. Lajoie, Matrigel: a complex protein mixture required for optimal growth of cell culture, *Proteomics*, 2010, **10**, 1886–1890.

262. N. Kohen, L. Little and K. Healy, Characterization of matrigel interfaces during defined human embryonic stem cell culture, *Biointerphases*, 2009, **4**, 69–79.

263. P. Hall, J. Lathia, M. Caldwell and C. ffrench-Constant, Laminin enhances the growth of human neural stem cells in defined culture media, *BMC Neurosci.*, 2008, **9**, 71.

264. K. Menezes, J. Ricardo Lacerda de Menezes, M. Assis Nascimento, R. de Siqueira Santos and T. Coelho-Sampaio, Polylaminin, a polymeric form of laminin, promotes regeneration after spinal cord injury, *FASEB J.*, 2010, **24**, 4513–4522.

265. Z. Wondimu, G. Gorfu, T. Kawataki, S. Smirnov, P. Yurchenco, K. Tryggvason and M. Patarroyo, Characterization of commercial laminin preparations from human placenta in comparison to recombinant laminins 2 α2β1γ1, 8 α4β1γ1, 10 α5β1γ1, *Matrix Biol.*, 2006, **25**, 89–93.

266. M. Paulsson, M. Aumailley, R. Deutzmann, R. Timpl, K. Beck and J. Engel, Laminin-nidogen complex. Extraction with chelating agents and structural characterization, *Eur. J. Biochem.*, 1987, **166**, 11–19.

267. Y. Kikkawa, N. Sanzen and K. Sekiguchi, Isolation and Characterization of laminin-10/11 secreted by human lung carcinoma cells: laminin-10/11 mediates cell adhesion through integrin α3β1, *J. Biol. Chem.*, 1998, **273**, 15854–15859.

268. I. C. Sroka, M. L. Chen and A. E. Cress, Simplified purification procedure of laminin-332 and laminin-511 from human cell lines, *Biochem. Biophys. Res. Commun.*, 2008, **375**, 410–413.

269. H. Fujiwara, Y. Kikkawa, N. Sanzen and K. Sekiguchi, Purification and characterization of human laminin-8: laminin-8 stimulates cell adhesion and migration through α3β1 and α6β1 integrin, *J. Biol. Chem.*, 2001, **276**, 17550–17558.

270. S. E. Baker, A. DiPasquale, E. L. Stock, G. Plopper, V. Quaranta, M. Fitchmun and J. C. R. Jones, Morphogenetic effects of soluble laminin-5 on cultured epithelial cells and tissue explants, *Exp. Cell Res.*, 1996, **228**, 262–270.

271. P. M. Carpenter, A. V. Dao, Z. S. Arain, M. K. Chang, H. P. Nguyen, S. Arain, J. Wang-Rodriguez, S.-Y. Kwon and S. P. Wilczynski, Motility induction in breast carcinoma by mammary epithelial laminin 332 (laminin 5), *Mol. Cancer Res.*, 2009, **7**, 462–475.

272. Y. Kariya, H. Sato, N. Katou, Y. Kariya and K. Miyazaki, Polymerized laminin-332 matrix supports rapid and tight adhesion of keratinocytes, suppressing cell migration, *PLoS One*, 2012, **7**, e35546.

273. E. Natarajan, J. D. Omobono II, Z. Guo, S. B. Hopkinson, A. Lazar, T. Brenn, J. C. R. Jones and J. G. Rheinwald, A keratinocyte hypermotility/growth arrest response involving laminin 5 and p16INK4A activated in wound healing and senescence, *Am. J. Pathol.*, 2006, **168**, 1821–1836.

274. S. Rodin, A. Domogatskaya, S. Strom, E. M. Hansson, K. R. Chien, J. Inzunza, O. Hovatta and K. Tryggvason, Long-term self-renewal of human pluripotent stem cells on human recombinant laminin-511, *Nat. Biotechnol.*, 2010, **28**, 611–615.

275. J. Gao, M. C. DeRouen, C.-H. Chen, M. Nguyen, N. T. Nguyen, H. Ido, K. Harada, K. Sekiguchi, B. A. Morgan, J. H. Miner, A. E. Oro and M. P. Marinkovich, Laminin-511 is an epithelial message promoting dermal papilla development and function during early hair morphogenesis, *Genes Dev.*, 2008, **22**, 2111–2124.

276. A. Croset, L. Delafosse, J.-P. Gaudry, C. Arod, L. Glez, C. Losberger, D. Begue, A. Krstanovic, F. Robert, F. Vilbois, L. Chevalet and B. Antonsson, Differences in the glycosylation of recombinant proteins expressed in HEK and CHO cells, *J. Biotech.*, 2012, **161**, 336–348.

277. S. B. Hopkinson, P. J. DeBiase, K. Kligys, K. Hamill and J. C. R. Jones, Fluorescently tagged laminin subunits facilitate analyses of the properties, assembly and processing of laminins in live and fixed lung epithelial cells and keratinocytes, *Matrix Biol.*, 2008, **27**, 640–647.

278. T. E. Sztal, C. Sonntag, T. E. Hall and P. D. Currie, Epistatic dissection of laminin–receptor interactions in dystrophic zebrafish muscle, *Hum. Mol. Genet.*, 2012, **21**, 4718–4731.

279. M. L. Weir, M. L. Oppizzi, M. D. Henry, A. Onishi, K. P. Campbell, M. J. Bissell and J. L. Muschler, Dystroglycan loss disrupts polarity and β-casein induction in mammary epithelial cells by perturbing laminin anchoring, *J. Cell Sci.*, 2006, **119**, 4047–4058.

280. O. A. Senkovich, J. Yin, C. Conant, J. Traylor, P. Adegboyega, D. J. McGee, R. E. Rhoads, S. Slepenkov and T. L. Testerman, Helicobacter pylori AlpA and AlpB bind host laminin and influence gastric inflammation in gerbils, *Infect. Immun.*, 2011, **79**, 3106–3116.

281. X. Wei, Y. Zhao, X. Dong, Y. Si, Z. Ma, C. Zhu and S. Pang, The interaction of laminin and its membrane receptor on mouse macrophage membrane studied by STM and FRAP (ast), *Cell Res.*, 1993, **3**, 21–26.
282. J. Graf, Y. Iwamoto, M. Sasaki, G. R. Martin, H. K. Kleinman, F. A. Robey and Y. Yamada, Identification of an amino acid sequence in laminin mediating cell attachment, chemotaxis, and receptor binding, *Cell*, 1987, **48**, 989–996.
283. Y. Yamada and H. K. Kleinman, Functional domains of cell adhesion molecules, *Curr. Opin. Cell Biol.*, 1992, **4**, 819–823.
284. M. L. Ponce and H. K. Kleinman, Identification of redundant angiogenic sites in laminin α1 and γ1 chains, *Exp. Cell Res.*, 2003, **285**, 189–195.
285. K. M. Malinda, M. Nomizu, M. Chung, M. Delgado, Y. Kuratomi, Y. Yamada, H. K. Kleinman and M. L. Ponce, Identification of laminin α1 and β1 chain peptides active for endothelial cell adhesion, tube formation, and aortic sprouting, *FASEB J.*, 1999, **13**, 53–62.
286. S.-K. Min, S.-C. Lee, S.-D. Hong, C.-P. Chung, W. H. Park and B.-M. Min, The effect of a laminin-5-derived peptide coated onto chitin microfibers on re-epithelialization in early-stage wound healing, *Biomaterials*, 2010, **31**, 4725–4730.
287. G. Damodaran, W. H. C. Tiong, R. Collighan, M. Griffin, H. Navsaria and A. Pandit, In vivo effects of tailored laminin-332 α3 conjugated scaffolds enhances wound healing: a histomorphometric analysis, *J. Biomed. Mater. Res., Part A*, 2013, **101**, 2788–2795.
288. P. Macchiarini, P. Jungebluth, T. Go, M. A. Asnaghi, L. E. Rees, T. A. Cogan, A. Dodson, J. Martorell, S. Bellini, P. P. Parnigotto, S. C. Dickinson, A. P. Hollander, S. Mantero, M. T. Conconi and M. A. Birchall, Clinical transplantation of a tissue-engineered airway, *Lancet*, 2008, **372**, 2023–2030.
289. H. C. Ott, T. S. Matthiesen, S.-K. Goh, L. D. Black, S. M. Kren, T. I. Netoff and D. A. Taylor, Perfusion-decellularized matrix: using nature's platform to engineer a bioartificial heart, *Nat. Med.*, 2008, **14**, 213–221.
290. H. C. Ott, B. Clippinger, C. Conrad, C. Schuetz, I. Pomerantseva, L. Ikomonou, D. Kotton and J. P. Vacanti, Regeneration and orthotopic transplantation of a bioartificial lung, *Nat. Med.*, 2010, **16**, 927–933.
291. T. H. Petersen, E. A. Calle, L. Zhao, E. J. Lee, L. Gui, M. B. Raredon, K. Gavrila, T. Yi, E. Herzog and L. E. Niklason, Tissue-engineered lungs for in vitro implantation, *Science*, 2010, **329**, 538–541.

292. B. E. Uygun, A. Soto-Gutierrez, H. Yagi, M.-L. Izamis, M. A. Guzzardi, C. Shulman, J. Milwid, N. Kobayashi, A. Tilles, F. Berthiaume, M. Hertl, Y. Nahmias, M. L. Yarmush and K. Uygun, Organ reengineering through development of a transplantable recellularized liver graft using decellularized liver matrix, *Nat. Med.*, 2010, **16**, 814–820.

293. A. Lichtenberg, I. Tudorache, S. Cebotari, M. Suprunov, G. Tudorache, H. Goerler, J.-K. Park, D. Hilfiker-Kleiner, S. Ringes-Lichtenberg, M. Karck, G. Brandes, A. Hilfiker and A. Haverich, Preclinical testing of tissue-engineered heart valves re-endothelialized under simulated physiological conditions, *Circulation*, 2006, **114**, I-559–I-565.

294. K. H. Nakayama, C. A. Batchelder, C. I. Lee and A. F. Tarantal, Decellularized rhesus monkey kidney as a three-dimensional scaffold for renal tissue engineering, *Tissue Eng., Part A*, 2010, **16**, 2207–2216.

295. P. B. Patil, P. B. Chougule, V. K. Kumar, S. Almström, H. Bäckdahl, D. Banerjee, G. Herlenius, M. Olausson and S. Sumitran-Holgersson, Recellularization of acellular human small intestine using bone marrow stem cells, *Stem Cells Transl. Med.*, 2013, **2**, 307–315.

296. C. C. Tate, D. A. Shear, M. C. Tate, D. R. Archer, D. G. Stein and M. C. LaPlaca, Laminin and fibronectin scaffolds enhance neural stem cell transplantation into the injured brain, *J. Tissue Eng. Regener. Med.*, 2009, **3**, 208–217.

297. K. Vats and D. S. W. Benoit, Dynamic manipulation of hydrogels to control cell behavior: a review, *Tissue Eng., Part B*, 2013, **19**, 455–469.

298. C. Deister, S. Aljabari and C. Scmidt, Effects of collagen 1, fibronectin, laminin, and hyaluronic acid concentration in multi-component gels on neurite extension, *J. Biomater. Sci., Polym. Ed.*, 2007, **18**, 983–997.

299. S. Suri and C. Schmidt, Cell-laden hydrogel constructs of hyaluronic acid, collagen, and laminin for neural tissue engineering, *Tissue Eng., Part A*, 2010, **16**, 1703–1716.

300. S. Ali, J. E. Saik, D. J. Gould, M. E. Dickinson and J. L. West, Immobilization of cell-adhesive laminin peptides in degradable PEGDA hydrogels influences endothelial cell tubulogenesis, *BioRes. Open Access*, 2013, **2**, 241–249.

301. A. T. Francisco, R. J. Mancino, R. D. Bowles, J. M. Brunger, D. M. Tainter, Y.-T. Chen, W. J. Richardson, F. Guilak and L. A. Setton, Injectable laminin-functionalized hydrogel for nucleus pulposus regeneration, *Biomaterials*, 2013, **34**, 7381–7388.

Part II

Biomaterials as Mimics of the ECM
A. Building Function Through Complexity

CHAPTER 5

Biomaterials: Incorporating ECM-Derived Molecular Features into Biomaterials

KRISTOPHER A. KILIAN*[a]

[a]Department of Materials Science and Engineering, University of Illinois at Urbana Champaign, 1304 West Green St, Urbana, IL 61801, USA
*E-mail: kakilian@illinois.edu

5.1 INTRODUCTION

The extracellular matrix (ECM) is a highly complex composite material, with hierarchical organization, many different organic and inorganic features, and dynamic self-assembly and disassembly processes that regulate tissue homeostasis and morphogenesis. The ECM differs significantly across tissue types, where the mechanical, topographical, biochemical and transport properties of the materials are dictated in a contextual fashion to guide specific cellular and tissue level functions. The presentation of these cues to cells is not a static process, but rather a dynamic interplay between individual cells, their neighbors and the ECM architecture. This spatiotemporal control of signaling patterns in the cellular microenvironment regulates diverse functions, including quiescence, migration, proliferation,

Mimicking the Extracellular Matrix: The Intersection of Matrix Biology and Biomaterials
Edited by Gregory A. Hudalla and William L. Murphy
© The Royal Society of Chemistry 2020
Published by the Royal Society of Chemistry, www.rsc.org

differentiation and apoptosis. Understanding how the properties of the ECM directs tissue form and function is important for fundamental biology research but also for establishing design criteria for next generation biomaterials that aim to recapitulate the structure and function of native extracellular matrices.

Mammalian cell culture is most often performed on rigid tissue culture plastic, where the surface is either pre-coated with defined ECM proteins or coated *in situ via* the physisorption of biomolecule components in the cell culture media. Both of these processes invariably lead to a heterogeneous deposition of physisorbed matrix components across the substrate. The composition of the initial coating can be defined to reflect aspects of the cultured cells' natural matrix; however, physisorption to rigid hydrophobic interfaces can cause denaturation of proteins and burying of adhesion cues, and the final surface composition is subject to competition from a large number of proteins in the media with different affinity for the substrate. Furthermore, most cell types will actively remodel their underlying ECM over time through secretion of matrix proteins and protease enzymes. Thus, ensuring that the cell culture surface presents a defined combination of matrix proteins to adherent cells is difficult at best and impossible over prolonged culture. This is important because a defined substrate that reflects the native composition of the microenvironment may be critical for many cell biology studies. In addition, there is considerable evidence that transferring a cell from the native *in vivo* environment to *in vitro* culture can lead to phenotypic changes that often result in senescence, differentiation, chromatin abnormalities and death.

In addition to the need for a defined ECM during cell culture, materials for implants, cell delivery formulations for regenerative medicine, and tissue engineering scaffolds require interfaces that promote the desired cellular processes and outcome. For instance, there are significant efforts to modify bone implant materials with bioactive ceramics, matrix proteins and morphogens that enhance osseointegration,[1,2] and to fabricate hydrogel materials that recapitulate the architecture and composition of hyaline cartilage.[3] Virtually every tissue presents a unique combination of matrix proteins and glycosaminoglycans (GAGs) that are assembled in a defined way to influence cell attachment, tissue mechanical properties, and diffusion and sequestration of growth factors; the organization of these components is dynamic, context dependent and ultimately critical to maintaining normal physiology.

In this chapter we explore the past, current and future prospects for integrating ECM-derived molecular features into biomaterials. Our primary objective is to describe the state of the art in modifying 2D and 3D materials with molecules derived from, or that mimic, the native ECM. Other important parameters including mechanical properties, cell adhesion mechanisms, sequestration of soluble factors, gradient formation, turnover and hierarchical organization will be the subject of later chapters. First we discuss efforts aimed at using natural components—derived from tissues or created through recombinant DNA techniques—to modify biomaterials. Next we focus on methodologies that incorporate small synthetic ECM fragments into biomaterials. Throughout the chapter we explore the future of integrating aspects of these approaches towards leveraging the optimal features of each for the design of ECM-mimetic biomaterials (Figure 5.1).

5.2 INCORPORATING NATURAL MOLECULES INTO BIOMATERIALS

When designing a 2D cell culture platform, implant or particle coating, or 3D scaffold material, often the most optimal strategy is to use the same molecules present in the host tissue. Natural biomaterials have the advantage of inherent biocompatibility and the potential to present many bioactive features by virtue of having a complete primary structure. Using natural materials from the ECM preserves many physiochemical features of native tissue with structural and biochemical aspects that promote new tissue development. Some disadvantages associated with using natural materials are limitations associated with production, immunogenicity from xenogenic materials and ill-defined structure and composition. In this section we will discuss the most common approaches to integrate natural molecules into biomaterials with a primary focus on ECM-derived proteins and polysaccharides. The inclusion of soluble factors, important considerations for both natural and synthetic biomaterials, will not be discussed here.

5.2.1 Bioconjugation of Polypeptides and Polysaccharides to Materials

The surface of proteins and polysaccharide are rich with functional groups that are amenable to covalent immobilization to synthetic biomaterials. Of the 20 amino acids, nearly half of these can be conjugated using simple methods under physiological conditions that

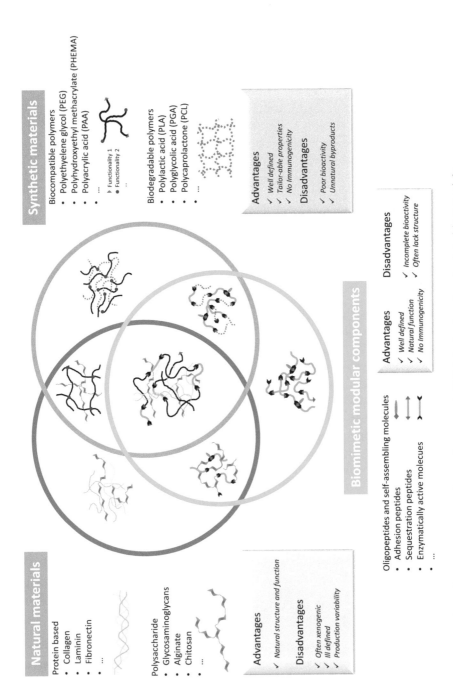

Figure 5.1   Overview of common approaches to incorporate ECM-derived molecules into biomaterials.

preserve biomolecule structure and function.[4] Important considerations when developing a method to conjugate a biomolecule to a surface includes solvent accessibility of the functional group and relative nucleophilicity. Of particular note are amino acids with ionizable side chains—aspartic acid, glutamic acid, lysine, arginine, cysteine, histidine, serine, threonine and tyrosine—where in the unprotonated state these side chains can act as potent nucleophiles for addition-type bioconjugation reactions (Figure 5.2a). Similarly, many polysaccharides have accessible functional groups for derivatization; in particular, free hydroxyl groups, carboxylic acids and amines appear in many ECM components (Figure 5.2b).

In order to immobilize the biomolecule, the surface of biomaterials often needs to be derivatized to present a complementary chemistry. Strategies to functionalize a surface largely depend on the type of material, and many different strategies have been developed over the years. Of particular note, plasma treatment can be used to change both the physical and chemical properties of a surface to modify the way in which it integrates with biological systems. For instance, oxygen plasma can be used to introduce hydroxyl and carboxylic acid groups onto the surface of many materials, including ceramics, metals and polymers.[5] This treatment can be used to increase the hydrophilicity of the material and further, provide reactive headgroups upon which polypeptides and polysaccharides may be conjugated (Figure 5.2c). Similarly, nitrogen plasma has been used to graft amino groups onto many different surfaces and implant materials,[6,7] which renders these interfaces amenable to coupling of activated carboxylates present on many proteins and polysaccharides. Another useful strategy for modifying the surface of many hard materials prior to conjugating a biomolecule of interest is to use short alkyl monolayers that react with the surface on one end and provide a functional headgroup on the other. Common monolayer modifications include alkyl silanes on glass, silicon and metal oxides,[8] alkyl phosphonates for modifying metal oxides,[9,10] and alkane thiolates for modifying metal surfaces[11] (Figure 5.2c). Using accessible amino acids on the protein or relevant functional group on the polysaccharide, and a complementary surface chemistry on the biomaterial, matrix proteins, polysaccharides and peptides can be conjugated with ease to study or direct a desired biological function.[12]

5.2.2 Protein-Based Biomaterials

Unsurprisingly, the inspiration for many protein-based biomaterials comes from the structure and composition of native tissues. Of the many different combinations of proteins that make up the ECM,

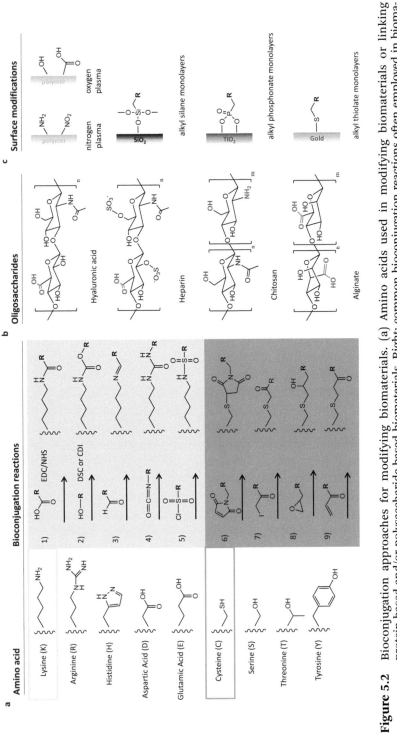

Figure 5.2 Bioconjugation approaches for modifying biomaterials. (a) Amino acids used in modifying biomaterials or linking protein-based and/or polysaccharide-based biomaterials. Right: common bioconjugation reactions often employed in biomaterials science. (b) Polysaccharides that are often used in modifying biomaterials; functional groups for bioconjugation are highlighted. (c) Surface modification schemes for incorporating distal reactivity for bioconjugation through routes depicted in (a) and (b).

the majority of protein-based scaffolds and hydrogels that are fabricated in the laboratory are composed of fibrin, collagen, elastin, gelatin, Matrigel or combinations thereof. Other commonly used protein-based biomaterials that will not be discussed here in depth include keratins[13,14] and silk proteins.[15]

Fibrin is a fibrous protein formed naturally during wound repair due to proteolytic processing of the serum protein fibrinogen, which aids polymerization and the formation of a 3D material during blood clotting.[16,17] The role of fibrin during wound healing is as temporary scaffolding that is replaced over time by the natural ECM. The polymerization mechanism of fibrin biomaterials makes it useful for biomedical applications, because the hydrogel network architecture can be varied through control of reaction conditions.[18] Fibrin biomaterials have found utility as a glue in clinical applications, where a patient's fibrinogen may be isolated from the blood and used as an autologous fixation material.[19–21] In addition to clinical applications, fibrin-based materials have proved useful as model soft materials for cell biology studies because fibrin contains natural cell adhesion sites (Arg-Gly-Asp (RGD) and Ala-Gly-Asp (AGD) peptide motifs)[22,23] and growth factor binding sites.[18] This is important because there is no requirement for further bioconjugation to mediate the attachment of cells. Fibrin gels have been used broadly for studies of cell differentiation and as a model for tissue engineering materials.[18] Because fibrin gels are very compliant, they have been particularly useful as materials to guide the differentiation of stem cells to "soft" lineages. For instance, embryonic stem cells and neural progenitor cells have been differentiated within fibrin matrices towards neurons and astrocytes.[24,25] Mesenchymal stem cells have been differentiated into osteoblasts in 3D fibrin gels;[26] however, high concentrations of fibrinogen and calcium were required, likely owing to the preference for high contractility during osteogenesis that is often fostered by stiff substrates. Fibrin has been used as a material to promote chondrogenesis[27] and as a scaffold material to support angiogenesis and neo-vascularization.[28,29] In a report by Liu *et al.*, fibrin gels were used to study tumorigenic cells and it was found that soft gels promote the selection and proliferation of a subset of melanoma cells that express markers associated with a cancer stem cell phenotype.[30]

Collagen is the most abundant fibrous protein, with 20 distinct forms found in native ECMs. Collagen is stabilized by a triple helical structure that is induced by a peptide repeat Gly–X–Y, where X and Y are proline or hydroxyproline.[31] The triple helices can form bundles which can further be assembled into microfibrils and larger structures through crosslinking of lysine residues by the enzyme lysyl oxidase. As the most abundant ECM protein, with good mechanical properties and biocompatibility, collagen is one of the most widely studied ECM components for biomaterials applications.[32] In addition to the large fibrillar forms of collagen, the fibril associated collagens with interrupted triple helices (FACIT) and basement membrane collagens are similarly important, particularly when recapitulating a microenvironment in which these forms are present. Monomeric and fibrous forms of collagen have been immobilized onto surfaces to mediate cell adhesion through physical adsorption,[4] crosslinking within synthetic materials[33] and *via* bioconjugation strategies as detailed in Section 5.2.1[34]. *In vitro* fibrillogenesis has been accomplished through careful control of conditions that lead to natural ECM architectures.[35] In contrast to fibrin gels, collagen can be processed in the laboratory to have a broader range of mechanical properties to suit the biophysical properties required for the application. This can be accomplished by varying the gel casting conditions but also through the use of crosslinking chemistry.[36] The mechanism of cell adhesion on collagen is mediated in part by RGD sequences[37] but also *via* alternative recognition sequences and sequestering of other adhesion molecules.[38] Collagen by itself and in co-polymer formulations has found broad utility as a cell culture substrate and in the fabrication of materials for tissue engineering and regenerative medicine.[39–43] Gelatin is a denatured form of collagen that can form a 3D structure upon cooling. Gelatin has found utility as a cell culture coating and as a 3D biomaterial, particularly when used with other ECM-derived and synthetic molecules.[44–47]

Matrigel is the trade name for an extract of basement membrane proteins from Engelbreth-Holm-Swarm mouse sarcoma and is one of the most widely used ECM-derived biomaterials for 2D coatings and 3D structures. Matrigel is primarily composed of laminin and collagen type IV but also contains entactin and other basement membrane constituents that can vary significantly across batches, which renders it inappropriate for studies of specific cell–protein interactions. A liquid at low temperature, Matrigel will self-assemble into a 3D network at physiological temperature and has been demonstrated as a viable matrix for studying endothelial and epithelial cell biology[48] and supporting stem cell growth and differentiation.[49,50]

5.2.3   Polysaccharide Biomaterials

Polysaccharides are polymers of saccharide monomers—derived from plants, animals and microorganisms in nature—that can be linear and branched and are connected *via* glycosidic linkages. The extent of branching, saccharide composition and pendant functional groups influences the physical and chemical properties of polysaccharides. Many polysaccharides contain functional groups that can be modified using bioconjugation strategies, described in Section 5.2.1, for immobilization to surfaces, labeling with other molecular species and for crosslinking to other polymer systems. Polysaccharide systems can form hydrogels and the mechanism of formation depends on the chemical functionality on the sugar rings with ionic interactions and hydrogen bonding dominating gelation. Polysaccharides in tissues serve a number of different functions, including providing compressive strength, sequestration of soluble factors and direct signaling to the cell surface, and are therefore of significant interest for incorporation into biomaterials. While there is a vast array of polysaccharide materials from many different biological sources, here we will focus on materials that have proven significantly useful in biomaterials applications, including GAGs, alginate and chitosan.

GAGs are linear polysaccharides that contain a repeating disaccharide unit of an amino sugar and uronic sugar, and are present in many ECMs, in particular connective tissue, and often in the presence of collagen matrices. Of the many different GAGs being considered for biomaterials design, hyaluronic acid (or hyaluronan) is one at the forefront due to its ubiquitous presence throughout mammalian tissue. Hyaluronic acid is the only non-sulfated GAG and plays an important role in many cellular processes, including migration, matrix assembly, angiogenesis and wound healing.[51] Hyaluronic acid contains carboxylic acid groups that allow it to form hydrogels by interacting with water and cations, and the volume in the solution state is approximately 1000-fold higher than the dry state. Hyaluronic acid interacts with specific cell surface receptors and has been proposed to augment integrin-mediated signaling.[52] Immobilization to surfaces has been reported using various bioconjugation techniques that typically involve modifying the polymer at the carboxylate or hydroxyl group for reaction with nucleophiles or oxidizing the vicinal hydroxyl groups to yield a reactive aldehyde, which can then be coupled to amines through a Schiff-base reaction.[53] Hyaluronic acid will readily form a hydrogel with mechanical properties of soft tissues and can be further crosslinked with itself and other polymers to modulate its viscoelastic properties.[53] Haylauronic acid biomaterials have been used as a support layer in cell culture, a coating for orthopedic implants[54,55] and as a 3D hydrogel material to promote stem cell growth and differentiation.[53] To further increase the complexity, heparin sulfate has been covalently grafted to hyaluronic acid in order to sequester growth factors *in situ*; this system was shown to enhance angiogenesis *in vivo*.[56]

Other GAGs of biomedical importance include heparin sulfate, keratin sulfate, chondroitin sulfate and dermatan sulfate, which differ from hyaluronic acid in the sulfation state of the pendant hydroxyl groups on the sugars (Figure 5.2b). These GAGs can undergo similar modifications to that of hyaluronic acid through activation of carboxylates or oxidation to yield aldehydes for Schiff-base reactions.[57]

Chitosan is a linear polysaccharide that is not present in significant quantities in natural systems. However, it is readily formed by deacetylation of chitin, a polysaccharide that is abundant in the exoskeletons of arthropods. After deacetylation of chitin, the amino groups enable the formation of hydrogel materials that are structurally similar to GAGs and when protonated have many unique properties that render it useful for biomedicine. For instance, chitosan shows antimicrobial activity,[58] mucoadhesive and hemostatic characteristics[59,60] and biodegradability.[61] Owing to its biocompatibility and many useful properties, chitosan has been approved by the US Food and Drug Administration (FDA) for use in wound dressings and has shown clinical promise for orthopedic biomaterials and synthetic skin.[62] After deacetylation, there are many amino and hydroxyl groups that can be functionalized and crosslinked to vary the physical and chemical properties of chitosan. Chitosan has been derivatized in various ways for covalent grafting to implant surfaces[63–65] and vascular grafts.[66] Chitosan hydrogels will naturally form through physical associations and electrostatic interactions with ions in solution. Furthermore, the polycationic nature renders it a good co-polymer system with proteins (*e.g.* gelatin, collagen, fibrin)[67,68] and anionic oligosaccharides (*e.g.* GAGs, alginate).[69,70] To further stabilize or enhance the mechanical and diffusional properties of chitogen hydrogels, chemical crosslinking schemes have been developed utilizing the amino and hydroxyl functionalities, including the formation of amides, esters and Schiff-base reactions.[71] Chitosan has been shown to undergo a sol–gel transition in the presence of β-glycerol phosphate,[72] and these gels have been demonstrated to support diverse differentiation outcomes, from neurogenesis[73] to osteogenesis.[74] Chitosan has emerged as one of the most promising materials for biopharmaceutical delivery, particularly when formulated as polyelectrolytes with other oligosaccharides.[75]

Alginate is an anionic polysaccharide that is harvested from seaweed and has attracted significant interest as a material for biomedical applications due to its biocompatibility, the presence of multiple functional groups throughout the material and simple gelation mechanism upon addition of divalent cations.[76] Importantly, alginate hydrogels display structural and mechanical aspects that mimic the macromolecular architectures in many tissues. Similarly to chitosan, there are also many favorable properties of alginate gels that make them useful for biomedical applications. For instance, the stimuli-responsive nature of alginate allows for dynamic studies where gelation can occur on demand to encapsulate cells in a defined way. In addition to ionic crosslinking, modulating the network architecture of alginate gels can be accomplished through other physical and covalent crosslinking approaches.[77] Alginate has found utility as a wound dressing,[78] for therapeutic delivery,[79] and alginate gels are one of the most commonly used model matrices for cell biology studies.[80] This is due to biocompatibility, resistance to protein adsorption and because there is no specific ligand present for mammalian cells, thus allowing covalent functionalization of the many carboxylic acids to control the incorporation of user-defined cell adhesion ligands.[81] Alginate materials functionalized with cell adhesion motifs will be covered in more depth in Section 5.3.

5.2.4 Natural and Synthetic Co-Polymer Biomaterials

The native extracellular matrix is not made up of either protein or polysaccharide materials alone, but rather is a complex architecture comprised of components from both categories. Towards the fabrication of scaffold materials that recapitulate the properties of native ECM, most commonly materials are sought that contain both protein and polysaccharide character. In addition, ECM molecules isolated from tissue often do not have the appropriate properties for biomaterials development and fabrication *ex vivo*. This can be related to mechanical properties, diffusional characteristics, biodegradation profile and porosity for cellular in-growth, amongst other requirements depending on the application. Synthetic materials harbor the advantage of being tailored in a defined way using careful control of chemistry and processing conditions to meet many of these requirements. Thus, the integration of natural biomaterials that contain desired cell signaling motifs into well-defined synthetic architectures has proved a popular strategy for many applications.

5.2.4.1 Protein and Polysaccharide Composite Materials. A drawback to using fibrin gels in many applications is the poor mechanical properties and the tendency for the material to be digested by proteolytic enzymes. The assembly kinetics and stability of fibrin gels has been guided in the laboratory by GAGs (hyaluronic acid and chondroitin sulfate), through specific peptide-mediated interactions of the oligosaccharides with fibrinogen during gelation.[82] Crosslinking hyaluronic acid during assembly of fibrin gels will lead to an inter-penetrating polymer network with tunable properties that expands the scope of using these materials across a larger array of tissues.[83] Composite hydrogel materials based on fibrin and GAGs have shown promise in cartilage and bone tissue engineering[84–88] and the inclusion of hyaluronic acid in fibrin gels has been shown to influence the rate of proliferation and matrix remodeling by endothelial cells for vascular grafts.[89] These examples demonstrate how incorporating multiple ECM-derived molecules into scaffolds can help maximize a preferred cellular outcome.

Collagen hydrogels have been integrated with GAGs to emulate the native ECM and have shown promise for engineering many tissues including skin, peripheral nerves, vasculature and orthopedic tissues.[90–93] Improved mechanical integrity that more closely corresponds to orthopedic tissues has been demonstrated using collagen I-chondroitin sulfate core shell architectures formed through lyophilization and dehydrothermal crosslinking.[94] Similarly, collagen I-hyaluronic acid scaffolds have been fabricated by freeze drying and crosslinking using carbodiimide chemistry and shown good resistance to collagenase activity.[95,96] Gelatin has also been mixed with GAGs to develop hybrid structures for tissue engineering. For instance, gelatin has been enzymatically crosslinked to hyaluronic acid and chondroitin sulfate for the culture of hepatocytes[97] and using azide–alkyne cycloaddition to crosslink the molecules by modifying the GAGs with an azido compound and the gelatin with propiolic acid.[98]

Chitosan and alginate have also proven attractive polysaccharides to fabricate hybrid scaffolds with protein biomaterials for improved properties. For instance, mixing different ratios of collagen and chitosan induces a pronounced difference in pore morphology and architecture[99] (Figure 5.3a). In this work, β-glycerophosphate was used to initiate gelation of both biopolymers and the resultant properties of the mixed scaffold favored an osteogenesis outcome in mesenchymal stem cells compared to collagen gels alone. Collagen–chitosan and gelatin–chitosan scaffolds have been prepared *via* various freeze drying and crosslinking approaches for tissue engineering applications in bone and cartilage,[100–103] skin[104] and smooth muscle constructs.[105] Collagen has been incorporated into alginate gels to tune physiochemical properties of scaffold materials[106] for tissue engineering of heart tissue,[107] vocal fold,[108] and orthopedic tissues.[109,110] The mechanical properties of collagen have also been improved by incorporating an inner alginate core to stabilize the porous collagen exterior.[111] To devise the optimal combination of protein and polysaccharide, a

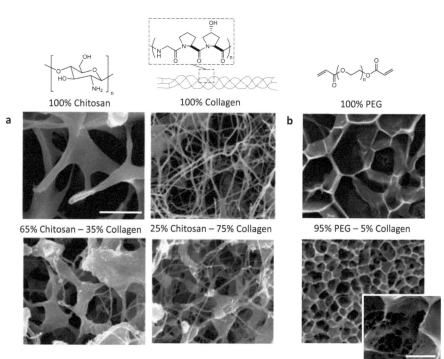

Figure 5.3 Mixing different biological and synthetic polymers can be used to tailor the structure of biomaterial scaffolds. (a) The effect on material architecture when the fraction of chitosan and collagen I are varied (adapted from ref. 99, copyright © 2010, with permission from Elsevier) and (b) when the fraction of poly(ethylene glycol) diacrylate hydrogels are mixed with collagen I (adapted from ref. 122, copyright © 2012, with permission from John Wiley & Sons, Inc.).

useful strategy is empirically screening different combinations. For instance: alginate, chitosan, collagen and gelatin were mixed together in different ratios to reveal an optimal combination of chitosan and collagen (or gelatin) for maintaining hepatocyte function.[112] Screening approaches have the advantage of using a functional read-out to determine the optimal material. However, in some instances the scaffold material may require mechanical or chemical attributes outside of the scope of available natural materials.

5.2.4.2 Integrating Natural Biomaterials with Synthetic Polymers. There are many viable options for integrating synthetic components with natural materials in order to optimize desired properties. Many of these simply involve mixing the natural and synthetic component under conditions in which a homogenous material is formed.[113–115] Alternatively, the rich complexity of functional groups on natural

biopolymers—coupled with the well-controlled chemistry and repro-
ducible fabrication of synthetic polymers—enables many different
covalent conjugation schemes to form biomaterials with diverse properties.[116] Specifically, the ability to tailor the ends and backbone of
synthetic polymers with functional groups provides exquisite control
over the assembly of complex scaffolds.

Polyethylene glycol (PEG) is a hydrophilic polymer that easily forms
a hydrogel through backbone hydrogen bonding with water, and can
be readily derivatized on either end with diverse functional groups.[117]
The use of di-functionalized polyethylene glycol has emerged as a
popular strategy for crafting hydrogel biomaterials, because PEG
forms a controlled hydrogel network that is highly tunable and biocompatible, and can be easily conjugated to a range of physiologically
important proteins.[118] As discussed in Section 5.2.2, one drawback of
many natural protein gels is the non-linear elasticity which for many
applications results in poor mechanical properties. Thus a common
strategy employed is modification of the hydrogel network with PEG
molecules in such a way that the biochemical integrity is maintained
(*e.g.* adhesion, sequestration sequences) and the physical properties
optimized. Seliktar and colleagues conjugated a poly(ethylene glycol)
diacrylate (PEGDA) molecule *via* Michael-type addition to the free thiols present on fibrinogen cysteine residues, followed by UV treatment
to form the hydrogel network.[119] This material supports cell adhesion
and importantly has highly tunable mechanical properties compared
to pure fibrin gels by varying the ratio of PEGDA to fibrinogen or, alternatively, varying the molecular weight of PEG used during polymerization. Fibrin–PEG hydrogels have been demonstrated to enhance
neuronal outgrowth[120] and increase vascularization potential.[121] PEG
has been used to direct the architecture of collagen self-assembly
into controlled fiber structure[122] (Figure 5.3b) and to modulate the
viscoelastic properties through interpenetrating polymer networks.
These strategies of self-assembled natural-synthetic co-polymers are
attractive because the synthetic component can be used to define the
structure while the natural self-assembly process is employed to create biomimetic materials. Covalent conjugation of PEGDA and collagen has the advantage of being able to precisely control mechanical
properties *in situ*[123] and for defining the density and presentation of
protein—and positioning of cells and aggregates—within the scaffold material.[124] Controlling the direct conjugation of collagen into
di-functional PEG networks has been shown to recapitulate aspects
of native tissue for directing capillary morphogenesis.[125] To increase
enzyme-mediated degradation gelatin, the denatured form of collagen,

is often used in hydrogel scaffolds. Covalently crosslinked PEGDA and thiolated gelatin were shown to increase the incidence of endothelial cell network formation compared to physically entrapped gelatin.[126]

To more closely emulate the biopolymeric architecture of native tissues, PEG hydrogels have also been fabricated to include many polysaccharides. Hyaluronic acid has been chemically modified in numerous ways to facilitate conjugation to synthetic polymers. For instance, hyaluronic acid has been derivatized to present free thiols, haloacetate groups, hydrazides, aldehydes, azides and alkynes for simple hydrogel formation when mixed with the appropriate molecule containing complementary functional groups.[53] Recent work demonstrated the clinical viability of photocrosslinking of hyaluronic acid and PEG transdermally for soft tissue reconstruction where the mechanics and volume could be modulated by varying the ratio of each component; interestingly this system is reversible by injection of hyaluonidase.[127] Importantly, the gelation kinetics and mechanical properties of GAG–PEG hydrogels can be modulated by changing the functional groups on the hyaluronan or PEG endgroups.[128,129] Inclusion of chondroiten sulfate and heparin sulfate into PEG hydrogels has been demonstrated to improve the behavior of vocal fold fibroblasts by emulating the structure and composition of the native tissue.[56] Hyaluronic acid and other GAGs are recognized by distinct cell surface receptors but will not promote cell adhesion *via* transmembrane integrin receptors on their own, thus requiring mixing and/or conjugation with cell adhesion proteins or peptides. Incorporating gelatin into hyaluronic acid and PEG co-polymer hydrogels has been developed for 2D and 3D cell culture materials and for tissue engineering applications.[130] Covalent modification of proteoglycans with short peptides to mediate cellular processes will be discussed in more detail in Section 5.3.

Another class of synthetic polymeric biomaterial of particular importance for applications in drug delivery and tissue engineering are the biodegradable polyesters. These materials can be tailored to have well-defined degradation profiles in physiological environments, and many are already approved for clinical use by regulatory bodies.[131] The inclusion of degradable polymers allows for an initial condition that can provide a framework for cellular connectivity and molecule encapsulation that changes over time as the ester backbone is hydrolyzed. This is important because providing a scaffold can direct cells to deposit their own matrix, which over time may replace the synthetic polyester. Encapsulation of bioactive molecules within the synthetic polymer will also enable the tuned release upon degradation to further promote a target biological process. Polysaccharides and polypeptides have been mixed with polydioxanone,[132] poly(β-hydroxy-butyrate),[133] poly(hydroxyvalerate),[134,135] polycaprolactone (PCL),[136] polyglycolic acid (PGA), polylactic acid (PLA) and the co-polymer polylactide-*co*-glycolide (PLGA).[137–140] Biodegradable polyesters are also useful to include within PEG-based materials to modulate degradation; for instance, PLA–PEG–PLA[141] and PEG–PCL[142] have been demonstrated as scaffold materials with tunable properties.

5.3   INCORPORATING SYNTHETIC ECM-DERIVED MOLECULES INTO BIOMATERIALS

Many of the proteins and polysaccharide biomaterials considered thus far are harvested from natural sources and so harbor the risk of immunogenicity, which questions whether these plant- and animal-derived components will be translated to clinical applications. Furthermore there are issues of lot to lot uniformity, consistency in post-translational modifications, packaging and sterilization, and industrial scalability. To ameliorate these concerns, recombinant DNA approaches or synthetic strategies to fabricate discreet entities derived from, or that mimic, the natural ECM (*e.g.* oligopeptides, oligosaccharides, inorganic particulates), are being actively explored for integration into biomaterials. In this section we will discuss how these biomimetic approaches are being translated into functional biomaterials.

5.3.1   Protein-Engineered and Self-Assembled Biomaterials

Recombinant DNA approaches can be used to express a protein or polypeptide of interest using nucleic acid vectors and a suitable host expression system (*e.g. Escherichia coli*, Baculovirus, mammalian cells). This strategy allows the selection—through computational, experimental or combination approaches—of generating modular protein or peptide domains that have a desired function without unwanted domains or problems associated with animal-derived products (Figure 5.4a). For instance, the recombinant production of polypeptides containing repeat units found in elastin, silk protein and resilin for the development of stimuli responsive materials.[143] Linking polypeptide domains together can be used to integrate multiple spatially defined protein functionalities into a single molecule. Silk-elastin-like proteins have been developed that display highly tunable properties by varying components of each protein's repeat primary structure.[144] Adding the tripeptide RGD motif from fibronectin to promote cell adhesion along with peptide domains that induce mineralization has been demonstrated.[145,146]

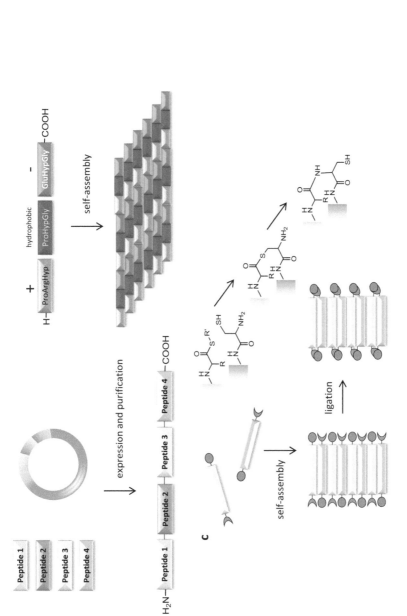
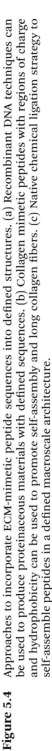

Figure 5.4 Approaches to incorporate ECM-mimetic peptide sequences into defined structures. (a) Recombinant DNA techniques can be used to produce proteinaceous materials with defined sequences. (b) Collagen mimetic peptides with regions of charge and hydrophobicity can be used to promote self-assembly and long collagen fibers. (c) Native chemical ligation strategy to self-assemble peptides in a defined macroscale architecture.

By combining the ideal mechanical properties engendered through the silk-elastin networks with cell adhesion motifs and mineralization domains, mesenchymal stem cells were shown to undergo osteogenesis on this material. Additional functionality can be included within the construct to cover a range of desirable functions. For instance, a resilin-like peptide was produced to include RGD peptides for cell adhesion, matrix metalloprotease (MMP) cleavage sequences to tune biodegradation and heparin-binding motifs for the sequestration of growth factors.[147] With the ease of which biomimetic modules can be assembled through plasmid construction, a host of functionalities may be included using protein-engineering approaches.

In contrast to the well-defined structural frameworks fabricated using motifs from elastin, silk and resilin, some desired proteins require post-translational modification that is incompatible with many scalable expression systems (*e.g.* collagen fiber assembly). Collagen polypeptides have been generated synthetically (ProHyp-Gly)$_{10}$ and assembled into a highly branched network; however, the resulting architecture does not resemble native collagen.[148] To form long collagen fibers, a strategy was developed wherein the collagen peptides self-assembled by virtue of a periodic cysteine residue that assembled overlapping peptides *via* intrastrand disulfides.[149] Other synthetic strategies to form collagenous biomaterials include self-assembly through stacking of electron-rich and electron-poor phenyl rings[150] and metal-triggered self-assembly.[151] Long collagen fibers with similar characteristics to those of native collagen were formed using the sequence (ProArgGly)$_4$–(ProHypGly)$_4$–(GluHypGly)$_4$ (Figure 5.4b).[152] While collagen materials generated using these approaches fail to fully recapitulate the structural and mechanical properties often desired, integrating these self-assembly strategies with other polymeric systems will surely lead to new and promising materials. Another strategy that has been shown as a useful method for fabricating complex protein structures from simple building blocks is native chemical ligation[153] (Figure 5.4c). This methodology has been used to entrap defined ECM motifs and mesenchymal stem cells within a modular PEG hydrogel material[154] and to modulate the mechanical properties of self-assembling β-sheet peptides for producing biomaterials that enhance endothelial cell proliferation.[155]

5.3.2 Synthetic Biomaterials Modified with Short Peptides

5.3.2.1 2D Model Systems. Truncation of the binding domain of fibronectin led to the discovery of the minimum sequence RGD that will facilitate cell adhesion through many integrin receptors.[156] This simple tripeptide has revolutionized the fabrication of synthetic bio-materials where grafting this sequence to a biomaterial surface will mediate cell adhesion and many other cellular processes.[157] While peptides do not contain all of the functionality of full length proteins, they offer the advantage of being well defined, synthesized through simple chemical means, and integrin specific.[158] This latter point in particular is critical to studies aimed at deciphering the precise inter-actions that mediate biochemical signaling from the matrix. This has led to the development of model substrates for studying diverse cel-lular processes, including cell adhesion, motility, proliferation, and differentiation.[159–163] One of the most widely employed model systems for cell biology studies are self-assembled monolayers (SAMs) of alkan-ethiolates on gold.[164] SAMs form well-ordered close-packed structures that mimic aspects of the ECM through simple mixing of terminally substituted alkanethiols. Peptide-modified SAMs have been used to show how the density of RGD peptides can influence adhesion and proliferation.[165,166] In addition to density, a linear and cyclic version of the RGD peptide were used to demonstrate how the affinity of peptide–integrin interaction can guide the differentiation of mesen-chymal stem cells.[162] Cells adherent to high affinity interfaces showed preferential expression of osteogenesis markers; decreasing the den-sity and affinity of the peptide ligands promoted the expression of markers associated with myogenesis and neurogenesis (Figure 5.5a). SAMs have proved quite versatile in exploring potential synergies between different adhesion motifs across different cell types and inte-grin classes.[167,168] The adhesion sequences RGD, IKVAV, YIGSR, and RETTAWA were found to direct changes in integrin expression during osteogenesis, adipogenesis, and chondrogenesis.[169] Thus modulating the incorporation of these simple adhesion sequences into biomateri-als at an appropriate density may prove a straight-forward method to promote specific differentiation outcomes in the absence of soluble media components.

Careful control of interfacial chemistry can enable multiple ECM derived cues to be presented on surfaces. For instance, a mixture of carboxyl and azide terminated alkanethiolates were recently used to immobilize the cell adhesion peptide RGDSP in combination with the heparin-binding sequence TYRSRKY *via* distinct orthogonal con-jugation reactions.[170] Using different conjugation schemes opens the way for interrogating multiple specific interactions. The use of com-bined adhesion peptides with sequences that bind proteoglycans and growth factors has demonstrated the importance of designing bio-materials beyond simple adhesion sequences, where sequestration of growth factors can influence adhesion, proliferation and soluble

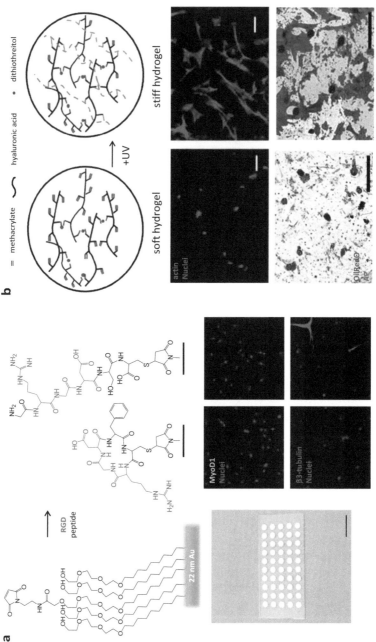

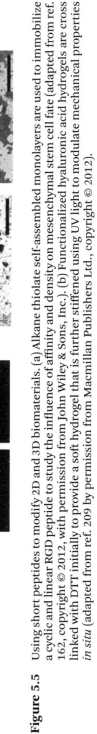

Figure 5.5   Using short peptides to modify 2D and 3D biomaterials. (a) Alkane thiolate self-assembled monolayers are used to immobilize a cyclic and linear RGD peptide to study the influence of affinity and density on mesenchymal stem cell fate (adapted from ref. 162, copyright © 2012, with permission from John Wiley & Sons, Inc.). (b) Functionalized hyaluronic acid hydrogels are cross linked with DTT initially to provide a soft hydrogel that is further stiffened using UV light to modulate mechanical properties *in situ* (adapted from ref. 209 by permission from Macmillan Publishers Ltd., copyright © 2012).

growth factor signaling.[171–174] In addition to adhesion peptides and those that sequester soluble factors, peptides directly derived from the binding sequences of growth factors and other morphogens have proved useful in modulating cell fate. For instance, SAMs presenting short peptides derived from a range of different proteins were used to discover unique surfaces that promote embryonic stem cell self-renewal.[175] These GAG-binding sequences were further developed to be integrated into hydrogel materials.[176] In addition to natural sequence selection, phage display technology based on the "biopanning and rapid analysis of selective interactive ligands" (BRASIL) technique has been used to identify non-natural short peptides that support embryonic stem cell growth.[177,178] Through combinations of naturally derived peptides and those discovered through these screening approaches, the literature contains a vast number of sequences with potential relevance to the development of synthetic biomaterials. Table 5.1 lists examples of short peptides derived from proteins that have shown utility in biomaterials science.

5.3.2.2 3D Synthetic Biomaterials. Many of the short peptide sequences studied using 2D model systems have proved translatable to more complex 3D biomaterials using di-functional PEG, GAGs and other synthetic hydrogel forming polymers.[36,179] In particular, hydrogel materials that do not contain endogenous cell instructive cues require modification, either through mixing with ECM proteins or by functionalization with oligopeptide motifs. Alginate has been modified using carbodiimide chemistry to conjugate RGD for regulation of skeletal muscle differentiation,[180,181] cardiac tissue engineering,[182] and for cellular delivery.[183] Both alginate and chitosan have been modified to present laminin-derived peptides for optimizing matrices to promote neurite outgrowth[184] and RGD peptides for promoting osteogenesis.[185,186] Similarly to chitosan and alginate, natural GAG materials can be easily derivatized with short peptides to enhance bioactivity. Hyaluronic acid interacts specifically with cell surface hyaluronan receptors but does not mediate appreciable cell adhesion on its own. Hyaluronic acid scaffolds have been functionalized with RGD peptides through sugar backbone oxidation,[187] carbodiimide activation of pendant carboxylates[188] or through Michael-type addition reactions with acrylate-modified hyaluronan.[189] Functional PEG hydrogels have been formed with collagen-mimetic synthetic peptides[190–192] and other ECM-derived short peptides.[193–196]

Besides ECM and morphogen-derived short peptide sequences, sequences that are amenable to enzymatic modification have emerged as another important class of peptides that mimic processes in the

Table 5.1   Short peptide sequences used in biomaterial fabrication.

Minimum sequence	Primary origin	References
Integrin binding		
GRGDS	Fibronectin	Pierschbacher *et al.*, *Nature*, 1984, **309**, 30
PHSRN	Fibronectin	Aota *et al.*, *J. Biol. Chem.*, 1994, **269**, 24756
REDV	Fibronectin	Hubbell *et al.*, *Bio/Technology*, 1991, **9**, 568
LDV	Fibronectin	Komoriya *et al.*, *J. Biol. Chem.*, 1991, **266**, 15075
YIGSR	Laminin	Graf *et al.*, *Biochemistry*, 1987, **26**, 6896
GIIFFL	Laminin	Graf *et al.*, *Cell*, 1987, **48**, 989
IKVAV	Laminin	Tashiro *et al.*, *J. Biol. Chem.*, 1989, **264**, 16174
PDGSR	Laminin	Kleinman *et al.*, *Arch. Biochem. Biophys.*, 1989, **272**
SIYITRF	Laminin	Nomizu *et al.*, *J. Biol. Chem.*, 1995, **270**, 20583
IAFQRN	Laminin	Nomizu *et al.*, *J. Biol. Chem.*, 1995, **270**, 20583
LQVQLSIR	Laminin	Nomizu *et al.*, *J. Biol. Chem.*, 1995, **270**, 20583
LRE	Laminin	Hunter *et al.*, *J. Neurosci.*, 1991, **11**, 3960
RNIAEIIKDI	Laminin	Hunter *et al.*, *Cell*, 1989, **59**, 905
SINNNR	Laminin	Richard *et al.*, *Exp. Cell Res.*, 1996, **228**, 98
YGYYGDALR	Laminin	Underwood *et al.*, *Biochem. J.*, 1995, **309**, 765
LGTIPG	Laminin	Mercham *et al.*, *J. Biol. Chem.*, 1989, **264**, 16652
DGEA	Collagen I	Staatz *et al.*, *J. Biol. Chem.*, 1991, **266**, 7363
FYFDLR	Collagen IV	Underwood *et al.*, *Biochem. J.*, 1995, **309**, 765
SVVYGLR	Osteopontin	Yokosaki *et al.*, *J. Biol. Chem.*, 1999, **274**, 36328
KRLDGS	Fibrinogen	Altieri *et al.*, *J. Biol. Chem.*, 1993, **268**, 1847
KQAGDV	Fibrinogen	Santoro *et al.*, *Cell*, 1987, **48**, 867
VTXG	Thorombospondin	Prater *et al.*, *J. Cell Biol.*, 1991, **112**, 1031
Proteoglycan binding		
KRSR	Vitronectin	Dee *et al.*, *J. Biomed. Mater. Res.*, 1998, **40**, 371
SHWSPWSS	Thrombospondin	Gou *et al.*, *Proc. Natl. Acad. Sci. U. S. A.*, 1992, **89**, 3040
FHRRIKA	Bone sialoprotein	Rezania *et al.*, *Biotechnol. Prog.*, 1999, **15**, 19
WQPPRARI	Fibronectin	Woods *et al.*, *Mol. Biol. Cell*, 1993, **4**, 605
TYRSRKY	FGF2	Lee *et al.*, *J. Biomed. Mater. Res.*, 2007, 970
FRHRNRKGY	Vitronectin	Dettin *et al.*, *J. Biomed. Mater. Res.*, 2002, **60**, 466
WSXW	Thorombospondin	Guo *et al.*, *Proc. Natl. Acad. Sci. U. S. A.*, 1992, **89**, 3040
EIKLLIS	Laminin	Tashiro *et al.*, *Biochem. J.*, 1994, **302**, 73

cellular microenvironment. These classes of materials are important because incorporating specificity into the degradation of a material—as opposed to the relatively non-specific degradation of many synthetic biodegradable polymers (Section 5.2.4.2)—can more closely recapitulate how tissue morphogenesis occurs *in vivo*. In Section 5.3.1 we presented recombinant DNA approaches to integrate MMP sequences into biomaterials to enable proteolytic degradation. There are a number of MMP degradable sequences, derived from natural ECM proteins and through screening of putative substrates, which are amenable to incorporation into scaffolds and hydrogel materials.[197,198] PEGDA has been directly conjugated with thiol end-functionalized

MMP sequences to promote enzymatic degradation and cell migration through the synthetic biomaterial.[199] The incorporation of adhesion peptides further supports interaction of cells with the synthetic matrix[200,201] and migration through the hydrogel network has been demonstrated to be modulated through MMP inhibition.[202] MMP and adhesion sequences have been incorporated into other synthetic and natural polymer systems including hyaluronic acid[203] and alginate.[204] The modular nature of the chemistry developed with these peptide hydrogels has led to the incorporation of growth factors for including protease-triggered release,[205,206] photoactive moieties for 3D patterning[207,208] and matrix stiffening[209] (Figure 5.5b).

In addition to homogenous hydrogel biomaterials, much of the ECM is organized with insoluble fibrous structures. Therefore, significant efforts have gone into developing strategies to fabricate nano- and micro-fiber structures using both natural materials and synthetic peptide-laden architectures. The most commonly used method for fabricating fiber structures uses electrospinning, in which viscoelastic polymer fibers are deposited randomly or in an aligned network.[210] One powerful aspect of electrospun structures is the ability to incorporate a rich diversity of composition into the material. Another approach that has benefited significantly from the design and development of bioactive short peptides is in the area of self-assembled peptide amphiphiles, which result in nanofibers of defined dimensions and hierarchical structures.[211]

5.4 CONCLUSION AND OUTLOOK

Designing materials that mimic the ECM is an important need in biomedicine. From the surfaces of metals, ceramics and polymers used in implants and fixative devices, to novel 3D architectures designed to emulate a tissues structure, controlling the spatiotemporal organization of ECM-derived features is key to directing a functional outcome. In this chapter we presented the state of the art in designing biomaterials to incorporate ECM-derived features. Natural materials have the advantage that they present bioactive motifs in a similar fashion to the native state and can be engineered *ex vivo* to form homogenous coatings and hydrogels as well as asymmetric fibrous scaffolds that reflect structural aspects of tissue. However, natural materials are commonly derived from plants and animals and thus carry a risk of immunogenicity. Furthermore, mimicking the spatiotemporal control of structure and biochemistry that is accomplished by nature is very challenging. To improve the control of biomaterials fabricated using natural molecules, integration with synthetic components is a powerful strategy because synthetic chemistry affords robust control over functionality, molecular weight, and structure. However, synthetic materials alone do not offer the breadth of activity present in naturally derived ECMs. Whether using naturally derived materials or synthetic polymer systems, careful control of bioconjugation within or between the materials—or at the interface of a biomaterials surface—can afford a vast range of structures and properties. Rather than choosing either naturally derived or synthetic systems alone, the emerging trend is to generate composite biomaterials systems that capitalize on the strength of both, *e.g.* the rich bioactivity of natural biopolymers and well-controlled synthetic architectures with defined mechanical properties.

Nature has spent millions of years developing robust materials composed of organic and inorganic molecules, with incredible diversity in structure and function. *In vivo*, materials present themselves to cells differentially across tissues, developmental stages, during normal tissue homeostasis and morphogenesis, and during pathological processes. Understanding how our cells and their microenvironments accomplish these feats—and translating this knowledge into functional biomaterials—is an incredibly challenging yet exciting task, as the development of biomimetic materials that recapitulate native tissue architecture has the potential to revolutionize medicine.

REFERENCES

1. K. Duan and R. Wang, *J. Mater. Chem.*, 2006, **16**, 2309–2321.
2. J. P. LeGeros, J. Wang, E. Garofalo, T. Salgado and R. Z. LeGeros, *Key Eng. Mater.*, 2008, **361–363**, 741–744.
3. M. Keeney, J. H. Lai and F. Yang, *Curr. Opin. Biotechnol.*, 2011, **22**, 734–740.
4. C. Wolf-Brandstetter and D. Scharnweber, *RSC Nanosci. Nanotechnol.*, 2012, **21**, 75–89.
5. P. K. Chu, J. Y. Chen, L. P. Wang and N. Huang, *Mater. Sci. Eng., R*, 2002, **R36**, 143–206.
6. P. Attri, B. Arora and E. H. Choi, *RSC Adv.*, 2013, **3**, 12540–12567.
7. A. A. Meyer-Plath, K. Schroder, B. Finke and A. Ohl, *Vacuum*, 2003, **71**, 391–406.
8. C. Haensch, S. Hoeppener and U. S. Schubert, *Chem. Soc. Rev.*, 2010, **39**, 2323–2334.
9. G. Guerrero, J. G. Alauzun, M. Granier, D. Laurencin and P. H. Mutin, *Dalton Trans.*, 2013, **42**, 12569–12585.

10. B. M. Silverman, K. A. Wieghaus and J. Schwartz, *Langmuir*, 2005, **21**, 225–228.
11. M. Mrksich, *Curr. Opin. Colloid Interface Sci.*, 1997, **2**, 83–88.
12. M. Mrksich and G. M. Whitesides, *Annu. Rev. Biophys. Biomol. Struct.*, 1996, **25**, 55–78.
13. A. Vasconcelos and A. Cavaco-Paulo, *Curr. Drug Targets*, 2013, **14**, 612–619.
14. S. Wang, F. Taraballi, L. P. Tan and K. W. Ng, *Cell Tissue Res.*, 2012, **347**, 795–802.
15. C. Vepari and D. L. Kaplan, *Prog. Polym. Sci.*, 2007, **32**, 991–1007.
16. S. L. Helgerson, T. Seelich, J. P. DiOrio, B. Tawil, K. Bittner and R. Spaethe, *Fibrin, Encyc. Biomat. Biomed. Eng.*, 2004, 603–610.
17. J. W. Weisel, *Adv. Protein Chem.*, 2005, **70**, 247–299.
18. P. A. Janmey, J. P. Winer and J. W. Weisel, *J. R. Soc., Interface*, 2009, **6**, 1–10.
19. C. Andree, B. I. J. Munder, P. Behrendt, S. Hellmann, W. Audretsch, M. Voigt, C. Reis, M. W. Beckmann, R. E. Horch and A. D. Bach, *Breast*, 2008, **17**, 492–498.
20. R. S. Martins, M. G. Siqueira, S. C. F. Da and J. P. P. Plese, *Surg. Neurol.*, 2005, **64**(suppl. 1), S10–S16, discussion S11:16.
21. R. Mittermayr, E. Wassermann, M. Thurnher, M. Simunek and H. Redl, *Burns*, 2006, **32**, 305–311.
22. J. Sanchez-Cortes and M. Mrksich, *Chem. Biol.*, 2009, **16**, 990–1000.
23. D. A. Cheresh, S. A. Berliner, V. Vicente and Z. M. Ruggeri, *Cell*, 1989, **58**, 945–953.
24. S. M. Willerth, K. J. Arendas, D. I. Gottlieb and S. E. Sakiyama-Elbert, *Biomaterials*, 2006, **27**, 5990–6003.
25. P. J. Johnson, A. Tatara, D. A. McCreedy, A. Shiu and S. E. Sakiyama-Elbert, *Soft Matter*, 2010, **6**, 5127–5137.
26. I. Catelas, N. Sese, B. M. Wu, J. C. Y. Dunn, S. Helgerson and B. Tawil, *Tissue Eng.*, 2006, **12**, 2385–2396.
27. D. Eyrich, F. Brandl, B. Appel, H. Wiese, G. Maier, M. Wenzel, R. Staudenmaier, A. Goepferich and T. Blunk, *Biomaterials*, 2007, **28**, 55–65.
28. R. R. Rao, A. W. Peterson, J. Ceccarelli, A. J. Putnam and J. P. Stegemann, *Angiogenesis*, 2012, **15**, 253–264.
29. J. S. Miller, K. R. Stevens, M. T. Yang, B. M. Baker, D.-H. T. Nguyen, D. M. Cohen, E. Toro, A. A. Chen, P. A. Galie, X. Yu, R. Chaturvedi, S. N. Bhatia and C. S. Chen, *Nat. Mater.*, 2012, **11**, 768–774.
30. J. Liu, Y. Tan, H. Zhang, Y. Zhang, P. Xu, J. Chen, Y.-C. Poh, K. Tang, N. Wang and B. Huang, *Nat. Mater.*, 2012, **11**, 734–741.

31. M. D. Shoulders and R. T. Raines, *Annu. Rev. Biochem.*, 2009, **78**, 929–958.

32. S.-T. Li, 2013.

33. A. J. Engler, S. Sen, H. L. Sweeney and D. E. Discher, *Cell*, 2006, **126**, 677–689.

34. J. Lee, A. A. Abdeen, D. Zhang and K. A. Kilian, *Biomaterials*, 2013, **34**, 8140–8148.

35. B. R. Williams, R. A. Gelman, D. C. Poppke and K. A. Piez, *J. Biol. Chem.*, 1978, **253**, 6578–6585.

36. J. Thiele, Y. Ma, S. M. C. Bruekers, S. Ma and W. T. S. Huck, *Adv. Mater.*, 2014, **26**, 125–148.

37. A. V. Taubenberger, M. A. Woodruff, H. Bai, D. J. Muller and D. W. Hutmacher, *Biomaterials*, 2010, **31**, 2827–2835.

38. J. Heino, *BioEssays*, 2007, **29**, 1001–1010.

39. W.-H. Zimmermann, I. Melnychenko and T. Eschenhagen, *Biomaterials*, 2004, **25**, 1639–1647.

40. A. M. Ferreira, P. Gentile, V. Chiono and G. Ciardelli, *Acta Biomater.*, 2012, **8**, 3191–3200.

41. J. Elisseeff, *PMSE Prepr.*, 2008, **98**, 726.

42. K. Y. Lee and D. J. Mooney, *Chem. Rev.*, 2001, **101**, 1869–1879.

43. J. R. Venugopal, M. P. Prabhakaran, S. Mukherjee, R. Ravichandran, K. Dan and S. Ramakrishna, *J. R. Soc., Interface*, 2012, **9**, 1–19.

44. Y. Huang, S. Onyeri, M. Siewe, A. Moshfeghian and S. V. Madihally, *Biomaterials*, 2005, **26**, 7616–7627.

45. G. Camci-Unal, D. Cuttica, N. Annabi, D. Demarchi and A. Khademhosseini, *Biomacromolecules*, 2013, **14**, 1085–1092.

46. Y.-C. Chen, R.-Z. Lin, H. Qi, Y. Yang, H. Bae, J. M. Melero-Martin and A. Khademhosseini, *Adv. Funct. Mater.*, 2012, **22**, 2027–2039.

47. C. B. Hutson, J. W. Nichol, H. Aubin, H. Bae, S. Yamanlar, S. Al-Haque, S. T. Koshy and A. Khademhosseini, *Tissue Eng., Part A*, 2011, **17**, 1713–1723.

48. I. Arnaoutova, J. George, H. K. Kleinman and G. Benton, *Angiogenesis*, 2009, **12**, 267–274.

49. L. G. Villa-Diaz, A. M. Ross, J. Lahann and P. H. Krebsbach, *Stem Cells*, 2013, **31**, 1–7.

50. C. Xu, M. S. Inokuma, J. Denham, K. Golds, P. Kundu, J. D. Gold and M. K. Carpenter, *Nat. Biotechnol.*, 2001, **19**, 971–974.

51. T. C. Laurent and J. R. E. Fraser, *FASEB J.*, 1992, **6**, 2397–2404.

52. A. Chopra, M. E. Murray, F. J. Byfield, M. G. Mendez, R. Halleluyan, D. J. Restle, A. D. Raz-Ben, P. A. Galie, K. Pogoda, R. Bucki, C. Marcinkiewicz, G. D. Prestwich, T. I. Zarembinski, C. S. Chen,

E. Pure, J. Y. Kresh and P. A. Janmey, *Biomaterials*, 2014, **35**, 71–82.

53. J. A. Burdick and G. D. Prestwich, *Adv. Mater.*, 2011, **23**, H41–H56.
54. L. G. Harris, L. M. Patterson, C. Bacon, I. ap Gwynn and R. G. Richards, *J. Biomed. Mater. Res., Part A*, 2005, **73A**, 12–20.
55. M. C. Schulz, P. Korn, B. Stadlinger, U. Range, S. Moeller, J. Becher, M. Schnabelrauch, R. Mai, D. Scharnweber, U. Eckelt and V. Hintze, *J. Mater. Sci.: Mater. Med.*, 2014, **25**, 247–258.
56. A. C. Jimenez-Vergara, D. J. Munoz-Pinto, S. Becerra-Bayona, B. Wang, A. Iacob and M. S. Hahn, *Acta Biomater.*, 2011, **7**, 3964–3972.
57. E. Gemma, O. Meyer, D. Uhrin and A. N. Hulme, *Mol. BioSyst.*, 2008, **4**, 481–495.
58. D. Campaniello and M. R. Corbo, 2010.
59. P. He, S. S. Davis and L. Illum, *Int. J. Pharm.*, 1998, **166**, 75–88.
60. S. B. Rao and C. P. Sharma, *J. Biomed. Mater. Res.*, 1997, **34**, 21–28.
61. I. Aranaz, M. Mengibar, R. Harris, I. Panos, B. Miralles, N. Acosta, G. Galed and A. Heras, *Curr. Chem. Biol.*, 2009, **3**, 203–230.
62. M. Rinaudo, *Prog. Polym. Sci.*, 2006, **31**, 603–632.
63. S. Y. Seo, S. H. Park, H. J. Lee, Y. Heo, H. N. Na, K. I. Kim, J. H. Han, Y. Ito and T. I. Son, *J. Appl. Polym. Sci.*, 2013, **128**, 4322–4326.
64. Z. Shi, K. G. Neoh, E. T. Kang, C. K. Poh and W. Wang, *Biomacromolecules*, 2009, **10**, 1603–1611.
65. Z. Shi, K. G. Neoh, E. T. Kang, C. Poh and W. Wang, *J. Biomed. Mater. Res., Part A*, 2008, **86A**, 865–872.
66. A. P. Zhu, Z. Ming and S. Jian, *Appl. Surf. Sci.*, 2005, **241**, 485–492.
67. J. S. Mao, Y. L. Cui, X. H. Wang, Y. Sun, Y. J. Yin, H. M. Zhao and K. D. Yao, *Biomaterials*, 2004, **25**, 3973–3981.
68. Y. Yin, Z. Li, Y. Sun and K. Yao, *J. Mater. Sci.*, 2005, **40**, 4649–4652.
69. A. V. Il'ina and V. P. Varlamov, *Appl. Biochem. Microbiol.*, 2005, **41**, 5–11.
70. S. I. Jeong, M. D. Krebs, C. A. Bonino, J. E. Samorezov, S. A. Khan and E. Alsberg, *Tissue Eng., Part A*, 2011, **17**, 59–70.
71. W. E. Hennink and N. C. F. van, *Adv. Drug Delivery Rev.*, 2002, **54**, 13–36.
72. J. Cho, M.-C. Heuzey, A. Begin and P. J. Carreau, *Biomacromolecules*, 2005, **6**, 3267–3275.
73. J. S. Kwon, G. H. Kim, D. Y. Kim, S. M. Yoon, H. W. Seo, J. H. Kim, B. H. Min and M. S. Kim, *Int. J. Biol. Macromol.*, 2012, **51**, 974–979.
74. K. S. Kim, J. H. Lee, H. H. Ahn, J. Y. Lee, G. Khang, B. Lee, H. B. Lee and M. S. Kim, *Biomaterials*, 2008, **29**, 4420–4428.

75. M.-C. Chen, F.-L. Mi, Z.-X. Liao, C.-W. Hsiao, K. Sonaje, M.-F. Chung, L.-W. Hsu and H.-W. Sung, *Adv. Drug Delivery Rev.*, 2013, **65**, 865–879.

76. K. Y. Lee and D. J. Mooney, *Prog. Polym. Sci.*, 2012, **37**, 106–126.

77. X. Zhao, N. Huebsch, D. J. Mooney and Z. Suo, *J. Appl. Phys.*, 2010, **107**, 063509/063501–063509/063505.

78. Y. Qin, *Polym. Adv. Technol.*, 2008, **19**, 6–14.

79. T. K. Giri, D. Thakur, A. Alexander, Ajazuddin, H. Badwaik and D. K. Tripathi, *Curr. Drug Delivery*, 2012, **9**, 539–555.

80. A. D. Augst, H. J. Kong and D. J. Mooney, *Macromol. Biosci.*, 2006, **6**, 623–633.

81. K. Y. Lee, H. J. Kong and D. J. Mooney, *Macromol. Biosci.*, 2008, **8**, 140–145.

82. R. D. LeBoeuf, R. R. Gregg, P. H. Weigel and G. M. Fuller, *Biochemistry*, 1987, **26**, 6052–6057.

83. F. Lee and M. Kurisawa, *Acta Biomater.*, 2013, **9**, 5143–5152.

84. S.-W. Kang, J.-S. Kim, K.-S. Park, B.-H. Cha, J.-H. Shim, J. Y. Kim, D.-W. Cho, J.-W. Rhie and S.-H. Lee, *Bone*, 2011, **48**, 298–306.

85. S.-H. Park, J. H. Cui, S. R. Park and B.-H. Min, *Artif. Organs*, 2009, **33**, 439–447.

86. R. C. Pereira, M. Scaranari, P. Castagnola, M. Grandizio, H. S. Azevedo, R. L. Reis, R. Cancedda and C. Gentili, *J. Tissue Eng. Regener. Med.*, 2009, **3**, 97–106.

87. M. Rampichova, E. Filova, F. Varga, A. Lytvynets, E. Prosecka, L. Kolacna, J. Motlik, A. Necas, L. Vajner, J. Uhlik and E. Amler, *ASAIO J.*, 2010, **56**, 563–568.

88. Y. Wei, Y. Hu, W. Hao, Y. Han, G. Meng, D. Zhang, Z. Wu and H. Wang, *J. Orthop. Res.*, 2008, **26**, 27–33.

89. P. Divya and L. K. Krishnan, *J. Tissue Eng. Regener. Med.*, 2009, **3**, 377–388.

90. I. V. Yannas, E. Lee, D. P. Orgill, E. M. Skrabut and G. F. Murphy, *Proc. Natl. Acad. Sci. U. S. A.*, 1989, **86**, 933–937.

91. B. A. Harley, M. H. Spilker, J. W. Wu, K. Asano, H. P. Hsu, M. Spector and I. V. Yannas, *Cells Tissues Organs*, 2004, **176**, 153–165.

92. B. A. C. Harley and L. J. Gibson, *Chem. Eng. J.*, 2008, **137**, 102–121.

93. C. Jungreuthmayer, S. W. Donahue, M. J. Jaasma, A. A. Al-Munajjed, J. Zanghellini, D. J. Kelly and F. J. O'Brien, *Tissue Eng., Part A*, 2009, **15**, 1141–1149.

94. S. R. Caliari, M. A. Ramirez and B. A. C. Harley, *Biomaterials*, 2011, **32**, 8990–8998.

95. N. Davidenko, J. J. Campbell, E. S. Thian, C. J. Watson and R. E. Cameron, *Acta Biomater.*, 2010, **6**, 3957–3968.

96. S.-N. Park, J.-C. Park, H. O. Kim, M. J. Song and H. Suh, *Biomaterials*, 2001, **23**, 1205–1212.
97. C. M. De, M. Massimi, A. Barbetta, R. B. L. Di, S. Nardecchia, D. L. Conti and M. Dentini, *Biomed. Mater.*, 2012, **7**, 055005–055013.
98. X. Hu, D. Li, F. Zhou and C. Gao, *Acta Biomater.*, 2011, **7**, 1618–1626.
99. L. Wang and J. P. Stegemann, *Biomaterials*, 2010, **31**, 3976–3985.
100. Z. Gong, H. Xiong, X. Long, L. Wei, J. Li, Y. Wu and Z. Lin, *Biomed. Mater.*, 2010, **5**, 055005/055001–055005/055009.
101. X. Yang, G. Han, X. Pang and M. Fan, *J. Biomed. Mater. Res A*, 2011, **00A**, 1–8.
102. Y. Zhang, X. Cheng, J. Wang, Y. Wang, B. Shi, C. Huang, X. Yang and T. Liu, *Biochem. Biophys. Res. Commun.*, 2006, **344**, 362–369.
103. Y. Zhang, J. Song, B. Shi, Y. Wang, X. Chen, C. Huang, X. Yang, D. Xu, X. Cheng and X. Chen, *Biomaterials*, 2007, **28**, 4635–4642.
104. S. D. Sarkar, B. L. Farrugia, T. R. Dargaville and S. Dhara, *J. Biomed. Mater. Res., Part A*, 2013, **101A**, 3482–3492.
105. E. Zakhem, S. Raghavan, R. R. Gilmont and K. N. Bitar, *Biomaterials*, 2012, **33**, 4810–4817.
106. L. Sang, X. Wang, Z. Chen, J. Lu, Z. Gu and X. Li, *Carbohydr. Polym.*, 2010, **82**, 1264–1270.
107. X. P. Bai, H. X. Zheng, R. Fang, T. R. Wang, X. L. Hou, Y. Li, X. B. Chen and W. M. Tian, *Biomed. Mater.*, 2011, **6**, 045002.
108. M. S. Hahn, B. A. Teply, M. M. Stevens, S. M. Zeitels and R. Langer, *Biomaterials*, 2006, **27**, 1104–1109.
109. H.-j. Lee, S.-H. Ahn and G. H. Kim, *Chem. Mater.*, 2012, **24**, 881–891.
110. X. Yang, L. Guo, Y. Fan and X. Zhang, *Int. J. Biol. Macromol.*, 2013, **61**, 487–493.
111. G. H. Kim, S. Ahn, Y. Y. Kim, Y. Cho and W. Chun, *J. Mater. Chem.*, 2011, **21**, 6165–6172.
112. K. Li, X. Qu, Y. Wang, Y. Tang, D. Qin, Y. Wang and M. Feng, *J. Biomed. Mater. Res., Part A*, 2005, **75A**, 268–274.
113. P. Giusti, L. Lazzeri, N. Barbani, L. Lelli, S. De Petris and M. G. Cascone, *Macromol. Symp.*, 1994, **78**, 285–297.
114. D. A. Rennerfeldt, A. N. Renth, Z. Talata, S. H. Gehrke and M. S. Detamore, *Biomaterials*, 2013, **34**, 8241–8257.
115. G. C. Ingavle, N. H. Dormer, S. H. Gehrke and M. S. Detamore, *J. Mater. Sci.: Mater. Med.*, 2012, **23**, 157–170.
116. S. C. Rizzi and J. A. Hubbell, *Biomacromolecules*, 2005, **6**, 1226–1238.
117. M. J. Roberts, M. D. Bentley and J. M. Harris, *Adv. Drug Delivery Rev.*, 2002, **54**, 459–476.

118. M. Gonen-Wadmany, L. Oss-Ronen and D. Seliktar, *Biomaterials*, 2007, **28**, 3876–3886.
119. L. Almany and D. Seliktar, *Biomaterials*, 2005, **26**, 2467–2477.
120. O. Sarig-Nadir and D. Seliktar, *Tissue Eng., Part A*, 2008, **14**, 401–411.
121. M. N. Mason and M. J. Mahoney, *J Biomed Mater Res A*, 2010, **95**, 283–293.
122. B. K. Chan, C. C. Wippich, C.-J. Wu, P. M. Sivasankar and G. Schmidt, *Macromol. Biosci.*, 2012, **12**, 1490–1501.
123. T. D. Sargeant, A. P. Desai, S. Banerjee, A. Agawu and J. B. Stopek, *Acta Biomater.*, 2012, **8**, 124–132.
124. K. Bott, Z. Upton, K. Schrobback, M. Ehrbar, J. A. Hubbell, M. P. Lutolf and S. C. Rizzi, *Biomaterials*, 2010, **31**, 8454–8464.
125. R. K. Singh, D. Seliktar and A. J. Putnam, *Biomaterials*, 2013, **34**, 9331–9340.
126. Y. Fu, K. Xu, X. Zheng, A. J. Giacomin, A. W. Mix and W. J. Kao, *Biomaterials*, 2012, **33**, 48–58.
127. A. T. Hillel, S. Unterman, Z. Nahas, B. Reid, J. M. Coburn, J. Axelman, J. J. Chae, Q. Guo, R. Trow, A. Thomas, Z. Hou, S. Lichtsteiner, D. Sutton, C. Matheson, P. Walker, N. David, S. Mori, J. M. Taube and J. H. Elisseeff, *Sci. Transl. Med.*, 2011, **3**, 93ra67.
128. J. L. Vanderhooft, B. K. Mann and G. D. Prestwich, *Biomacromolecules*, 2007, **8**, 2883–2889.
129. R. Jin, L. S. M. Teixeira, A. Krouwels, P. J. Dijkstra, B. C. A. van, M. Karperien and J. Feijen, *Acta Biomater.*, 2010, **6**, 1968–1977.
130. X. Z. Shu, Y. Liu, F. Palumbo and G. D. Prestwich, *Biomaterials*, 2003, **24**, 3825–3834.
131. D. Eglin, D. Mortisen and M. Alini, *Soft Matter*, 2009, **5**, 938–947.
132. M. Gao, Y. Cai, W. Wu, Y. Shi and Z. Fei, *Biomed. Mater.*, 2013, **8**, 045003–045010.
133. L. N. Novikova, J. Pettersson, M. Brohlin, M. Wiberg and L. N. Novikov, *Biomaterials*, 2008, **29**, 1198–1206.
134. W. Meng, S.-Y. Kim, J. Yuan, J. C. Kim, O. H. Kwon, N. Kawazoe, G. Chen, Y. Ito and I.-K. Kang, *J. Biomater. Sci., Polym. Ed.*, 2007, **18**, 81–94.
135. G. T. Koese, F. Korkusuz, P. Korkusuz and V. Hasirci, *Tissue Eng.*, 2004, **10**, 1234–1250.
136. D. Feingold-Leitman, E. Zussman and D. Seliktar, *J. Bionanosci.*, 2009, **3**, 45–57.
137. K. Haberstroh, K. Ritter, J. Kuschnierz, K.-H. Bormann, C. Kaps, C. Carvalho, R. Muelhaupt, M. Sittinger and N.-C. Gellrich, *J. Biomed. Mater. Res., Part B*, 2010, **93B**, 520–530.

138. P. Y. Lee, E. Cobain, J. Huard and L. Huang, *Mol. Ther.*, 2007, **15**, 1189–1194.
139. H. Tan, J. Wu, L. Lao and C. Gao, *Acta Biomater.*, 2009, **5**, 328–337.
140. H. Fan, H. Tao, Y. Wu, Y. Hu, Y. Yan and Z. Luo, *J Biomed Mater Res A*, 2010, **95**, 982–992.
141. A. S. Sawhney, C. P. Pathak and J. A. Hubbell, *Macromolecules*, 1993, **26**, 581–587.
142. M. A. Rice and K. S. Anseth, *Tissue Eng.*, 2007, **13**, 683–691.
143. R. L. DiMarco and S. C. Heilshorn, *Adv. Mater.*, 2012, **24**, 3923–3940.
144. R. Dandu and H. Ghandehari, *Prog. Polym. Sci.*, 2007, **32**, 1008–1030.
145. C. W. P. Foo, S. V. Patwardhan, D. J. Belton, B. Kitchel, D. Anastasiades, J. Huang, R. R. Naik, C. C. Perry and D. L. Kaplan, *Proc. Natl. Acad. Sci. U. S. A.*, 2006, **103**, 9428–9433.
146. A. J. Mieszawska, L. D. Nadkarni, C. C. Perry and D. L. Kaplan, *Chem. Mater.*, 2010, **22**, 5780–5785.
147. M. B. Charati, J. L. Ifkovits, J. A. Burdick, J. G. Linhardt and K. L. Kiick, *Soft Matter*, 2009, **5**, 3412–3416.
148. K. Kar, P. Amin, M. A. Bryan, A. V. Persikov, A. Mohs, Y.-H. Wang and B. Brodsky, *J. Biol. Chem.*, 2006, **281**, 33283–33290.
149. F. W. Kotch and R. T. Raines, *Proc. Natl. Acad. Sci. U. S. A.*, 2006, **103**, 3028–3033.
150. M. A. Cejas, W. A. Kinney, C. Chen, G. C. Leo, B. A. Tounge, J. G. Vinter, P. P. Joshi and B. E. Maryanoff, *J. Am. Chem. Soc.*, 2007, **129**, 2202–2203.
151. D. E. Przybyla and J. Chmielewski, *J. Am. Chem. Soc.*, 2008, **130**, 12610–12611.
152. S. Rele, Y. Song, R. P. Apkarian, Z. Qu, V. P. Conticello and E. L. Chaikof, *J. Am. Chem. Soc.*, 2007, **129**, 14780–14787.
153. P. E. Dawson, T. W. Muir, I. Clark-Lewis and S. B. H. Kent, *Science*, 1994, **266**, 776–779.
154. J. P. Jung, A. J. Sprangers, J. R. Byce, J. Su, J. M. Squirrell, P. B. Messersmith, K. W. Eliceiri and B. M. Ogle, *Biomacromolecules*, 2013, **14**, 3102–3111.
155. J. P. Jung, J. L. Jones, S. A. Cronier and J. H. Collier, *Biomaterials*, 2008, **29**, 2143–2151.
156. M. D. Pierschbacher and E. Ruoslahti, *Nature*, 1984, **309**, 30–33.
157. U. Hersel, C. Dahmen and H. Kessler, *Biomaterials*, 2003, **24**, 4385–4415.
158. E. Ruoslahti, *Annu. Rev. Cell Dev. Biol.*, 1996, **12**, 697–715.
159. D. L. Hern and J. A. Hubbell, *J. Biomed. Mater. Res.*, 1998, **39**, 266–276.

160. M. H. Fittkau, P. Zilla, D. Bezuidenhout, M. P. Lutolf, P. Human, J. A. Hubbell and N. Davies, *Biomaterials*, 2004, **26**, 167–174.
161. M. J. Cooke, T. Zahir, S. R. Phillips, D. S. H. Shah, D. Athey, J. H. Lakey, M. S. Shoichet and S. A. Przyborski, *J. Biomed. Mater. Res., Part A*, 2010, **93A**, 824–832.
162. K. A. Kilian and M. Mrksich, *Angew. Chem., Int. Ed.*, 2012, **51**, 4891–4895.
163. S. G. Le, A. Magenau, K. Gunaratnam, K. A. Kilian, T. Boecking, J. J. Gooding and K. Gaus, *Biophys. J.*, 2011, **101**, 764–773.
164. M. Mrksich, *Acta Biomater.*, 2009, **5**, 832–841.
165. G. A. Hudalla and W. L. Murphy, *Langmuir*, 2009, **25**, 5737–5746.
166. C. Roberts, C. S. Chen, M. Mrksich, V. Martichonok, D. E. Ingber and G. M. Whitesides, *J. Am. Chem. Soc.*, 1998, **120**, 6548–6555.
167. Y. Feng and M. Mrksich, *Biochemistry*, 2004, **43**, 15811–15821.
168. J. Sanchez-Cortes and M. Mrksich, *ACS Chem. Biol.*, 2011, **6**, 1078–1086.
169. J. E. Frith, R. J. Mills, J. E. Hudson and J. J. Cooper-White, *Stem Cells Dev.*, 2012, **21**, 2442–2456.
170. G. A. Hudalla and W. L. Murphy, *Langmuir*, 2010, **26**, 6449–6456.
171. G. A. Hudalla, J. T. Koepsel and W. L. Murphy, *Adv. Mater.*, 2011, **23**, 5415–5418.
172. G. A. Hudalla, N. A. Kouris, J. T. Koepsel, B. M. Ogle and W. L. Murphy, *Integr. Biol.*, 2011, **3**, 832–842.
173. A. Rezania and K. E. Healy, *Biotechnol. Prog.*, 1999, **15**, 19–32.
174. L. Li, J. R. Klim, R. Derda, A. H. Courtney and L. L. Kiessling, *Proc. Natl. Acad. Sci. U. S. A.*, 2011, **108**, 11745–11750.
175. J. R. Klim, L. Li, P. J. Wrighton, M. S. Piekarczyk and L. L. Kiessling, *Nat. Methods*, 2010, **7**, 989–994.
176. S. Musah, S. A. Morin, P. J. Wrighton, D. B. Zwick, S. Jin and L. L. Kiessling, *ACS Nano*, 2012, **6**, 10168–10177.
177. R. Derda, L. Li, B. P. Orner, R. L. Lewis, J. A. Thomson and L. L. Kiessling, *ACS Chem. Biol.*, 2007, **2**, 347–355.
178. R. Derda, S. Musah, B. P. Orner, J. R. Klim, L. Li and L. L. Kiessling, *J. Am. Chem. Soc.*, 2010, **132**, 1289–1295.
179. M. P. Lutolf and J. A. Hubbell, *Nat. Biotechnol.*, 2005, **23**, 47–55.
180. J. A. Rowley, Z. Sun, D. Goldman and D. J. Mooney, *Adv. Mater.*, 2002, **14**, 886–889.
181. J. A. Rowley, G. Madlambayan and D. J. Mooneyb, *Biomaterials*, 1998, **20**, 45–53.
182. M. Shachar, O. Tsur-Gang, T. Dvir, J. Leor and S. Cohen, *Acta Biomater.*, 2011, **7**, 152–162.

183. S. J. Bidarra, C. C. Barrias, K. B. Fonseca, M. A. Barbosa, R. A. Soares and P. L. Granja, *Biomaterials*, 2011, **32**, 7897–7904.

184. Y. Yamada, K. Hozumi, F. Katagiri, Y. Kikkawa and M. Nomizu, *Biopolymers*, 2010, **94**, 711–720.

185. M.-H. Ho, D.-M. Wang, H.-J. Hsieh, H.-C. Liu, T.-Y. Hsien, J.-Y. Lai and L.-T. Hou, *Biomaterials*, 2005, **26**, 3197–3206.

186. X. Liu, W. Peng, Y. Wang, M. Zhu, T. Sun, Q. Peng, Y. Zeng, B. Feng, X. Lu, J. Weng and J. Wang, *J. Mater. Chem. B*, 2013, **1**, 4484–4492.

187. J. R. Glass, K. T. Dickerson, K. Stecker and J. W. Polarek, *Biomaterials*, 1996, **17**, 1101–1108.

188. F. Z. Cui, W. M. Tian, S. P. Hou, Q. Y. Xu and I. S. Lee, *J. Mater. Sci.: Mater. Med.*, 2006, **17**, 1393–1401.

189. J. Lam and T. Segura, *Biomaterials*, 2013, **34**, 3938–3947.

190. H. J. Lee, J.-S. Lee, T. Chansakul, C. Yu, J. H. Elisseeff and S. M. Yu, *Biomaterials*, 2006, **27**, 5268–5276.

191. P. J. Stahl, N. H. Romano, D. Wirtz and S. M. Yu, *Biomacromolecules*, 2010, **11**, 2336–2344.

192. P. J. Stahl and S. M. Yu, *Soft Matter*, 2012, **8**, 10409–10418.

193. D. R. Schmidt and W. J. Kao, *J. Biomed. Mater. Res., Part A*, 2007, **83A**, 617–625.

194. A. L. Gonzalez, A. S. Gobin, J. L. West, L. V. McIntire and C. W. Smith, *Tissue Eng.*, 2004, **10**, 1775–1786.

195. W. J. Seeto, Y. Tian and E. A. Lipke, *Acta Biomater.*, 2013, **9**, 8279–8289.

196. F. Yang, C. G. Williams, D.-a. Wang, H. Lee, P. N. Manson and J. Elisseeff, *Biomaterials*, 2005, **26**, 5991–5998.

197. H. Nagase and G. B. Fields, *Biopolymers*, 1996, **40**, 399–416.

198. J. Patterson and J. A. Hubbell, *Biomaterials*, 2010, **31**, 7836–7845.

199. J. L. West and J. A. Hubbell, *Macromolecules*, 1999, **32**, 241–244.

200. M. P. Lutolf, J. L. Lauer-Fields, H. G. Schmoekel, A. T. Metters, F. E. Weber, G. B. Fields and J. A. Hubbell, *Proc. Natl. Acad. Sci. U. S. A.*, 2003, **100**, 5413–5418.

201. P. J. Yang, M. E. Levenston and J. S. Temenoff, *Tissue Eng., Part A*, 2012, **18**, 2365–2375.

202. G. P. Raeber, M. P. Lutolf and J. A. Hubbell, *Biophys. J.*, 2005, **89**, 1374–1388.

203. S. Khetan and J. A. Burdick, *Biomaterials*, 2010, **31**, 8228–8234.

204. K. B. Fonseca, D. B. Gomes, K. Lee, S. G. Santos, A. Sousa, E. A. Silva, D. J. Mooney, P. L. Granja and C. C. Barrias, *Biomacromolecules*, 2014, **15**, 380–390.

205. S. E. Sakiyama-Elbert, A. Panitch and J. A. Hubbell, *FASEB J.*, 2001, **15**, 1300–1302.

206. A. H. Zisch, M. P. Lutolf, M. Ehrbar, G. P. Raeber, S. C. Rizzi, N. Davies, H. Schmoekel, D. Bezuidenhout, V. Djonov, P. Zilla and J. A. Hubbell, *FASEB J.*, 2003, **17**, 2260–2262.
207. P. M. Kharkar, K. L. Kiick and A. M. Kloxin, *Chem. Soc. Rev.*, 2013, **42**, 7335–7372.
208. A. M. Kloxin, A. M. Kasko, C. N. Salinas and K. S. Anseth, *Science*, 2009, **324**, 59–63.
209. M. Guvendiren and J. A. Burdick, *Nat. Commun.*, 2012, **3**, 1792/1791–1792/1799.
210. S. Agarwal, J. H. Wendorff and A. Greiner, *Adv. Mater.*, 2009, **21**, 3343–3351.
211. J. H. Collier, J. S. Rudra, J. Z. Gasiorowski and J.-W. P. Jung, *Chem. Soc. Rev.*, 2010, **39**, 3413–3424.

CHAPTER 6

Biomaterials: Modulating and Tuning Synthetic Extracellular Matrix Mechanics

ELIZABETH JIN[*a] AND WAN-JU LI[*a]

[a]Department of Orthopedics and Rehabilitation, Department of Biomedical Engineering, University of Wisconsin-Madison, Madison, WI 53705, USA
*E-mail: ekirk@wisc.edu, li@ortho.wisc.edu

6.1 INTRODUCTION

The extracellular matrix (ECM) constituting the microenvironment of a tissue is a collection of proteins, glycoproteins, and proteoglycans. Composed of varying ratios of these molecules, the ECM in tissues or organs possesses different chemical and physical properties. In the ECM microenvironment of a tissue or organ, cell regulation is coordinated by both chemical and physical cues of the surrounding ECM to direct cell activities, such as adhesion, migration, proliferation, and differentiation.[1,2] In particular, physical cues including geometry, topography, and mechanics have been shown to direct tissue formation during development and maintain tissue homeostasis throughout the postnatal life.[1,3] For example, during embryo development, ECM mechanics plays an important role in germ layer organization[4] and

Mimicking the Extracellular Matrix: The Intersection of Matrix Biology and Biomaterials
Edited by Gregory A. Hudalla and William L. Murphy
© The Royal Society of Chemistry 2020
Published by the Royal Society of Chemistry, www.rsc.org

gastrulation.[5] In the postnatal life, cells in a healthy tissue continue to receive mechanical cues from the ECM to regulate tissue homeostasis whereas during tumor formation or disease progression, abnormal ECM deposition or cross-linking alters stiffness of the ECM, which in turn changes mechanical cues to disrupt tissue homeostasis, ultimately leading to malignancy.[6] Recently, emerging research findings of stem cell studies have demonstrated that lineage differentiation of embryonic stem cells (ESCs) is affected by mechanical properties of their surrounding environment.[7] Stiffness of the ECM, independently of biochemical cues, was identified as a critical environmental factor capable of directing lineage-specific differentiation of adult stem cells.[8] These research findings collectively suggest that ECM mechanics plays an important role in regulation of cell activities.

In recent years, considerable research effort has been devoted to understanding the effect of ECM mechanics on cell behavior, particularly in the fields of cancer biology and regenerative medicine. In this chapter, we would like to discuss the theme of modulation of synthetic ECM mechanics through reviewing previously published literature that report a variety of engineering approaches for fabrication of synthetic ECM with tunable mechanical properties. Further, we would like to discuss how the synthetic ECM with tunable mechanical properties in two-dimensional (2D) substrates and three-dimensional (3D) matrices are used to study the role of ECM mechanics in regulation of cell behavior.

6.2 STIFFNESS OF THE ECM REGULATES CELL BEHAVIOR IN TISSUES

Stiffness, also referring to the rigidity of a substrate, is defined as the extent to which a substrate resists deformation in response to an applied force.[9] Stiffness is commonly used to determine whether a material is compliant (soft) or rigid (hard). In biology, stiffness has been used to represent collective mechanical properties of a biological substrate. Tissues in our body are composed of different ECM molecules and have specific stiffness for their physiological needs. For example, the tissue stiffness of brain is around 50 Pa, whereas that of bone is about 2–4 GPa. In addition, stiffness of a tissue is often changed as a result of tissue pathogenesis.[10,11] In a healthy tissue, the ECM is constantly remodeled through processes of matrix production and degradation. However, diseases, such as cancer, often involve disruption of balanced matrix homeostasis and/or uncontrolled matrix cross-linking, which results in alteration of ECM stiffness and subsequent abnormal cell behavior that ultimately leads to tissue malignancy.

Owing to the critical role of ECM stiffness on regulation of cell response, it is important to gain a thorough understanding of the mechanisms behind how stiffness influences cell behavior, normal development, and disease progression. As the mechanisms become more understood, it will be possible to enhance tissue formation for regenerative medicine and to control metastasis for cancer treatment if approaches to manipulate ECM stiffness to achieve preferred cell behavior are successfully developed. Hydrogels and nanofibers are two material structures that closely resemble the ground substance and fibrous component, respectively, of the ECM and are commonly used to study cell responses to ECM stiffness. In addition, these two scaffold types are easy to be modified in a controllable manner to achieve the desired mechanical properties.

6.3 HYDROGELS AS THE SYNTHETIC ECM

Hydrogels have become a leading material used for drug delivery[12] and cell encapsulation applications[13,14] owing to their propensity for modification. The mechanical properties of a hydrogel determine the potential success of the scaffold for its intended applications. Hydrogels are commonly used for cell encapsulation partly because they allow for adequate diffusion of molecules and can be made into structures with tunable stiffness. However, generally diffusivity and stiffness of a hydrogel are inversely correlated[15] and these two properties determine cell viability in the hydrogel. For example, hydrogels that are too stiff may decrease cell viability[14] because it is more difficult for oxygen and nutrients to reach encapsulated cells in a stiffer hydrogel.[13] In order for hydrogels to be applicable for cell encapsulation and drug delivery as well as other applications, it is necessary to develop engineering approaches to tune the mechanical properties of a hydrogel in a controllable manner, and would be ideal to decouple the mechanical properties from other material properties.

6.4 APPROACHES TO MODULATE STIFFNESS IN HYDROGELS

To alter the stiffness of a hydrogel, modifications in the pre-gelled components,[14,16–21] the hydrogel cross-linking process,[22,23] or the fabrication method[13,24–31] have been applied. A thoughtfully engineered hydrogel can provide mechanical properties which make it more appropriate for applications such as cell culture, cell encapsulation, and drug delivery systems.[32] However, the approaches used to tune hydrogel stiffness can also alter other properties such as swelling ratio, and diffusivity, which may impact the response of cells that are interacting with the hydrogel.[33,34] This section will cover the approaches used to tune hydrogel stiffness as well as several methods used to decouple stiffness from other hydrogel properties.

6.4.1 Polymer Composition

The simplest approach used to alter hydrogel stiffness is to alter the concentration of one of the pre-gelled components. This can include altering the concentration of the polymer[17,19,20] or altering the concentration of the cross-linking agent.[16,18,35] An increase in either polymer concentration or cross-linker concentration leads to an increase in hydrogel mechanical properties due to an increase in cross-linking density.[17] A major advantage of changing the concentration to enhance mechanical properties is that the surface chemistry of the hydrogel remains unchanged without introducing any new chemical or material into the system.[36] However, a disadvantage is that changes in concentration alter not only hydrogel stiffness but also other hydrogel properties such as swelling, permeability, and pore size. Hydrogel stiffness must be decoupled from other material properties in order to avoid introducing additional variables into the system.[34] One attempt to decouple stiffness from diffusivity involved the use of a hydrophilic cross-linking agent. Typically, hydrogel diffusivity decreases with increasing concentration and increasing cross-link density. However, with increasing concentrations of the hydrophilic cross-linking agent, the elastic modulus increases while diffusion decreases only moderately. Hydrophilic cross-linkers are kinetically favorable for osmotic water entry, encouraging diffusion transfer even as cross-linking density increases.[15,35]

Another approach commonly used to tune hydrogel stiffness is to change polymer molecular weight.[14,21] With this approach, the overall polymer concentration does not vary but the size of the polymer chain is altered, which then impacts cross-linking and gel stiffness. In some cases, decreasing polymer molecular weight produces a hydrogel with increased mechanical strength. For example, as alginate molecular weight decreases, hydrogel stiffness increases due to an increase in cross-linking density.[14] However, it should be noted that the effect of molecular weight on hydrogel stiffness is dependent on the segment of the polymer which is capable of forming crosslinks. To decrease molecular weight, long polymer chains are cleaved. If the site of cleavage occurs in a segment of the polymer which is essential for cross-linking, then the stiffness will instead decrease. Complicating things further, changes in molecular weight do not always correlate linearly to changes in stiffness. For example, when polyethylene glycol (PEG) hydrogels are formed using polymers with molecular weights of 860, 3900, and 9300, the two highest molecular weight polymers produce hydrogels with similar mechanical properties while the lowest molecular weight polymer produces a significantly stiffer hydrogel.[21] Therefore, stiffness is largely controlled by the effect that changes in polymer molecular weight have on the ability of the polymer to form crosslinks. By understanding this concept, it is possible to alter the molecular weight of a polymer without altering the stiffness at all. For example, through cleavage of specific sites on the alginate polymer, molecular weight is reduced to half of the original molecular weight but the resulting hydrogel demonstrates similar mechanical properties compared to the hydrogel with the full-length alginate sequence. Alginate is cleaved so that the gluronic block concentration and function are left unchanged, allowing for similar crosslinking to occur despite the decrease in molecular weight.[37] Therefore, it is important to understand which polymer components are crucial for crosslinking in order to be able to decrease molecular weight as a way to tune hydrogel mechanical properties.

6.4.2   Polymerization Technique

Another technique used to control hydrogel stiffness involves photopolymerization. Photopolymerization uses photoinitiators such as Irgacure 2959, methacrylate, and acrylate as well as UV light to encourage the formation of crosslinks.[31,38] Mechanical properties can be tuned based on the type of photoinitiator that is used. Three photoinitiators—Irgacure 12959, I184, and I651—were used to polymerize a hydrogel. It was found that the hydrogels formed with the I651 photoinitiator had much lower strength than hydrogels formed with the other two photoinitiators. I651 has a much higher molar extinction coefficient, meaning that free radicals are formed much more quickly, leading to less efficient crosslinking.[39] Therefore, careful photoinitiator selection can improve the overall hydrogel mechanical properties. Regardless of the crosslinking agent that is used, photoinitiators can induce the formation of crosslinks in a time-dependent manner. The length of time during which a gel is exposed to UV light dictates the hydrogel stiffness, with a longer duration of UV exposure resulting in an increased concentration of crosslinks and therefore increased stiffness.[22,23] Careful selection of a photoinitiator and control over UV light exposure allows for tuning of hydrogel stiffness.

6.4.3 Fabrication Method

Hydrogel stiffness can also be patterned through manipulation of photopolymerization. In order to pattern stiffness, a photomask is required. A photomask blocks UV light, which induces the formation of crosslinks, from reaching specific areas on the gel in order to generate stiffness patterns (Figure 6.1). The mask can be printed on a transparency with black ink and the dark portion placed over areas in which polymerization is not desired so that crosslinking is restricted, resulting in patterned stiffness.[40] One particularly useful stiffness pattern for the study of ECM mechanics is the stiffness gradient. *In vivo*, stiffness gradients occur naturally between two tissue types or within the same tissue. One way to form a stiffness gradient is to use a sliding photomask instead of a stationary photomask. As with the stationary photomask, a sliding mask involves a transparency with a dark region which inhibits light from reaching the gel. However, with a sliding mask, the whole polymer is initially exposed to light and the darkened sheet gradually covers the gel, moving in a linear fashion, starting from one direction until it blocks light from reaching the entire gel. This method takes advantage of the time-dependent crosslinking process that occurs with photopolymerization. The section of hydrogel covered first is the most elastic while the section covered last receives the greatest UV light exposure and is the most stiff.[31] While the use of photomasks provides a simple approach for patterning hydrogel stiffness, a more precise approach is sometimes required. Laser scanning

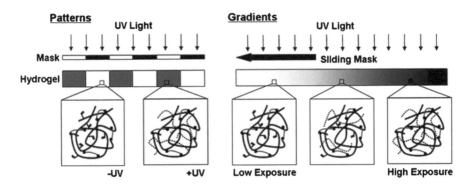

Figure 6.1   Hydrogel patterning through the use of a photomask. A photomask containing UV blocking and transparent regions is applied to a UV-sensitive polymer. With the application of UV light, the hydrogel obtains a stiffness pattern, since polymerization is dependent on UV light exposure. Stiffness gradients can also be produced through the use of a sliding photomask. (Reproduced from ref. 31 with permission from The Royal Society of Chemistry.)

photolithograph techniques provide a method for patterning three-dimensional hydrogels through the controlled localization of light. A pre-crosslinked hydrogel is submerged in a solution and positioned on a stage under a confocal microscope with a two-photon laser. The laser light can be precisely directed to a target area to induce photo-polymerization, thus providing stiffness patterning for the hydrogel.[30] By controlling light exposure through the use of a mask or laser scanning photolithography techniques, hydrogels can be endowed with stiffness patterns and gradients.

When used in concert with previously mentioned approaches, microfluidic channels provide a spatially specific method to control hydrogel stiffness. Burdick *et al.* designed a hydrogel system which uses the photopolymerization technique and a microfluidic gradient generator consisting of multiple inlet areas and one outlet area. Two monomer solutions are injected into the microfluidic channels which repeatedly split, recombine, and mix the solutions.[24] Mixing occurs in such a way as to create a gradient in a perpendicular direction to the fluid flow.[27] After mixing, polymer solution exits the microfluidic device and UV light crosslinks the polymer and sets the gradient. By using a microfluidic gradient generator, it is possible to form a hydrogel with a stiffness gradient using different concentrations of photoinitiator or polymer. Another research group created a hydrogel gradient using UV light and microfluidics by controlling the duration of UV light exposure.[41] Polymer solution can be injected sequentially into neighboring microfluidic channels. After the first channel is filled, it is exposed to UV light prior to the filling of the second channel. After the initial exposure for the first channel, the second channel is filled and the original channel and second channel are both exposed to UV light. This process is repeated until the whole hydrogel has been polymerized. Polymerization of hydrogels using microfluidic channels provides enhanced control over hydrogel stiffness and can be used to create gradients. Stiffness gradients created using microfluidics have a more precise rate of change in stiffness within the gradient than that of a gradient created using other methods. Microfluidic chambers are useful in the manufacture of hydrogels due to their precision when tuning hydrogel mechanical properties.

Another approach to tune hydrogel stiffness requires the formation of interpenetrating polymer networks (IPNs) or 'double-network hydrogels'. This technique involves the combination of two different polymers into one gel (Figure 6.2). The combinations of polymers that can form IPNs are virtually endless and it is even possible to combine synthetic and natural polymers.[13,25,42] By carefully selecting the two

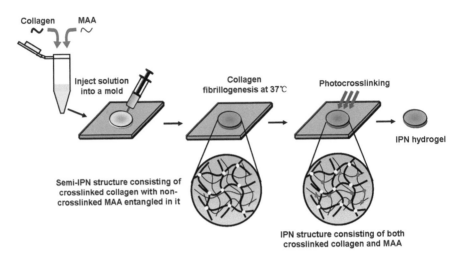

Figure 6.2   IPN hydrogel formation. Two polymers of different types are combined. One polymer is crosslinked first, then the other polymer through a different mechanism. This produces a hydrogel with enhanced mechanical properties due to the entangled networks between the two polymers. (Reproduced from ref. 42 with permission from The Royal Society of Chemistry.)

materials used to form an IPN, it is possible to form a hydrogel with tunable mechanical properties. For example, by combining two different polymers, a hydrogel 20 times stronger than a hydrogel made of only one of those materials can be manufactured.[26] To form an IPN hydrogel, a two-step sequential network formation technique is used. Two different polymers are mixed, then one polymer is polymerized. Once the first polymer is polymerized, the second polymer is polymerized using a different polymerizing mechanism. This method promotes network entanglement between the two different polymers to increase the hydrogel stiffness. It should also be noted that the mechanical properties depend not only on the polymer combinations but also on the ratio of the first polymer network to the second network.[26] IPNs with a loosely crosslinked second network have higher mechanical properties than IPNs with more tightly crosslinked second networks. In general, hydrogels made of only one polymer tend to crack if applied forces are over the limit of their mechanical strength. However, in IPN hydrogels, an entangled and loosely crosslinked secondary network is able to enhance the strength while also reducing cracking. This results in a hydrogel with high toughness.[26] The IPN approach provides an opportunity for tuning mechanical properties of a hydrogel through the use of a variety of polymer combinations.

Mechanical strength and toughness can be concurrently improved within these gels, making them attractive synthetic ECM options.

6.5 STRATEGIES THAT CONTROL HYDROGEL STIFFNESS OVER TIME

As mentioned earlier, mechanical properties of a tissue change due to aging, injury, and disease.[43-45] The approaches for tuning hydrogel stiffness previously discussed can be used to mimic mechanical properties of a tissue affected by these pathological problems, and the properties are often dynamically changed over time. A considerable amount of research has been devoted to development of hydrogels with mechanical properties that can be dynamically altered.[28,46-51] In this section, we will discuss different approaches used to tune the mechanical properties of hydrogels temporally so that the material can stiffen or soften in a controllable manner.

6.5.1 Softening of Hydrogels

Polymers with degradable crosslinks can form a hydrogel with tunable mechanical properties.[28] These gels will soften rapidly when exposed to light due to crosslink cleavage. Additionally, by applying light to one side of the gel, a stiffness gradient is formed, since crosslinks which are closer to the light are cleaved more quickly than those which are further away.[29] Another approach which provides temporal control over time-dependent softening of hydrogels involves the use of surfactants. The mechanical properties of gelatin hydrogels gradually decrease when bound to anionic, cationic, and non-ionic surfactants. As surfactant concentrations increase, the gelatin hydrogel undergoes gradual softening. All three types of surfactant exert a similar softening effect on the gel. An advantage to using this technique is that the diffusivity remains unchanged as gel stiffness decreases, decoupling diffusivity and stiffness.[46] The use of surfactant to gradually soften a hydrogel allows for the tuning of hydrogel mechanical properties.

6.5.2 Stiffening of Hydrogel

In vivo, cells often experience the impact of matrix stiffening caused by development, injury, or disease.[43-45] Therefore, it is useful to develop a hydrogel which is able to stiffen over time. For instance, hydrogel mechanical properties can be manipulated so that it mimics the stiffening of the myocardium during development. To do this, the polymer is functionalized with diacrylate, which initiates a Michael-type addition reaction. The dynamics of this reaction are controlled by polymer molecular weight. Higher molecular weight leads to a more rapid stiffening of the hydrogel. However, degradation of this hydrogel eventually outpaces crosslinking, allowing the mechanical weakening and clearance of the hydrogel.[51] In the case where degradation is not desired, hydrogels can also be endowed with long-term stability as well as fast dynamic changes to mechanical properties. In this case, temporal control over hydrogel stiffness is provided through the use of multiple polymerization steps. Previously, we discussed the tuning of hydrogel mechanics through the application of UV light. Here, using UV light, gels are exposed at multiple time points, stiffening the gel in steps since stiffness is dependent on the duration of UV exposure. In this way, hydrogel stiffness increases over a period of time as UV light is occasionally applied to the hydrogel at short time increments.[47] By inducing gelation in stages or by combining two polymers, hydrogel mechanical properties can be increased over time.

6.5.3   Reversibility of Hydrogel Mechanics

The development of hydrogels that can stiffen and soften reversibly provides even greater control over hydrogel mechanics. In order for mechanical properties to be reversibly altered, there must be a way to both induce and degrade the same crosslink. One way to do this is to use a hydrogel made from a biological material so that the processes to form and degrade the gel mimics reversible processes occurring *in vivo*. For example, DNA hydrogels have the controllable property of stiffness that can be reversibly altered through the manipulation of DNA strand formation. To form the gels, two DNA side chains, a crosslinking strand of DNA and its complementary strand of DNA, are synthesized. To crosslink the hydrogel, the crosslinking DNA strand is delivered to the gel. This strand forms a crosslink between the two side chains. For degradation, the reversing DNA strand is delivered. Since the reversing strand is complimentary to the crosslinking strand, the crosslinking strand detaches from the two side chains and binds to the reversing strand. The addition or removal of DNA crosslinks within the hydrogel dynamically controls the mechanical properties.[49] The use of DNA hydrogels to tune mechanical properties is advantageous because the crosslinks within a gel allow changes in stiffness to occur without changing temperature, pH, light, or other environmental factors.[48] Additionally, stiffness of DNA hydrogels can be easily modulated *in situ*.[50] However, it is challenging to alter stiffness of the gel without affecting its swelling. No work has yet been done to decouple these two properties of DNA gels.[49] Another hydrogel featuring reversible stiffness is a mixture of two naturally derived ECM components, collagen and alginate. The collagen is gelled first and then the alginate is gelled with the addition of calcium chloride and uncrosslinked using sodium citrate, a chelator of calcium to manipulate stiffness of the hydrogel complex.[52] Both the DNA hydrogel and the collagen/alginate hydrogel complex are designed as a viable approach to study reversible changes in ECM stiffness and their biological impact.

6.6 NANOFIBERS AS THE SYNTHETIC ECM

Nanofibers are another option of the synthetic ECM for examining the effect of stiffness on various biological responses. Nanofibers are formed either from an electrospinning process[53] or through self-assembly.[54,55] Electrospinning tends to produce fibers ranging from several microns down to tens of nanometers in diameter[56–58] while self-assembly can produce a hydrogel composed of fibers which are only a few nanometers in diameter.[59] It is relatively easy to modulate mechanical properties of an electrospun nanofibrous structure by controlling fiber diameter, and several approaches have been developed to control fiber diameter.[60–64] In contrast, tuning mechanical properties of self-assembling nanofibers is more challenging due to the complex chemistry that induces fiber formation. However, previous studies have shown it is possible to manipulate the chemistry of self-assembly to modulate mechanical properties of nanofiber hydrogels.[65–67]

6.7 ELECTROSPUN NANOFIBERS

6.7.1 Nanofiber Diameter

One of the approaches to control stiffness of a nanofiber is to modulate the nanofiber diameter.[68] Typical approaches for altering fiber diameter include altering the viscosity,[56,62,63,69–73] electrostatic forces,[57,62,64,74,75] solvent selection,[58,61,76,77] environmental conditions,[57,58,60,75,78,79] or fiber alignment.[76,80] The diverse array of options for tuning fiber diameter makes electrospinning a popular method for scaffold fabrication. As the effect of electrospinning parameters on fiber diameter becomes more understood, they will be utilized to tune stiffness of a nanofibrous structure. In the following sections, we will discuss how these fabrication parameters affect fiber diameter.

6.7.1.1 Viscosity. The most explored approach used to alter nanofiber diameter is to change solution viscosity by either altering the solubilized polymer concentration or polymer molecular weight.[56,60,62,70,81,82]

As polymer concentration is increased, the viscosity increases as well.[60,62] Similarly, an increase in the molecular weight of the polymer leads to an increase in solution viscosity due to an increase in polymer chain entanglements.[83] By simply increasing polymer solution viscosity, nanofiber diameters become larger.[56,63,69,72] Viscosity is important in determining fiber morphology because a more viscous solution is composed of more densely concentrated polymer molecules. The organic solvent used to solubilize the polymer molecules evaporates during electrospinning. When a high concentration of polymer solution is used, a small amount of organic solvent quickly evaporates before a polymer jet is split into small diameter fibers and collected onto a plate, thus resulting in large diameter fibers. The approach of increasing polymer molecular weight or concentration in order to increase fiber diameter can be applied to a wide range of polymers. Previous studies have demonstrated that fiber diameter can be controlled for synthetic polymers such as polystyrene, poly-lactic acid (PLA), and polyurethane as well as naturally derived polymers including collagen and chitosan by simply altering polymer molecular weight or concentration of polymer solution.[63,69–71,73] The versatility and simplicity of altering solution viscosity by changing polymer concentration or molecular weight has contributed to the popularity of this approach for tuning fiber diameter.[68]

6.7.1.2 Electrostatic Forces. Fiber diameter can also be tuned by altering the applied electrostatic forces. The approach of altering electrostatic forces to manipulate fiber diameter is more complex than the simple method of altering the viscosity of a polymer solution. To control the electrostatic forces, either the voltage or the distance of the nozzle to the collection plate can be adjusted.[57,62,64,84,85] An increase in applied voltage results in an increase in fiber diameter. This is due to an increase in voltage creating an electrostatic field that is able to draw a greater amount of polymer solution from the nozzle than an electrostatic field with a lower applied voltage. Alternatively, increasing the distance of the nozzle to the collection plate can produce fibers with smaller diameter than if a shorter distance is used. As distance increases, the electrostatic field strength is lessened, resulting in lower electric potential and less solution being drawn through the nozzle.[62] While adjustments in the electric field strength at first seems straightforward, it is important to note that increasing the electric potential can sometimes cause the jet of solution to split into multiple smaller jets.[74] This occurs when the electrostatic field strength becomes stronger than the surface tension of a polymer solution, reducing the fiber diameter significantly. Adjusting fiber diameter by altering the electric potential is a viable approach for producing a nanofibrous structure with desired mechanical properties. Another approach that alters fiber diameter is to control the ionic concentration in a polymer solution with the addition of salt.[57,64,75] An increase in salt results in an increase in charge density. The impact of the addition of salt on nanofiber thickness, however, is convoluted. Increasing NaCl concentration from 1% to 5% results in a larger fiber diameter. An increase in fiber diameter is caused by the balance between electrostatic forces and coulombic repulsion forces. An increase in salt concentration results in an increase in coulombic repulsion forces which ordinarily lead to smaller fibers. However, increasing the concentration of NaCl also increases electrostatic and viscoelastic forces, which offset the coulombic repulsion forces and produces a faster fluid flow rate. Therefore, despite the increase in coulombic repulsion forces, fiber diameter is increased.[57] At lower salt concentrations, the balance in forces is reversed. When NaCl solution concentrations are increased from 0% to 1%, fiber diameter decreases. The addition of salt is known to increase the charge density of the solution, making it more conductive. This increase in conductivity is responsible for the smaller fiber diameter.[64,86] Further, fiber diameter can also be controlled by the molecular composition of the salt as well as the salt concentration. A solution containing 1% NaCl produces fibers with a smaller diameter than a solution containing 1% KH_2PO_4. Ions with a smaller atomic radius have a higher charge density under an applied electric field, resulting in the production of fibers with a larger diameter.[64] To tune fiber diameter through the manipulation of applied electric forces or solution conductivity, many variables, including surface tension, electrical potential, and coulombic repulsion forces, must be considered.

6.7.1.3 Solvent Selection. The types of solvent and the combinations of solvent used in the electrospinning process can also affect fiber diameter. There are a number of solvents that are capable of dissolving polymers for electrospinning. Diameters of the resulting fibers can be affected by the dielectric point and boiling point of solvents.[58,61,76,77] For example, when dimethylformamide (DMF), a common solvent used for electrospinning, is added to a polymer solution to lower the solution viscosity and increase conductivity, the resulting fibers are thinner than fibers spun from a solution without DMF.[76]

The impact of viscosity and solution conductivity on fiber diameter has previously been discussed so we will focus on other mechanisms through which different solvents can alter fiber diameter. Selection of the right solvent with an appropriate boiling point is a convenient strategy through which fiber diameter can be altered. Solvents with lower boiling points evaporate faster. The faster the solvent evaporates, the less the jet is stretched, resulting in a fiber with a larger diameter.[58] Further tuning of fiber diameter is obtained by combining two solvents with different boiling points. For example, a polymer can be dissolved in either tetrahydrofuran (THF) or DMF. THF has a significantly lower boiling point than DMF, resulting in fibers with a larger diameter than those spun with DMF. However, if the polymer is dissolved in a combination of THF and DMF, the resulting fibers have a diameter which is somewhere in between the diameters of fibers spun from THF or DMF. The resulting fiber diameter is dependent on the ratio of THF to DMF, allowing for increased diameter tunability.[61] Careful solvent selection allows for increased control over fiber diameter, thus tuning the mechanical properties without changing polymer concentration or altering electrostatic forces.

6.7.1.4 Environmental Conditions. Temperature is a key environmental factor which can be used to tune fiber diameter during the electrospinning process. In general, as temperature increases, fiber diameter decreases.[57,79] An elevation in temperature results in a decrease in solution viscosity, which in turn decreases fiber diameter. However, a rise in temperature increases the evaporation rate, which has been shown to generate larger fiber diameter.[58] Since viscosity is a more dominant factor in determining fiber diameter than evaporation rate, fiber diameter tends to decrease with increasing temperature despite the increase in the rate of evaporation. The simplicity of controlling fiber diameter through the alteration of temperature has led this to be a popular tuning approach during scaffold fabrication.

Changes in relative environmental humidity are also known to have an impact on both nanofiber diameter and nanofiber mechanical properties. An increase in relative humidity results in either an increase in fiber diameter[62,78] or a decrease in fiber diameter[79] depending on polymer chemistry. For example, the diameter of cellulose acetate (CA) fibers increases with increasing humidity while the diameter of poly(vinylpyrrolidone) (PVP) fibers decreases under identical conditions.[79] These differences in the effect that increasing humidity has on fiber diameter are correlated with variations in chemical and molecular interactions between different polymers. Additionally, water tolerance of a polymer contributes to fiber diameter and uniformity. As relative humidity increases, both poly(acrylonitrile) (PAN) and polysulfone (PSU) nanofibers demonstrate an increase in average fiber diameter. However, at high relative humidity, electrospinning PAN yields fibers with uniform morphology while electrospinning PSU produces both large and small fibers due to phase separation. PSU has a lower water tolerance than PAN and therefore undergoes faster phase separation.[78] Studies comparing the effects of humidity on fiber diameter suggest that it is important to consider both the properties of the polymer selected for electrospinning as well as environmental conditions such as relatively humidity in order to produce nanofibers with optimal diameters.

6.7.2 Decoupling Parameters

Cell behavior can be regulated by either fiber diameter or stiffness of a nanofibrous structure. However, stiffness of a nanofibrous structure is closely related to the fiber diameter. To understand how cell behavior is regulated by stiffness of a nanofibrous structure, it is necessary to decouple stiffness from fiber diameter. A new fabrication technique, coaxial electrospinning, can decouple fiber diameter and fiber mechanical properties. For coaxial electrospinning, two solutions with different concentrations of the same polymer are simultaneously electrospun to produce a fiber with an inner core and an outer cylinder. The inner core diameter, which is tuned by altering the feeding rate of the polymers, determines the mechanical stiffness of fibers.[80] The outer core is able to even out differences in inner core diameter so that all nanofibers are of a uniform size, providing a unique opportunity to study cell behavior based solely on the effect of stiffness.

6.7.3 Nanofiber Alignment Within a Scaffold

Stiffness of a nanofibrous structure can be altered by changing fiber alignment. Electrospinning typically produces fibers which are oriented randomly. However, by controlling fiber orientation, it is possible to alter the mechanical properties of a nanofibrous structure. The most common approach used to align nanofibers requires the use of a rotating drum.[87] By electrospinning fibers onto the drum, the orientation can be controlled. Fiber orientation along a specific direction is increased with increasing rotational speed of the drum. As alignment increases, Young's modulus of the structure increases in the direction parallel to the fiber orientation but decreases in the perpendicular direction.[53] Another approach used to align fibers uses a collector with two parallel conductive strips and an insulating substance between the two strips. Nanofibers spun onto the collector are perpendicular to the strips and are stretched between them.[88] This approach allows for the orientation of fibers to be controlled without the use of a moving collection plate. As stated previously, aligned fibers demonstrate increased strength in the direction of fiber alignment but decreased strength in the direction perpendicular to fiber alignment. To increase strength in the perpendicular direction, multiple bundles of aligned fibers can be twisted into an integrated structure. In addition, stiffness of a twisted nanofibrous structure can be tuned by controlling the twist angle. It has been shown that an increase in twist angle improves the strength of a twisted structure.[89] Another approach to increase the strength of an aligned nanofibrous structure is through the approach of fiber braiding. A braided structure made of three aligned nanofiber bundles, compared to those made of four or five nanofiber bundles, has higher Young's modulus, yield stress, and ultimate stress.[90] Through altering fiber alignment and organization, the mechanics of the overall nanofiber structure can be tuned.

6.8   SELF-ASSEMBLING NANOFIBERS

Self-assembling nanofibers have gained much attention for their application as a synthetic ECM structure due to their high water content and tunability. Self-assembling nanofibers have qualities of both hydrogels and electrospun nanofibers. They have a high water content, similar to a hydrogel, and also form nanometer-scale fibers for favorable biological interactions with cells. Self-assembly can be induced through changes in divalent ion concentration,[54,55,66] amino acid sequence,[54,67,91] or pH.[55,65,92] By tuning the factors that induce self-assembly, it is possible to alter the mechanics of a self-assembling nanofibrous structure. This section will discuss several key approaches that have been implemented to control the stiffness of self-assembling nanofibers.

6.8.1   Ion Concentration

One approach used to control stiffness of a self-assembling nanofibrous structure during self-assembly is the manipulation of ion concentration. This approach takes advantage of the chemistry involved in a self-assembly process. Peptides that form self-assembling nanofibers contain repeats of hydrophobic and hydrophilic amino acids. Due to the amphiphilic nature of these peptides, beta-sheet structures are formed during self-assembly. The formation of a beta-sheet structure is a result of molecular interactions through hydrogen bonding and van der Waals forces.[77] Since ion concentration effects molecular interactions and self-assembly, this provides a mechanism to tune mechanical properties of a self-assembling nanofibrous structure.[55] In general, increasing the ion concentration results in an increase in strength of a self-assembling structure. However, it has been reported that an overdose of ions may weaken the mechanical properties of the structure.[93] Ions, such as calcium, magnesium, iron, copper, or salt, can be used to modulate mechanical properties of a self-assembling structure. The use of different ions as well as various concentrations of ions allows researchers to tune stiffness of a self-assembling structure.[54,55,66]

6.8.2 Amino Acid Sequence

The stability of a chemical bond can be altered by changing the peptide chemistry through the alteration of an amino acid sequence. For example, by changing the positions of two amino acids, valine and alanine, within a peptide sequence, stiffness of a self-assembling structure can be altered. In addition, it has been found that the addition of valine increases stiffness of the structure while the addition of alanine decreases the stiffness.[91] The variation in mechanical properties can be attributed to hydrogen bonding. Controlling hydrogen bond formation is key to tuning mechanical properties of a self-assembling nanofibrous structure. Self-assembling nanofibers with stronger mechanical properties are those forming beta-sheets with a lower amount of twisting and with more stable hydrogen bonding. Another approach to altering hydrogen bonding and thus altering mechanical properties of self-assembling nanofibers is through methylation of an amino acid sequence. Methylation is a chemical process that adds methyl groups to a peptide, which decreases its ability to form stable hydrogen bonds. Therefore, increasing methylation of a peptide decreases the strength of self-assembling nanofibers. Additionally, methylation occurring at certain key locations in an amino acid sequence can inhibit the formation of all hydrogen bonds.[67] Disulfide bonds also play an important role in regulating the process of self-assembling. For example, the replacement of certain amino acid residues with cysteine residues in a peptide can induce the formation of additional disulfide bonds.[54] Since disulfide bonds are much stronger than hydrogen bonds, the addition or removal of a disulfide bond alters the mechanical properties of a self-assembling structure to a greater extent than the addition or removal of a hydrogen bond. By modifying an amino acid sequence, the arrangement of hydrogen bonds and disulfide bonds can be altered in order to tune mechanical properties of nanofibers.

6.8.3 Alteration of pH

Mechanical properties of self-assembling nanofibers can also be tuned by changing the pH of a solution in which self-assembly occurs. When the pH of a solution is changed, chemical interactions between peptides are altered, resulting in the formation of self-assembling nanofibers with different mechanical properties. It has been shown that as the pH increases, deprotonation of functional groups occurs, which in turn increases the solubility, hydration, and self-assembly of peptides.[55] Notably, the process of controlling the pH of a solution to manipulate a self-assembling process is often reversible. The self-assembled nanofibers can be disassembled and reassembled through deprotonation or protonation of functional groups,[65] providing a convenient approach for controlling mechanical properties of a self-assembling nanofibrous structure.

6.9 CONCLUSION

Hydrogels and nanofibers are considered viable choices among all synthetic ECM for research and translational applications due to tunability of their mechanical properties and versatility of the approaches used to fabricate them. Mechanical properties of electrospun nanofibrous structures can be tuned through changes in solution viscosity, electrostatic force, solvent selection, environmental condition, or fiber orientation. Mechanical properties of hydrogels can be tuned through changes in polymer concentration or molecular weight, photopolymerization techniques, formation of IPNs, or the use of masks and microfluidics during fabrication. Fabrication of synthetic ECM with tunable mechanical properties is to develop a viable system that can be used to study how cells are regulated by mechanical cues of the surrounding environment. While much progress has been made towards achieving this aim, critical strategies, such as how to decouple stiffness from other material properties are required for researchers to study cellular response to ECM mechanics.

REFERENCES

1. A. Berrier and K. Yamada, *J. Cell. Physiol.*, 2007, **213**, 565–573.
2. R. Hynes, *Science*, 2009, **326**, 1216–1219.
3. B. Geiger, J. Spatz and A. Bershadsky, *Nat. Rev. Mol. Cell Biol.*, 2009, **10**, 21–33.
4. M. Krieg, Y. Arboleda-Estudillo, P. H. Puech, J. Kafer, F. Graner, D. J. Muller and C. P. Heisenberg, *Nat. Cell Biol.*, 2008, **10**, 429–436.
5. R. Keller, L. A. Davidson and D. R. Shook, *Differentiation*, 2003, **71**, 171–205.
6. M. J. Paszek, N. Zahir, K. R. Johnson, J. N. Lakins, G. I. Rozenberg, A. Gefen, C. A. Reinhart-King, S. S. Margulies, M. Dembo, D. Boettiger, D. A. Hammer and V. M. Weaver, *Cancer Cell*, 2005, **8**, 241–254.
7. N. D. Evans, C. Minelli, E. Gentleman, V. LaPointe, S. N. Patankar, M. Kallivretaki, X. Chen, C. J. Roberts and M. M. Stevens, *Eur. Cells Mater.*, 2009, **18**, 1–13, discussion 13–14.
8. A. J. Engler, S. Sen, H. L. Sweeney and D. E. Discher, *Cell*, 2006, **126**, 677–689.
9. E. Baumgart, *Injury*, 2000, **31**(suppl. 2), S-B14-23.
10. A. Pathak and S. Kumar, *Proc. Natl. Acad. Sci. U. S. A.*, 2012, **109**, 10334–10339.
11. M. Zaman, L. Trapani, A. Sieminski, D. MacKellar, H. Gong, R. Kamm, A. Wells, D. Lauffenburger and P. Matsudaira, *Proc. Natl. Acad. Sci. U. S. A.*, 2006, **103**, 10889–10894.
12. T. Hoare and D. Kohane, *Polymer*, 2008, **49**, 1993–2007.
13. H. Geckil, F. Xu, X. Zhang, S. Moon and U. Demirici, *Nanomedicine*, 2010, **5**, 469–484.
14. H. Kong, D. Kaigler, K. Kim and D. Mooney, *Biomacromolecules*, 2004, **5**, 1720–1727.
15. C. Cha, J. Jeong, J. Shim and H. Kong, *Acta Biomater.*, 2011, **7**, 3719–3728.
16. A. Banerjee, M. Arha, S. Choudhary, R. Ashton, S. Bhatia, D. Schaffer and R. Kane, *Biomaterials*, 2009, **30**, 4695–4699.
17. S. Bryant and K. Anseth, *J. Biomed. Mater. Res.*, 2002, **59**, 63–72.
18. A. Engler, S. Sen, H. Sweeney and D. Discher, *Cell*, 2006, **126**, 677–689.
19. M. LeRoux, F. Guilak and L. Setton, *J. Biomater. Res.*, 1999, **47**, 46–53.
20. A. Martinez-Ruvalcaba, E. Chornet and D. Rodrigue, *Carbohydr. Polym.*, 2007, **67**, 586–595.
21. J. Temenoff, K. Athanasiou, R. Lebaron and A. Mikos, *J. Biomater. Res.*, 2002, **59**, 429–437.

22. C. Decker, *J. Coat. Technol. Res.*, 1987, **59**, 97–106.
23. K. Nguyen and J. West, *Biomaterials*, 2002, **23**, 4307–4314.
24. J. Burdick, A. Khademhosseini and R. Langer, *Langmuir*, 2004, **20**, 5153–5156.
25. N. Desai, A. Sojomihardjo, Z. Yao, N. Ron and P. Soon-Shiong, *J. Microencapsulation*, 2000, **17**, 677–690.
26. J. Gong, Y. Katsuyama, T. Kurokawa and Y. Osada, *Adv. Mater.*, 2003, **15**, 1155–1158.
27. N. Jeon, S. Drtinger, D. Chiu, I. Choi, A. Stroock and G. Whitesides, *Langmuir*, 2000, **16**, 8311–8316.
28. A. Kloxin, A. Kasko, C. Salinas and K. Anseth, *Science*, 2009, **34**, 59–63.
29. A. Kloxin, M. Tibbitt, A. Kasko, J. Fairbairn and K. Anseth, *Adv. Mater.*, 2010, **22**, 61–66.
30. S. Lee, J. Moon and J. West, *Biomaterials*, 2008, **29**, 2962–2968.
31. R. Marklein and J. Burdick, *Soft Matter*, 2009, **6**, 136–143.
32. S. Khetan and J. Burdick, *J. Visualized Exp.*, 2009, (32), DOI: 10.3791/1590.
33. G. Nicodems and S. Bryant, *Tissue Eng., Part B*, 2008, **14**, 149–165.
34. S. Zustiak, H. Boukari and J. Leach, *Soft Matter*, 2010, **6**, 3609–3618.
35. C. Cha, R. Kohman and H. Kong, *Adv. Funct. Mater.*, 2009, **19**, 3056–3062.
36. J. Tse and A. Engler, *Current Protocols in Cell Biology*, Extracellular Matrix, 2010, ch. 10, vol. 47, unit 10.16.
37. H. Kong, M. Smith and D. Mooney, *Biomaterials*, 2003, **24**, 4023–4029.
38. R. Sunyer, A. Jin, R. Nossal and D. Sackett, *PLoS One*, 2010, 7, e46107.
39. I. Mironi-Harpaz, D. Wang, S. Venkatraman and D. Seiktar, *Acta Biomater.*, 2010, **8**, 1838–1848.
40. S. Nemir, H. Hayenga and J. West, *Biotechnol. Bioeng.*, 2009, **105**, 636–644.
41. Y. Cheung, E. Azeloglu, D. Shiovitz, K. Costa, D. Seliktar and S. Sia, *Angew. Chem., Int. Ed.*, 2009, **48**, 7188–7192.
42. J. Sun, W. Xiao, Y. Tang, K. Li and H. Fan, *Soft Matter*, 2012, **8**, 2398–2404.
43. J. Clark, J. Cheng and K. Leung, *Burns*, 1996, **22**, 443–446.
44. A. Guerin, G. London, S. Marchais and F. Metivier, *Nephrol., Dial., Transplant.*, 2000, **15**, 1014–1021.
45. J. Kim, H. Yim, S. Kim, J. Ahn, Y. Jung, M. Joo, S. Kim, J. Kim, Y. Seo, J. Yeon, H. Lee, S. Um, S. Lee, K. Byun, J. Choi and H. Ryu, *Korean J. Gastroenterol.*, 2009, **54**, 155–161.
46. M. A. H. Bohidar, *Eur. Polym. J.*, 2005, **41**, 2395–2405.
47. M. Guvendiren and J. Burdick, *Nature*, 2012, 3, 792.

48. D. Lin, B. Yurke and N. Langrana, *J. Mater. Res.*, 2005, **20**, 1456–1464.
49. M. Previtera, K. Trout, D. Verma, U. Chippanda, R. Schloss and N. Langrana, *Ann. Biomed. Eng.*, 2012, **40**, 1061–1072.
50. S. Um, J. Lee, N. Park, S. Kwon, C. Umbach and D. Luo, *Nat. Mater.*, 2006, **5**, 797–801.
51. J. Young and A. Engler, *Biomaterials*, 2011, **32**, 1002–1009.
52. B. Gillette, J. Jensen, M. Wang, J. Tchao and S. Sia, *Adv. Mater.*, 2010, **22**, 686–691.
53. W. Li, R. Mauck, J. Cooper, X. Yuan and R. Tuan, *J. Biomech.*, 2007, **40**, 1686–1693.
54. L. Aulisa, H. Dong and J. Hartgerink, *Biomacromolecules*, 2009, **10**, 2694–2698.
55. J. Stendahl, M. Rao, M. Guler and S. Stupp, *Adv. Funct. Mater.*, 2006, **16**, 499–508.
56. H. Lin, Y. Kuo, S. Chang and T. Ni, *Biomed. Mater.*, 2013, **8**, 2.
57. P. Supaphol, C. Mit-Uppatham and M. Nithitanakul, *J. Polym. Sci., Part B: Polym. Phys.*, 2005, **43**, 3699–3712.
58. L. Wannatong, A. Sirivat and P. Supaphol, *Polym. Int.*, 2004, **53**, 1851–1859.
59. S. Zhang, F. Gelain and X. Zhao, *Semin. Cancer Biol.*, 2005, **15**, 413–420.
60. Y. Ji, B. Li, S. Ge, J. Sokolov and M. Rafailovich, *Langmuir*, 2006, **22**, 1321–1328.
61. K. Lee, H. Kim, Y. La, D. Lee and N. Sung, *J. Polym. Sci., Part B: Polym. Phys.*, 2002, **40**, 2259–2268.
62. C. Thompson, G. Chase, A. Yarin and D. Reneker, *Polymer*, 2007, **48**, 6913–6922.
63. J. Zeng, H. Haoqing, A. Schaper, J. Wendorff and A. Greiner, *e-Polymers*, 2003, **3**, 102–110.
64. X. Zong, K. Kim, D. Fang, S. Ran, B. Hsiao and B. Chu, *Polymer*, 2002, **43**, 4403–4412.
65. H. Cui, M. Webber and S. Stupp, *Pept. Sci.*, 2010, **94**, 1–18.
66. B. Ozbas, J. Kretsinger, K. Rajagopal, J. Schneider and D. Pochan, *Macromolecules*, 2004, **37**, 7331–7337.
67. S. Paramonov, H. Jun and J. Hartgerink, *J. Am. Chem. Soc.*, 2006, **128**, 7291–7298.
68. E. Tan and C. Lim, *Compos. Sci. Technol.*, 2006, **66**, 1102–1111.
69. M. Demir, I. Yilgor, E. Yilgor and B. Erman, *Polymer*, 2002, **43**, 3303–3309.
70. T. Jarusuwannapoom, W. Hongrojjanawiwat, S. Jitjaicham, L. Wannatong, M. Nithitanakul, C. Pattamaprom, P. Koombhongse, R. Rangkupan and P. Supaphol, *Eur. Polym. J.*, 2005, **41**, 409–421.

71. J. Matthews, G. Wnek, D. Simpson and G. Bowlin, *Biomacromolecules*, 2002, **3**, 232–238.
72. M. McKee, G. Wilkes, R. Colby and T. Long, *Macromolecules*, 2004, **37**, 1760–1767.
73. K. Ohkawa, D. Cha, H. Kim, A. Nishida and H. Yamamoto, *Macrocolecular Rapid Commun.*, 2004, **25**, 1600–1605.
74. A. Frenot and I. Chronakis, *Curr. Opin. Colloid Interface Sci.*, 2003, **8**, 64–75.
75. C. Mit-uppatham, M. Nithitanakul and P. Supaphol, *Macromol. Chem. Phys.*, 2004, **205**, 2327–2338.
76. K. Lee, H. Kim, M. Khil, Y. Ra and D. Lee, *Polymer*, 2003, **44**, 1287–1294.
77. F. Zhang, B. Zuo and L. Bai, *J. Mater. Sci.*, 2009, **44**, 5682–5687.
78. L. Huang, N. Bui, S. Manickam and J. McCutcheon, *Polym. Phys.*, 2011, **49**, 1734–1744.
79. S. Vrieze, T. Van Camp, A. Nelvig, B. Hagstrom, P. Westbroek and K. De Clerck, *J. Mater. Sci.*, 2009, **44**, 1357–1362.
80. J. Drexler and H. Powell, *Acta Biomater.*, 2011, **7**, 1133–1139.
81. A. Arinstein and E. Zussman, *Polym. Phys.*, 2011, **49**, 691–707.
82. P. Gupta, C. Elkins, T. Long and G. Wilkes, *Polymer*, 2005, **46**, 4799–4810.
83. S. Tan, R. Inai, M. Kotaki and S. Ramakrishna, *Polymer*, 2005, **46**, 6128–6134.
84. X. Geng, O. Kwon and J. Jang, *Biomaterials*, 2005, **26**, 5427–5432.
85. Z. Huang, Y. Zhang, M. Kotaki and S. Ramakrishna, *Compos. Sci. Technol.*, 2003, **63**, 2223–2253.
86. S. Han, J. Youk, K. Min, Y. Kang and W. Park, *Mater. Lett.*, 2008, **62**, 759–762.
87. P. Katta, M. Alessandro, R. Ramsier and G. Chase, *Nano Lett.*, 2004, **4**, 2215–2218.
88. D. Li, Y. Wang and Y. Xia, *Adv. Mater.*, 2004, **16**, 361–366.
89. S. Fennessey and R. Farris, *Polymer*, 2004, **45**, 4217–4225.
90. J. Barber, A. Handorf, T. Allee and W. Li, *Tissue Eng., Part A*, 2013, **19**, 1265–1274.
91. E. Pashuck, H. Cui and S. Stupp, *J. Am. Chem. Soc.*, 2010, **132**, 6041–6046.
92. M. Greenfield, J. Hoffman, M. Cruz and S. Stuff, *Langmuir*, 2010, **26**, 3641–3647.
93. H. Jun, S. Paramonov and J. Hartgerink, *Soft Matter*, 2006, **2**, 177–181.

B. Gradients and Patterns Within the Extracellular Matrix

CHAPTER 7

Biomaterials: Protein Interactions with Glycosaminoglycan-Based Biomaterials for Tissue Engineering

MELISSA C. GOUDE[†a], TOBIAS MILLER[†a], TODD C. McDEVITT[a,b], AND JOHNNA S. TEMENOFF[*a,b]

[a]Wallace H. Coulter Department of Biomedical Engineering, Georgia Institute of Technology and Emory University, 313 Ferst Drive, Atlanta, Georgia 30332, USA; [b]Parker H. Petit Institute for Bioengineering and Bioscience, Georgia Institute of Technology, Atlanta, Georgia 30332, USA
*E-mail: johnna.temenoff@bme.gatech.edu

7.1 INTRODUCTION

Glycosaminoglycans (GAGs) are a class of linear polysaccharides that are ubiquitous in the human body and possess multiple biological functions essential for life.[1] Such functions consist of: (1) osmotically attracting water and therefore maintaining hydrostatic pressure to confer mechanical stability in connective tissues such as cartilage;[2–5] (2) covalent attachment to proteoglycans that regulate cell function;[6] and (3) acting in conjunction with proteins on cell surfaces *via* receptors or co-receptors to modulate the local biological environment.[7]

[†]Equal contributions to this work.

Mimicking the Extracellular Matrix: The Intersection of Matrix Biology and Biomaterials
Edited by Gregory A. Hudalla and William L. Murphy
© The Royal Society of Chemistry 2020
Published by the Royal Society of Chemistry, www.rsc.org

Based on their numerous biological functions, GAGs have been extensively explored as biomaterials for controlled protein delivery.[8–11] In addition, the low toxicity and high biocompatibility of GAGs allows for use in a variety of tissue engineering applications.[12]

Many of their biological functions are conferred by the unique chemical structure of GAGs, consisting of repeating disaccharide units that are specific for each GAG species. Sulfated GAG species such as chondroitin sulfate (CS), heparin, heparan sulfate (HS), dermatan sulfate (DS), and keratan sulfate (KS) bear negative charges that vary in density and position within the disaccharide units.[13] In addition to sulfated GAGs, hyaluronic acid (HA) is non-sulfated and therefore is the GAG with the least net negative charge.[14] Based on this negative net charge, GAGs attract positively charged proteins.[15–17] For protein delivery applications, a number of GAG-based approaches have been developed that mimic the interactions that occur naturally between GAGs in the extracellular matrix (ECM) and growth factor binding partners.

In order to form tissue-engineered constructs, GAGs can be easily chemically modified for assembly into carriers of cells and biomolecules.[18] These generally include functionalization to form three-dimensional scaffolds for a wide range of applications, from neural to bone tissue regeneration.[19,20] However, chemical modifications can have effects on protein binding[21] and affect degradation processes,[22,23] which, in turn, influence molecular release characteristics. Therefore, a thorough understanding of the chemical properties of GAGs, both native and modified, is a key factor for successful implementation of GAG-based biomaterial strategies in tissue engineering and drug delivery applications. To that end, this chapter begins with a summary of GAG structure and protein-binding properties followed by mechanisms of GAG degradation. A better understanding of GAG chemistry will lead to enhanced predictability of GAG-based scaffold degradation (and protein release if applicable), and thus the ability to design more efficacious strategies for harnessing the innate properties of GAGs for a broad range of regenerative medicine applications, some of which are highlighted in the last section of this chapter.

7.2 GAG STRUCTURE AND PROTEIN BINDING

It is generally believed that net negative charge is primarily responsible for mediating GAG interactions with oppositely charged proteins, but polyelectrolyte complexation does not fully explain protein affinity to GAGs. Therefore, this section reviews the primary structure of

GAGs, which is determined by carbohydrate repeat units with specific sulfation patterns that can affect binding and resulting bioactivity of complexed proteins. A general overview of carbohydrate structure is followed by structural information for each major class of GAGs in order to provide background for the reader to better understand how altering the chemical structure for inclusion in scaffolds for drug delivery or tissue engineering may affect key biological activity of these biomolecules.

7.2.1 Carbohydrate Structures and Nomenclature

For the reader's reference, this section summarizes the most important monosaccharide structures, conformations and configurations of GAG subunits in order to better understand the specific epitopes presented in the following sections. Recommendations on sugar nomenclature rules and determination of conformation were published by the International Union of Pure and Applied Chemistry (IUPAC).[24,25] Monosaccharide units of relevance for GAGs are uronic acid and amino sugars (Figure 7.1). Such monosaccharides can acquire different conformations in solution (Figure 7.2A). Among the most well-known conformations are chair (*C*), boat (*B*), and envelope (*E*). The chair conformations and intermediate conformation between chair and boat, skew-boat (*S*), plays an important role for antithrombin III (AT-III) binding with heparin (Figure 7.2A). In solution, each carbohydrate is in equilibrium with its different conformations. Furthermore, the α/β- and D,L nomenclature is also used to distinguish between different configurations (Figure 7.2B).

7.2.2 Structure of Heparin and Heparan Sulfate

The carbohydrate composition for heparin and HS is similar but differs in monosaccharide ratios and sulfation pattern distribution. The most prominent disaccharide repeat unit in heparin consists of 2-*O* sulfated L-iduronic acid (IduA2S, α-1,4) and a mixture of either *N*- and 6-*O* sulfated (GlcNS6S) or *N*-acetylated D-glucosamine (GlcNAc, α-1,4). In HS, instead of heparin's IduA2S, the majority of uronic acid residues are D-glucuronic acid (GlcA, β-1,4). These repeat units are connected in a complex pattern including other residues with additional *O*- and *N*-sulfated groups: GlcNAc can be additionally 6-*O* sulfated (GlcNAc6S), or GlcNS less commonly 3-*O* sulfated (GlcNS3·6S).[26] Unfractionated heparin has a molar mass between 3 and 30 kDa (15 kDa average),[27] whereas HS, *e.g.* from human liver, was found to

Important monosaccharide components of GAGs

Uronic acid residues

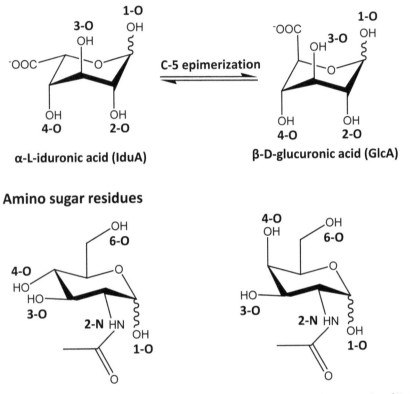

α-L-iduronic acid (IduA)   **β-D-glucuronic acid (GlcA)**

C-5 epimerization

Amino sugar residues

N-acetyl-α -D-glucosamine (GlcNAc)   **N-acetyl-α -D-galactosamine (GalNAc)**

Figure 7.1   The most prominent monosaccharides present in GAGs. Uronic acid sugars possess a carboxyl function connected to C5 of the ring atom, whereas amino sugars have an amino function at position C2. This amino moiety may exist as a free amine (rare), acetylated (shown above) or sulfated (see Figure 7.3) within GAG polysaccharides. Reprinted from: T. Miller, M. C. Goude, T. C. McDevitt and J. S. Temenoff, Molecular engineering of glycosaminoglycan chemistry for biomolecule delivery, *Acta Biomater.*, 2014, **10**, 1705–1719, with permission from Elsevier.

have a molar mass around 24 kDa.[28] HS is a key component of proteo-glycans (PGs) secreted into the ECM, such as perlecan,[29] or agrin,[30] but HS transmembrane PGs can also serve as receptors or co-receptors (*e.g.* syndecans).[31] Consequently, HS is present in many tissues and can serve multiple functions, while the presence of heparin in humans is limited to very few tissues. The best-described occurrence of heparin

A. Selected monosaccharide conformations

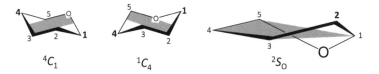

$^4C_1$ $^1C_4$ ^2S_O

B. Definition of C_1/C_5 stereochemistry and D,L configuration

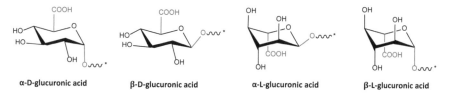

α-D-glucuronic acid   β-D-glucuronic acid   α-L-glucuronic acid   β-L-glucuronic acid

Figure 7.2   (A) Carbohydrate conformations are shown as chair (C) and skew-boat (S). In the chair conformations, the gray, imaginary planes connect ring carbons on parallel sites. Ring carbons which are out-of-plane, determine the conformations. In this nomenclature, the carbon above the plane is noted with a superscript to the C, while the carbon below becomes a subscript. For the skew-boat conformation, the gray plane connects carbons 1, 3, 4, 5. Carbons above this plane are indicated with superscripts, while those below become subscripts. As no carbons are below the plane in this conformation, the subscript is O. (B) Carbohydrate stereochemistry and D,L configuration with glucuronic acid as an example. The α/β nomenclature describes the configuration of the stereo-centers at C1 and C5, whereas D,L nomenclature is related to the substituent position in the Fischer-projection (will not be further explained here – please see[24] for details). Reprinted from: T. Miller, M. C. Goude, T. C. McDevitt and J. S. Temenoff, Molecular engineering of glycosaminoglycan chemistry for biomolecule delivery, *Acta Biomater.*, 2014, **10**, 1705–1719, with permission from Elsevier.

is in mast cell granules, where its function and evolutionary role is still not fully understood.[32,33]

    Besides tissue distribution and function, there are general structural differences between HS and heparin. Specifically, these species differ in the overall charge distributions along the polymeric chain: HS exhibits sulfate-rich, or "S-rich," regions[34] that are separated by disaccharide units that contain mainly unsulfated, acetylated glucosamine and GlcA (NA-regions[35]). Interestingly, the number of HS sulfate clusters change during cell differentiation, leading to more sulfated regions with greater differentiation, whereas stem cells exhibit fewer amounts of sulfate clusters.[36,37] The combination of sulfated and non-sulfated regions in HS leads to a very flexible conformational structure because the alternating structure consisting of regions with high and low sulfation may cause HS to bind and "wrap" around a variety of proteins through non-carbohydrate sequence-specific inter-actions.[38,39] Although HS–protein interactions may not always be sequence specific, different cell types produce HS derivatives with various repeating monosaccharide patterns in the sulfated regions that potentially account for some protein specificity in certain tissues, but the exact physiological role of tissue-dependent HS compositions remains unknown.[40]

In contrast to non-sequence-specific interactions, carbohydrate sequence-specific GAG–protein interactions have been elucidated for heparin/HS. The most well-investigated example is basic fibro-blast growth factor (FGF-2) binding to heparin/HS, for which spec-ificities and effects have been studied since the 1980s.[41] The FGF family consists of 22 distinct isoforms that are sub-divided into seven sub-families.[42] The transmembrane tyrosine kinase receptor for FGF is activated by heparin or HS as a co-factor, which induces FGF-dimerization and enhances FGF signaling.[43] From a series of studies on this topic, a minimal pentasaccharide sequence[44] from heparin was found to be responsible for FGF-2 pairing. Moreover, a 6-*O* desulfated (glucosamine residue) heparin was deemed to promote FGF-2 bind-ing, and 2-*O* sulfate groups (iduronic acid residue) were found to be essential for interaction. Besides FGF-2 attraction, 2-*O*-sulfation of the iduronic acid was found to be essential for binding to proteins such as human glial cell neurotropic factor,[45] and endostatin.[46] However, in other studies, a decasaccharide unit, fully *N*-sulfated and partially 2-*O*- (IduA residue) and 6-*O*- (GlcNS residue) sulfated, was determined to be the minimum binding motif for FGF-1 and FGF-4.[47] In addition, the carboxyl function of the uronic acid residues (Figure 7.1) was found to be necessary for binding to FGF-2.

The results described above highlight the inconsistencies in the minimal binding sequence lengths and sulfation patterns thought to be required for FGF binding.[48–51] In order to explain the different results for the FGF superfamily, one important point to consider is the heterogeneity of the FGF protein family, as well as conformational changes of the polysaccharide itself. Conformational aspects are under current investigation using molecular modeling[52] and analytical tools such as nuclear magnetic resonance (NMR) spectroscopy.[53,54] Gen-erally, pyranose sugars (six-membered ring system consisting of five ring carbons and one oxygen) such as iduronic acid are present in dif-ferent conformations simultaneously: chair conformations ($^4C_1$, $^1C_4$) or skew-boat conformation ($^2S_0$) (Figure 7.2A). These conformations are in equilibrium to each other when sugars are in solution and the glycosidic bonds confer a small degree of freedom to switch between conformations.[55] In this context, the role of 2-*O* sulfation (IduA) for binding of AT-III for anticoagulation purposes is important, as the 2-*O*-sulfation (IduA) as well as the 3-*O*-sulfation (GlcNS3·6S) instigate a conformational change to the skew-boat conformation upon binding of heparin to AT-III.[56] Once bound to AT-III, heparin potentiates the anticoagulant activity of AT-III by factor 2000.[57] However, the AT-III binding region[55,56,58,59] lies outside of the active binding region for FGFs (FGF-1/2),[60] so the relative contribution of GAG conformation to interaction with FGF remains unknown. The carbohydrate conformations and sulfate group activities that are relevant for protein binding are summarized in Figure 7.3.

Overall, heparin and HS exhibit structural similarities but differ mainly in the distribution and quantity of sulfate groups per polymer chain. The sulfate cluster structure of HS allows for a variety of

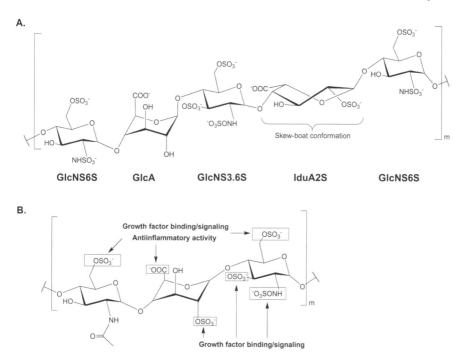

Figure 7.3 Conformation of carbohydrates in heparin and heparan sulfate: (A) pentasaccharide binding sequence of heparin for AT-III; (B) important binding sites for growth factors to a heparin/HS trisaccharide sample sequence. Position of sulfate groups is labeled with their biological activity. The blue boxes indicate necessity of presence for protein binding. Reprinted from: T. Miller, M. C. Goude, T. C. McDevitt and J. S. Temenoff, Molecular engineering of glycosaminoglycan chemistry for biomolecule delivery, *Acta Biomater.*, 2014, **10**, 1705–1719, with permission from Elsevier.

non-carbohydrate sequence-specific protein interactions, whereas heparin possesses a defined AT-III carbohydrate binding sequence. Moreover, the fact that the polyelectrolyte nature of the GAGs alone might lead to "random" binding based solely on electrostatic interactions is well characterized for HS[39,61] and can likely be assumed for heparin as well.

7.2.3   Structure of Chondroitin Sulfate and Dermatan Sulfate

CS and DS are structurally related and often found jointly in PGs. The monosaccharide components of the most prominent repeating disaccharide units in CS are D-glucuronic acid (GlcA, β-1,4) and N-acetyl-D-galactosamine (GalNAc, β-1,3).[62] Epimerization at C-5 (Figure 7.1) during DS biosynthesis of some D-glucuronic acid residues to L-iduronic acid (IduA) converts CS to DS, which is done *in vivo* by specific epimerases.[63,64] Generally, the IduA residues can only be found adjacent to GalNAc-4-sulfate (GalNAc4S).[65] Although molar masses of CS and DS are similar (23 kDa[66] and 25 kDa,[67] respectively), the sulfation patterns of both GAGs are very diverse, and a separate nomenclature for different CS species has been established. Generally, CS polymers are referred to as CS-A (GalNAc4S), CS-C (GalNAc6S), CS-D (GlcA2S, GalNAc6S) and CS-E (GalNAc4,6S) depending on the combination of sulfate positions. For DS, the most frequent sulfation pattern is the GalNAc4S associated with IduA.[68] Functions of the sulfate groups for CS and DS are summarized in Figure 7.4.

Interaction of growth factors with CS and DS occurs in a PG-bound state, as neither CS nor DS is available in its free state in blood or tissue. PGs often contain mixtures of CS and DS that are specific for the tissue in which they were synthesized. CS-rich PGs such as versican (connective tissues) and aggrecan (cartilage) and DS-rich PGs such as decorin and biglycan (both found in connective tissues)[69,70] are important co-factors for receptor function.[71]

For DS, the IduA residue, similar to heparin and HS, is able to exist in multiple conformations including the $^4C_1$, $^1C_4$ and $^2S_0$.[62] In heparin, besides the carbohydrate conformations mentioned in Section 7.2.2, a specific pentasaccharide carbohydrate sequence is essential to bind to AT-III and inhibit thrombin (Figure 7.3A).[56] Although it does not contain this specific carbohydrate sequence, DS was found to be an alternative anticoagulant,[72,73] whose anticoagulative properties are linked to the content of IduA2S. In contrast to regular DS, highly sulfated DS derivatives with high IduA2S were found to bind heparin co-factor II in a less specific manner than DS with lower sulfation[74]

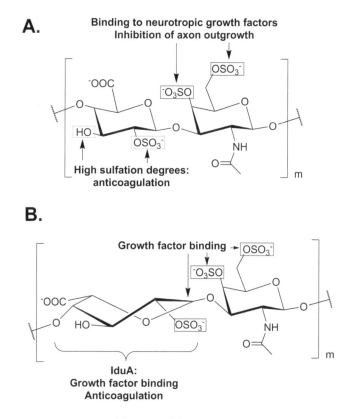

Figure 7.4 Structures of CS (A) and DS (B). Blue boxes indicate necessity for protein binding. Reprinted from: T. Miller, M. C. Goude, T. C. McDevitt and J. S. Temenoff, Molecular engineering of glycosaminoglycan chemistry for biomolecule delivery, *Acta Biomater.*, 2014, **10**, 1705–1719, with permission from Elsevier.

and, in tests of platelet aggregation, was found to bind also to AT-III.[75] A 4,6-disulfated DS is marketed for this purpose[76] with, however, a different mode of action compared to heparin. Heparin anticoagulation effects are mostly caused by the specific interaction with AT-III, whereas DS activates heparin co-factor II, which then inhibits thrombin.[77,78] Similarly, for chemically oversulfated CS,[79] a change of the GlcA residue from the $^4C_1$ to $^1C_4$ conformation mimics the stereochemistry of the 2-O sulfated IduA of heparin (a part of the specific AT-III binding sequence), making oversulfated CS more active toward heparin co-factor II with similar potential compared to DS.

Carbohydrate sequence and sulfation specificity of growth factor to CS and DS has been researched extensively and, especially for neurotropic growth factors, there is strong evidence for a carbohydrate sequence specificity that triggers protein interactions and amplifies neuronal regeneration.[80] However, for many other growth factors, a clear carbohydrate sequence specificity has not consistently been reported, and therefore pure electrostatic interactions between growth factors and CS/DS are possible.

CS and DS have received much attention as components of PGs that can affect neural outgrowth and neural stem cell differentiation.[81,82] Depending on the presentation of the C4,6S epitope, it was found to either inhibit (as part of a PG mimic)[83] or promote (as part of a tetra-saccharide)[80] neuronal outgrowth. Neuronal growth factors such as midkine and brain-derived neurotropic factor bind with high affinity to C4,6S, and it has been suggested that C4,6S recognition motifs are present in receptors or co-receptors in the neural environment for such growth factors.[84] Besides effects in neural regeneration and scarring, CS is able to interact with growth factors in other tissues as well. The 6-O sulfate group was deemed to be essential for binding TGFβ1,[85] which is important in cartilaginous tissue formation. In another study,[86] CS and sulfated HA were compared to each other in terms of binding affinity to bone morphogenetic protein 4 (BMP-4), which has a variety of targets in regenerating bone and cartilage tissue.[87,88] CS and HA share one carbohydrate unit, GlcA, and C4S was compared to HA that was sulfated at position 6-O (HA) of the GalNAc. C4S was discovered to have a lower BMP-4 affinity compared to 6-O sulfated HA and it was hypothesized that, besides other conformational differences, the 6-O sulfate in GalNAc promoted binding of BMP-4. However, the dual 4,6-sulfation pattern on CS was found to be important for binding of BMP-4 to stimulate osteoblast differentiation and mineralization,[89] which leads to the assumption that C4,6S might be a strong promoter of growth factor activity outside the neuronal environment.

For soluble DS–PGs secreted into wounds, interaction with FGF-2 after injury was found to have a tremendously positive effect on wound healing.[90] When CS/DS mixed polymers were isolated from the proteoglycan decorin, a trisulfated pentasaccharide FGF-2 binding motif within CS/DS was suggested containing two potential sulfation patterns: 2-O sulfated IduA with 4,6-disulfated GalNAc or 2-O sulfated IduA and GlcA combined with 4-O sulfated GalNAc.[91] In other studies, an octasaccharide binding sequence for FGF-2 and a decasaccharide sequence for FGF-7 was discovered and it was found that 4-O sulfation of GalNAc and the presence of IduA were essential for effective binding.[92] In addition, a decasaccharide sequence containing IduA was elucidated as the minimal binding motif for FGF-10.[93] The examples above indicate the importance of specific oligosaccharide carbohydrate binding sequences that often demonstrate high activity with a 4,6-sulfation pattern for CS and 4-O sulfation for DS. It also appears that epimerization at C-5 to form IduA from GlcA, the most important structural difference between CS and DS, changes the growth factor selectivity and affinity between CS and DS.

CS and DS are closely related GAGs found in a wide range of PGs. In contrast to the common non-sequence-specific binding of HS to many proteins, there is strong evidence that the 4,6-sulfation epitope of CS enhances growth factor binding and signaling. This has been found to play an important role in the central nervous system by regulating neurite outgrowth.[83] In addition to sulfation pattern, carbohydrate conformation is a key aspect in DS for both growth factor interactions as well as promoting anticoagulant activities.

7.2.4   Structure of Keratan Sulfate

KS primarily consists of the repeating units D-galactose (Gal, β-1,3) and *N*-acetylglucosamine (GlcNAc) that occur in many combinations within the polymer chain.[94] Both monosaccharides were found to have a sulfate group at position 6-O.[95] KS is an important component of PGs (*e.g.* aggrecan), where it is attached to CS and DS. A specific nomenclature of different types of KS based on the type of attachment to the PG core protein has developed, referring to KS as type I (N-linked to Asn residue in PG), II (O-linked to Ser/Thr residues in PG), or III (mannose linked to Ser residue in PG).[96] Typical molar masses of KS range from 7 to 22 kDa[97] and KS PGs are present in high quantities especially in the cornea,[98] with lower concentrations in cartilage (*e.g.* in aggrecan)[99] and many other tissues, such as brain[100] and intervertebral disc.[101]

The sulfation pattern of monosaccharides is not equally distributed over the KS chains, leading to regions of high and low sulfation.[102] Generally, unsulfated regions confer a conformational flexibility that can result in the formation of helical structure in solution, as determined by X-ray diffraction data,[103] which makes it conformationally similar to CS and DS.

Relatively little is known about the influence of the sulfation pattern of KS on binding to specific proteins. The sulfation pattern of KS was shown to be important for the binding affinity to galectin proteins,[104] a class of mainly intracellular proteins possessing a carbohydrate-binding motif that can be secreted into the ECM and regulate cell adhesion.[105] By examining KS tri- to pentasaccharides with various sulfation patterns, the unsulfated 6-OH group of the Gal residue was found to be essential for galectin binding, whereas 6-O sulfation of the GlcNAc residue did not increase binding of galectins[104] (Figure 7.5).

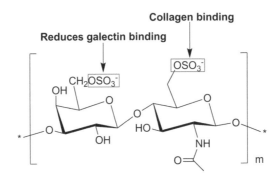

Figure 7.5 Primary sequence of keratan sulfate: 6-*O* sulfate D-galactosamine and 6-*O* sulfate *N*-acetylglucosamine. The blue box indicates necessity for growth factor binding, whereas the red box indicates reduction of galectin binding. Reprinted from: T. Miller, M. C. Goude, T. C. McDevitt and J. S. Temenoff, Molecular engineering of glycosaminoglycan chemistry for biomolecule delivery, *Acta Biomater.*, 2014, **10**, 1705–1719, with permission from Elsevier.

Furthermore, desulfated KS polysaccharides (containing galactose) exhibited binding to galectins whereas the sulfated species did not exhibit any detectable effects, suggesting that for this specific class of proteins, 6-*O* sulfation of Gal residues inhibits binding and that interaction of galectins with GAGs are primarily mediated by hydrophobic and/or van der Waals forces instead of electrostatic interactions.[104]

KS is jointly found with CS and DS in PGs, but the effect of protein binding to single KS polymer chains within PGs has not been a focus of current research and therefore knowledge remains limited on this topic. However, what is known about KS-protein binding (with galectins) demonstrates that the unsulfated Gal residue is essential for binding through a non-electrostatic route. This behavior contrasts with what is known about heparin/HS and CS/DS, where sulfation primarily determines affinity to growth factors, and suggests that strategies for protein binding may need to be tailored based on the class of GAG to be incorporated in the scaffold or drug delivery vehicle.

7.2.5   Structure of Hyaluronan

HA is an unusual member of the GAG family compared to the aforementioned species because it is the only one that is not sulfated. HA polymers consist of the disaccharide units D-GlcA (β-1,4) and D-GlcNAc (β-1,3) and are naturally found with polymerization degrees ranging between 2000 and 25 000 disaccharide units (1–10 MDa)[106] (Figure 7.6). The size of HA assemblies can increase further by aggregation that

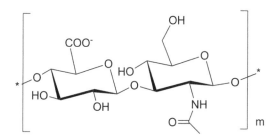

Figure 7.6   Disaccharide units of HA: D-glucuronic acid and D-*N*-acetylglucos-
amine. Reprinted from: T. Miller, M. C. Goude, T. C. McDevitt and J. S.
Temenoff, Molecular engineering of glycosaminoglycan chemistry for
biomolecule delivery, *Acta Biomater.*, 2014, **10**, 1705–1719, with permission
from Elsevier.

occurs in tissues such as cartilage, where HA forms a ternary complex with aggrecan and a link protein.[107] HA can be found in every
connective tissue and with high concentrations in the vitreous humor
of the eye.[108] Carbohydrate conformation within the polymer can be
assumed to be in the $^4C_1$ conformation, which is stabilized intramolec-
ularly by strong hydrogen bonding between the carbohydrates.[109] This
hydrogen bonding is in rapid exchange with water molecules, which is
one explanation for the remarkably high water binding capacity even
though HA is unsulfated and therefore has diminished electrostatic
interactions with water.[108]

HA is known to act intracellularly (*e.g.* intracellular Rhamm-
protein)[110] and extracellularly, where it serves as a direct messenger
molecule by binding to the cell-surface receptor CD-44.[111] In addition to CD-44, HA is known to interact with other cell-surface receptors, such as extracellular Rhamm[112] and ICAM-1.[113] The anticipated
mode of interaction and binding to proteins/receptors is based on
carboxyl-amino group interactions. The interaction of basic amino
acids of CD-44 with HA was revealed *via* scrambling of amino acid
sequences of soluble CD-44.[114] However, in a rather recent study an
octasaccharide binding sequence of HA to CD-44 was found where the
interactions between both species were mainly governed by hydrogen
bonding and less pronounced by electrostatic interactions.[115]

HA is able to induce signaling *via* extracellular receptors (*e.g.* CD-44,
Rhamm) as well as *via* binding to intracellular proteins to regulate
cellular functions. This is in strong contrast to other GAGs that act
through growth factor binding to induce signal transduction. However, promotion of growth factor binding to HA can be achieved by
chemically sulfating the polymer.[85,86,116]

Overall, GAGs have been found to bind to growth factors electrostatically in a carbohydrate sequence-specific or non-sequence-specific manner. Therefore, depending on the nature of the particular interactions, it is quite possible that functionalization of these biopolymers for improved crosslinking or degradation in specific tissue engineering/drug delivery applications may alter affinity toward the loaded protein and thus affect release kinetics, making it imperative to quantify release for each new modification. However, since material degradation and payload release are often coupled in these materials, the following section focuses more specifically on GAG degradation mechanisms.

7.3 DEGRADATION OF GAGs AND THEIR DERIVATIVES

As summarized in the previous section, due to high affinity binding between positively charged proteins and many GAGs, degradation plays a key role in effecting release of loaded growth factors as well as promoting tissue development within scaffolds. Since some chemical modifications of GAGs reduce the degradability by enzymes that are often responsible for *in-vitro* and *in-vivo* degradation, more specifics on how GAG modification can affect degradation are presented in this section in order to help guide rational design of GAG-containing formulations for a range of applications. Degradation of GAG molecules can occur by: (1) chemically induced degradation; and (2) enzymatically induced degradation (see Figure 7.7). These methods are similar insofar as they are able to cleave either glycosidic bonds within the polysaccharide backbone or bonds of chemical crosslinkers. However, differences in their cleavage mechanisms lead to varying degradation products.

7.3.1 Chemical Degradation Mechanisms

Generally, GAGs have been found to be very stable against chemical degradation (*i.e.* CS and KS were seen to be stable in 2 N NaOH at room temperature for more than 120 h[117]). Consequently, if degradation of GAG-based scaffolds occur, it is generally through degradation at functionalization sites introduced prior to loading of drugs or cells. In this vein, a rather non-specific degradation mechanism is hydrolytic cleavage of ester bonds that were previously added on GAG molecules, *e.g.* by esterification with carbodiimide chemistry. For example, in a recent study,[118] photopolymerized, methacrylamide modified alginate–heparin hydrogels degraded in water primarily through this mechanism, showing significant (70%) mass loss over 8 weeks.

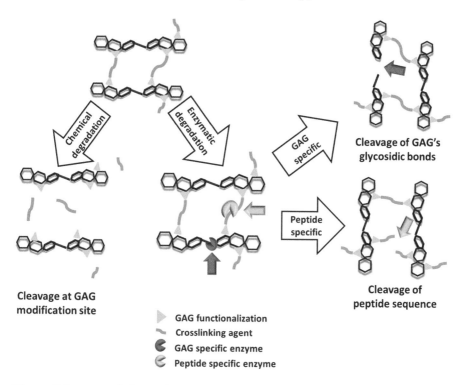

Figure 7.7 Degradation principles of GAGs. Chemical degradation focuses on degradation at GAG modification sites as the polymers themselves are chemically quite stable. Enzymatic mechanisms consist of GAG-specific enzymes that cleave glycosidic bonds, whereas peptide sequence-specific bonds can be cleaved by enzymes if a specific sequence is introduced into the drug delivery system. Reprinted from: T. Miller, M. C. Goude, T. C. McDevitt and J. S. Temenoff, Molecular engineering of glycosaminoglycan chemistry for biomolecule delivery, *Acta Biomater.*, 2014, **10**, 1705–1719, with permission from Elsevier.

Another example of dissociation through reactions at functional groups is the cleavage of disulfide bonds by glutathione (GSH). Disulfide bonds can be formed between modified GAGs, in one example hyaluronan, and another molecule with thiol groups [*e.g.* thiolated polyethylene glycol (PEG)] to form cross-linked hydrogels that can be degraded through the addition of GSH.[119] After GSH treatment, hydrogels were found to undergo bulk degradation, whereas treatment with hyaluronidase led to surface degradation.

Overall, chemical degradation of GAG-based delivery systems or scaffolds under physiological conditions results in the release of cargo (if loaded) and full-length modified GAG molecules. For drug delivery it should be noted that non-enzyme-based degradation occurs through a bulk mechanism that makes cargo release dependent on diffusion length and can therefore be controlled by the geometry of the system (*e.g.* thickness, shape).

7.3.2 Enzymatic Degradation Mechanisms

For enzymatic degradation reactions, the varieties of conditions as well as GAG cleavage mechanisms have been thoroughly reviewed.[120,121] Therefore, in this section we will only briefly describe the enzymatic modes of action and highlight potential effects of GAG chemical derivatization on their digestion by enzymes. Generally, heparin/ HS-cleaving enzymes can be categorized into two groups that cleave the glycosidic bond between two sugars, but have different modes of action: lyases (bacterial in origin, prefer heparin) and hydrolases (mammalian, prefer HS) (Figure 7.8). Lyases are either endo- or exoglycosidases that cleave glycosidic bonds by an eliminative mechanism

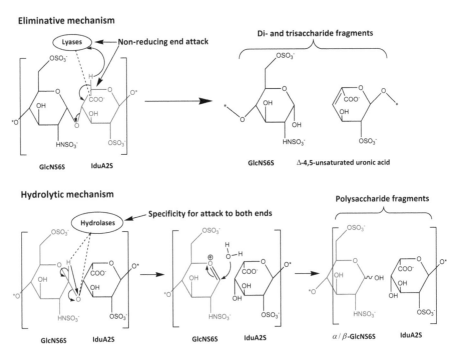

Figure 7.8   Mechanisms of enzymatic heparin/HS degradation. Lyases cause eliminative cleavage whereas hydrolases lead to hydrolytic degradation. Red color indicates the center of action in the scheme. Reprinted from: T. Miller, M. C. Goude, T. C. McDevitt and J. S. Temenoff, Molecular engineering of glycosaminoglycan chemistry for biomolecule delivery, *Acta Biomater.*, 2014, **10**, 1705–1719, with permission from Elsevier.

involving uronic acid residues by abstracting the proton at C-5 position with formation of a double bond between C-4 and C-5. The carboxyl function of the uronic acid residue is essential for the cleavage mechanism.[122] The group of GAG lyases is comprised of heparinases, chondroitinases, and hyaluronidases, each with varying subtypes requiring individual reaction conditions and exhibiting their own reaction kinetics.

In contrast, hydrolases are endoglycosidases that form an intermediate oxonium by delivering a proton to the glycosidic oxygen followed by water attack (hydroxyl addition removes the double bond). For hydrolases, the resulting cleaved residues are completely saturated. The group of hydrolases consists of heparin hydrolases, keratanases, and hyaluronan hydrolases. Depending on the tissue origin of the hydrolases (*e.g.* platelets, tumors, fibroblasts), enzyme activity can depend on sulfation or acetylation of carbohydrates and therefore sulfation pattern and adjacent carbohydrate clusters are important in obtaining complete degradation.[120]

Comparing these two classes of enzymes, for heparin hydrolases in general, the *O*-sulfation pattern and glucuronic acid content of the polymer is important for enzymatic cleavage, whereas, as previously mentioned, lyases require a carboxyl function and are active against glucuronic and iduronic acid residues adjacent to glucosamino sugars. An additional difference is that lyases are only able to cleave bonds on the non-reductive side of sugars, whereas hydrolases can be specific for both reducing or non-reducing sites.[120] Each heparin lyase and hydrolase enzyme isoform has its own carbohydrate cleaving sequence and therefore we refer the reader to published literature on this topic.[120,123–126]

Chemical GAG modifications, such as the commonly performed carboxyl modifications of the uronic acid residues of heparin, affect enzymatic digestion. Since lyases require a free carboxyl function for their action, heparin-based scaffolds cannot be degraded *via* this mechanism with high degrees of modifications at this position. However, hydrolases, which have specific preferences for *O*-sulfate groups and specific carbohydrates adjacent to them (*e.g.* glucuronic acid for heparin hydrolases), will be sensitive to sulfation pattern and compositional changes of the GAG molecule.

Generally, lyases and hydrolases degrade GAGs to a different extent: lyases degrade GAGs into oligosaccharides,[127] but hydrolases produce smaller, still polymeric GAG fragments of lower molecular weight compared to the starting material. In this context, an example for heparin hydrolases was carried out with poly(vinyl alcohol) (PVA)/heparin crosslinked hydrogels where platelet extract, a known source of heparin hydrolases, was used for degradation.[128] By analyzing soluble heparin prior to hydrogel incorporation, it was found that heparin was degraded by the enzyme into fragments with approximately half the molar mass of the original molecule (17–19 kDa heparin into 8.6 kDa fragments). As heparin hydrolases represent the enzyme class that is relevant in mammals, these degradation properties are assumed to be relevant to *in-vivo* conditions for crosslinked hydrogels. Moreover, this finding highlights the fact that even if *in-vitro* degradation (often tested with lyases) fails, owing to different modes of action between lyases and hydrolases, *in-vivo* degradation may still be achieved.

Instead of relying on GAG-degrading enzymes, another approach to achieve enzyme-specific degradable drug delivery systems is *via* incorporation of specific peptide sequences into the crosslinking chemistry or within the GAG backbone. For example, a peptide-functionalized PEG containing a matrix metalloprotease (MMP) cleavable sequence[129] was crosslinked with heparin and loaded with vascular endothelial growth factor (VEGF), which is known to promote the proliferation of human umbilical vein endothelial cells (HUVECs). A comparison between MMP-degradable and non-degradable hydrogels indicated that the number of HUVECs increased significantly and spread in three dimensions after 1 week only in the MMP-degradable hydrogel group. Thus, it was concluded that VEGF release was promoted by the degradability of the hydrogels.

In comparing the degradation mechanisms, chemically degraded gels (hydrolysis) will undergo primarily bulk degradation, whereas for enzymatically degraded hydrogels, surface *vs.* bulk degradation is dependent on the enzymatic reaction rate and enzyme diffusion rate through the gel.[130] In general, enzymatic degradation of GAG-based biomaterials seems to be a beneficial approach over chemical degradation because of the much milder conditions and greater *in-vivo* relevance, but several caveats remain. For example, the commonly used lyases are commercially available but do not represent the class of enzymes present *in vivo* in mammals. This drawback can be circumvented by including peptide-based enzyme-responsive sequences, which may lead to much more controlled degradation profiles in response to enzymes specifically upregulated in diseased tissues. For GAG-based materials used for sustained protein release, it should also be noted that, due to electrostatic attraction, cargo release is often dependent on GAG release (assuming GAG-binding sites exceed cargo concentration). Regardless of chemical *vs.* enzymatic degradation mechanisms, it is thus very plausible to assume the release of GAG–cargo complexes, which may have additional effects on biological activity of the released cargo.[131]

7.4 GAG-BASED SCAFFOLDS FOR TISSUE ENGINEERING

Depending on the application, GAGs can be included in tissue engineering scaffolds to improve mechanical properties,[132] or to increase bioactivity due the GAGs themselves[133,134] and/or their binding capacity for growth factors.[135,136] In the latter case, GAG-based scaffolds are loaded with the bioactive protein *via* electrostatic interactions, often at the time of fabrication, and some form of degradation is designed into the scaffold to facilitate growth factor release and space for tissue ingrowth or development.[137,138] As such, the protein-binding properties as well as means of GAG degradation as discussed in previous sections are very important to the development of GAG-based scaffolds for a myriad of regenerative medicine applications. A few examples of these applications are highlighted in this section to give the reader an idea of the variety of uses for GAGs in tissue engineering, as well as to present differences in how major classes of GAGs are utilized in current tissue engineering research.

7.4.1 Bulk Hydrogel Scaffolds

This section focuses on the incorporation of GAGs in bulk hydrogel scaffolds. Hydrogels as discussed here are crosslinked, hydrophilic, polymeric constructs that have a high affinity for water[139] and have not undergone specific processing to create macropores or extensive interconnected porous structures. Owing to the wide range of properties and applications of GAG-based hydrogels, examples are organized by each major GAG species, starting with the most heavily sulfated.

7.4.1.1 Heparin and Heparan Sulfate. Because of the high degree of sulfation along the GAG backbone, heparin-based scaffolds are often used in tissue engineering to sequester growth factors, although heparin is only present in a few tissues natively.[140] In hydrogel scaffolds, heparin (or HS) has often been combined with another polymer, such as linear or branched PEG, to improve handling or tune protein-release properties.[135,141] For crosslinking, heparin has been functionalized with moieties such as thiols, maleimide and tyramine to react with PEG.[135,142,143] Owing to the versatility of these crosslinking schemes, heparin can be used as a part of strategies to regenerate target tissues with very different biochemical and mechanical properties.[144]

Heparin–PEG scaffolds can be used in "soft" and water-rich tissues such as liver (tissue modulus <1 kPa)[145] and brain (tissue modulus 0.5–1 kPa).[146] Heparin has been thiolated and crosslinked with PEG diacrylate (PEG-DA) to encapsulate rat hepatocytes and hepatocyte growth factor (HGF) for liver tissue engineering.[135] The heparin hydrogel demonstrated a smaller burst release of HGF (10%) and a slower release profile over 30 days in comparison to the PEG-DA control, which underwent a 50% burst release and reached ~100% release by day 15.[135] Degradation of this heparin-based hydrogel has been shown to occur through hydrolytic cleavage of ester bonds in the PEG-DA.[21] Compared to PEG-DA controls, the heparin-based scaffolds yielded higher cell viability of primary rat hepatocytes, albumin and urea secretion and maintained a hepatocyte phenotype for 20 days.[135]

Neuronal tissue replacement has also been investigated by the delivery of mouse mesencephalic neural stem cells in a construct of heparin and amine-functionalized starPEG crosslinked with carbodiimide chemistry.[136] The FGF-release profile from the heparin–starPEG construct showed an initial burst at 6 hours and a sustained continuous release of ~20% of the loaded growth factor over 1 week.[136] The construct demonstrated little degradation in the presence of collagenase type IV over 8 hours.[147] Addition of FGF-2 and the RGD (sequence of amino acids arginine-glycine-aspartic acid) adhesion peptide to the scaffold increased cell number and colony outgrowth while minimizing differentiation within 7 days.[136]

In separate studies, a hydrogel system was fabricated consisting of HS and other biopolymers (chitosan, gelatin, HA) where each component was selected to serve a specific purpose in the construct: HA to promote angiogenesis, HS to facilitate FGF-2 signaling and chitosan and gelatin for structural support.[19] Enzymatic degradation of the HS-based construct in lysozyme and collagenase type I led to ~85% weight loss after 3 days, but the degradation rate slowed by day 5. Rat embryonic neural stem and progenitor cells encapsulated in the scaffold containing 50:50:1:1 chitosan:gelatin:HA:HS volume ratio maintained a more mature neuronal phenotype, characterized by increased neuron-specific class III beta-tubulin (Tuj1)-positive cells and neurite outgrowth, over 7 days of culture.[19]

Other heparin–PEG systems have been utilized for engineering tissues that require higher elastic moduli, such as in myocardial tissue (tissue modulus 20–500 kPa),[142] cartilage (tissue modulus ~1000 kPa)[143] and cortical bone (tissue modulus 1.5×10^7 kPa).[148] However, the moduli of the heparin–PEG hydrogels discussed here are still significantly lower than that of the native tissue, and thus they are may be not useful for load-bearing applications without additional mechanical support.[149] The delivery of human aortic adventitial fibroblasts and basic fibroblast growth factor (bFGF) in a maleimide–thiol-based heparin-containing PEG hydrogel has been evaluated for cardiovascular tissues. The addition of PEG–thiol at a higher concentration increased the elastic modulus by seven-fold (2.8 kPa), which improved cell adhesion in combination with fibronectin.[142] Maleimide–thiol adducts have been shown to undergo degradation at physiological pH and temperature due to retro reactions in the presence of other thiol compounds.[137]

For repair of connective tissues, bovine chondrocytes have been delivered for cartilage repair in dextran–heparin hydrogels. In these studies, the increase in dextran content resulted in scaffolds of higher storage modulus up to a maximum of 48 kPa.[143] However, scaffolds containing 25 : 75 weight ratio of dextran and heparin demonstrated less susceptibility to hydrolytic degradation due to the stability of the urethane bond against hydrolysis[150] and enhanced collagen II and aggrecan gene expression and collagen production in chondrocytes compared to 50 : 50 or 75 : 25 dextran–heparin scaffolds.[143] Another heparin-based hydrogel was prepared from photocrosslinking of heparin methacrylate and PEG dimethacrylate (PEG-DMA) for the osteogenic differentiation of human mesenchymal stem cells (hMSCs).[151] The heparin–PEG scaffold has demonstrated sustained release of bFGF over 5 weeks relative to PEG-DMA controls, which released 100% of the growth factor within 8 hours.[152] The bioactivity of the released bFGF was demonstrated with the increase in hMSC proliferation compared to MSCs without bFGF treatment (a high of 46% increase in cell number) due to the ability of heparin in binding growth factors.[152]

In an innovative use of this material, in follow-on studies hMSCs were stimulated toward the osteoblastic lineage (with concomitant increase in BMP production) with fluvastatin. When seeded on these heparin-containing hydrogels, which should sequester cell-secreted BMP, and stimulated with fluvastatin, osteogenic differentiation, as measured by RUNX2 and osteopontin (OPN) gene expression by day 28, was higher compared to stimulated MSCs on hydrogels comprised of PEG and poly(lactic) acid, but without incorporating heparin.[153]

In a similar vein, oligo-(poly-(ethylene glycol)-fumarate) (OPF) was incorporated into a heparin-based construct to promote osteogenic differentiation in human mesenchymal stem cells in co-culture.[148] After co-culture with osteoblasts for 21 days, increased calcium accumulation, alkaline phosphatase activity and mineralization were observed in gels with 10% heparin by weight as compared to PEG-based gels with no heparin. Because no soluble factors were added to induce osteogenic differentiation in this case, it was concluded that the heparin aided in sequestering signals from nearby osteoblasts to promote osteogenic differentiation of encapsulated MSCs.[148]

Overall, the ease in functionalizing heparin/HS for crosslinking, as well as its high level of negative charge, has promoted its use in many tissue-engineering constructs as carriers for both cells and growth factors or other proteins. In recent studies, heparin-based hydrogels have also been explored to enhance cell-derived (autocrine) signals or paracrine signaling between two cell types in order to improve differentiation of encapsulated progenitor cells. Thus, the electrostatic interactions between heparin/HS and proteins described in previous sections can be employed in several paradigms in tissue engineering scaffolds as a part of regenerative approaches to a wide range of tissues.

7.4.1.2 Chondroitin Sulfate. As reviewed earlier in this chapter, CS also has the ability to bind to positively charged proteins, although it has fewer sulfate groups along the backbone than heparin. Since CS is found mostly in connective tissues,[154] its tissue engineering application has been primarily directed towards the regeneration of cartilage and bone.[20,155,156] CS is commonly paired with PEG for similar reasons to heparin/HS.[20,155] To form hydrogels, CS is usually methacrylated for chemically crosslinking *via* free radical polymerization with either thermal[20,157] or UV light initiation.[138,155] CS can also be crosslinked with carbodiimide chemistry.[156,158]

CS-based scaffolds have been designed to either: (1) induce differentiation of stem cells such as MSCs; or (2) promote survival and function of differentiated cells. In contrast to heparin-based scaffolds, CS gels are not often used as stand-alone drug delivery depots, but, rather the focus is often on enhancement of biochemical properties in the presence of growth factors, such as TGF-β, in cell culture media.[20,155] CS in combination with various PEG crosslinkers has been used toward regeneration of cartilage by promoting chondrogenic differentiation of seeded hMSCs.[20,155] In one study, enzymatic degradation of the modified CS polymer and PEG-DA with chondroitinase ABC was possible, but did not occur due to the limited amount of the enzyme secreted by the hMSCs *in vitro*, which resulted in low amounts of ECM deposition after 21 days. In addition, this study focused on the effects of desulfated CS on TGF-β sequestration from the media and hMSC chondrogenic differentiation. The results showed that aggrecan and collagen II gene expression increased in the desulfated scaffolds compared to those containing natively sulfated CS.[20] In contrast, in another study, a cell-laden CS methacrylate/PEG-DA scaffold experienced degradation due to cell-secreted enzymes that allowed caprine MSCs (cMSCs) to form cartilaginous aggregates after 3 weeks with enhanced GAG and collagen production compared to a PEG-DA-only scaffold.[155]

To increase control over degradation in CS-based gels, a MMP-sensitive sequence was added to a CS–PEG-DMA hydrogel, which resulted in a scaffold that mimicked the superficial zone of cartilage. The addition of CS increased the swelling ratio and the equilibrium water content of the hydrogels relative to PEG-DMA scaffolds. Enzymatic degradation of the CS–PEG components in chondroitinase resulted in ~80% weight loss over 21 days, whereas the MMP-PEG components in collagenase experienced ~50% weight loss, compared to PEG-DMA controls that underwent minimal degradation.[138] The CS–PEG-MMP peptide hydrogel promoted high collagen II synthesis in the encapsulated cMSCs and low GAG production levels after 6 weeks.[138]

CS-based hydrogels have also been utilized to promote function of differentiated cells, such as chondrocytes, for cartilage tissue engineering. In order to increase swelling ratio while maintaining high modulus in the hydrogel scaffold, CS methacrylate was photocrosslinked with PEG-DMA and used to encapsulate bovine articular chondrocytes.[132] Compared to a PEG-DMA scaffold of similar swelling ratio, the incorporation of 40% CS increased the compressive modulus of the copolymer construct by four-fold.[132] After 14-day culture, the bovine chondrocytes in CS–PEG hydrogels exhibited higher collagen II gene expression than pure CS constructs.[132] A decrease in GAG content was observed at week 4 in CS–PEG hydrogels compared to PEG-DMA controls, which may indicate possible degradation by cell-secreted enzymes.[132]

Other CS-based hydrogel scaffolds function not only as an artificial ECM to promote cellular activity, but also act as a tissue adhesive.[158–160] The articular cartilage tissue adhesive was first formulated with methacrylated CS and PEG-DA, which could be photocrosslinked *in situ* and has the capability to bond other biomaterials containing vinyl groups.[160] In another formulation, CS was conjugated to *N*-hydroxysuccinimide (NHS) to react with PEG–amine through the carboxyl group or with the primary amines found in the proteins of the tissue.[158] The multifunctional CS–PEG hydrogel can be polymerized *in situ* and is degradable enzymatically by chondroitinase ABC (70% weight loss in 30 hours),[161] which may better promote *in-vivo* tissue integration. The same CS–PEG system has also been used to encapsulate rabbit corneal cells, keratocytes, epithelial cells, and endothelial cells as a corneal adhesive for repairing ocular defects.[162] Similarly, CS–NHS was reacted with the amines in proteins of bone marrow (BM) aspirate to encapsulate bovine meniscal fibrochondrocytes.[159] *In vitro*, the CS–BM adhesive has been shown to promote total collagen production in encapsulated bovine chondrocytes and collagen I production in meniscus fibrochondrocytes. Bovine meniscus bonded by the CS–BM adhesive was subcutaneously implanted in athymic rats for 12 weeks. At 12 weeks, the menisci remained fused only in the group treated with CS–BM (30 : 70 volume ratio) adhesive.[159] However, no CS–BM adhesive was observed at week 12, which may be due to degradation.[159]

CS-based hydrogels or tissue adhesives have been primarily explored to date to engineer musculoskeletal tissues, particularly cartilage/fibrocartilage. Rather than the controlled delivery of growth factors seen with heparin, CS has mainly been studied as a means to enhance the function of cells in the presence of exogenous cues as well as improve the stiffness of hydrogels without sacrificing the high water content needed for cell encapsulation. Thus, much like its native function in cartilage, CS has been employed for both biochemical and structural reasons towards regeneration of orthopedic tissues.

7.4.1.3 Hyaluronan.   Unlike other GAGs, HA is non-sulfated and therefore its major biological activity is conferred through binding to specific cell-surface receptors, rather than interactions with growth factors.[108] HA is a component of every connective tissue and possesses extremely high water binding capacity,[108] which makes it a strong candidate for engineering cardiac, cartilaginous, and even osteoid tissues.[163] Similar to CS, HA is often incorporated in hydrogel scaffolds to achieve both high swelling ratios and compressive moduli[164–166] as well as impart biological activity that can be related to the molecular weight of the HA employed.[133,134] HA is often methacrylated to allow for covalent crosslinking into hydrogels.[167,168]

HA's high water binding combined with the ability to form low modulus tissue constructs can be harnessed for neural tissue engineering. In one HA-based hydrogel, the methacrylation of HA (37%, 87%, and 160% degrees of substitution per disaccharide) was tuned by the synthesis process,[169] which resulted in a range of compressive moduli (1.5, 2.6, and 7.2 kPa, respectively), and after photocrosslinking that spanned those of embryonic and mature central nervous system tissues. Hydrogels with highest methacrylation experienced the slowest rate of enzymatic degradation in hyaluronidase and did not fully degrade until 24 hours, compared to the least methacrylated HA hydrogel that completely degraded within 6 hours.[169] Murine neural progenitor cells encapsulated in the (softest) hydrogel with mechanical properties closest to that of neonatal brain exhibited the most mature neuronal phenotype indicated by long branched processes with positive β-III tubulin staining after 3 weeks.[169]

Higher elastic moduli are required of scaffolds used to repair myocardial tissues (20–500 kPa).[170] Therefore HA of different methacrylation degree (30 or 60%) was crosslinked *in situ via* radical initiation into a scaffold to examine effects of crosslinking on hydrogel mechanical properties.[168] The hydrogel containing the highly methacrylated HA (60%) exhibited an enhanced modulus (~43 kPa) and experienced little hydrolytic degradation *in vitro* (less than 25% weight loss after 20 weeks) and *in vivo*, allowing it to attenuate dilation of the left ventricle after infarction after 8 weeks in an ovine model.[168] A separate PEG-DMA–HA methacrylate construct was used to encapsulate porcine valvular interstitial cells. Increases in the PEG-DMA content of the scaffold could be used to slow the rate of enzymatic degradation in hyaluronidase with the hydrogel containing the lowest amount of PEG–DMA (1.3 wt%) reaching ~100% mass loss within 15 days. The degradability of the 1 wt% PEG–DMA scaffold and the subsequent release of cleaved HA over 14 days improved cell proliferation and increased cellular elastin production compared to non-degradable constructs.[167]

HA has also been incorporated for engineering of orthopedic tissues. Chondrogenic differentiation of cMSCs has been investigated in PEG–HA scaffolds over 6 weeks.[171] The presence of HA suppressed GAG deposition in the cMSCs and induced lower levels of SOX9, aggrecan, and collagen II gene expression than PEG-DA controls and PEG-collagen scaffolds, which indicated a lack of chondrogenic phenotype.[171] However, the MSCs in the PEG–HA hydrogel stained strongly with Alizarin Red and promoted the highest calcium accumulation compared to the PEG–DA and PEG–collagen hydrogels.[171] Degradation of the PEG–HA scaffold was not specifically discussed in the study.[171] From another laboratory, scaffolds containing modified (sulfated) HA and collagen I were prepared *via* fibrillogenesis in an acetic acid buffer and used to encapsulate hMSCs for osteogenic differentiation. Derivatives of HA were sulfated with sulfur trioxide/dimethylformamide (SO_3–DMF) complexes with or without tetrabutylammonium salt.[172] Hydrolytic degradation of the collagen–HA occurred with a small loss of the collagen content (~8%) over 8 days and significant loss of HA derivatives within the first hour (50–65%).[116] The incorporation of sulfated HA resulted in increased calcium phosphate deposition, alkaline phosphatase activity, and gene expression from encapsulated cells.[172]

Like CS, HA-based hydrogel constructs have been employed due to both their high swelling capacity as well as the potential for bioactivity of HA degradation products through interaction with specific cell-surface receptors. However, because HA is found natively in many tissues, the range of applications for which HA hydrogels have been explored is greater than for CS. Like the other GAGs discussed in this section, the ability to tune levels of HA modification during synthesis allows the fabrication of crosslinked constructs with a range of mechanical properties and degradation times that can be tailored for particular regenerative medicine applications.

7.4.2 Porous Scaffolds

In addition to hydrogels, GAGs have been incorporated as a part of tissue engineering scaffolds that have macroporosity and an interconnected pore architecture to promote cell growth, penetration, and ECM production.[173,174] GAG-based porous scaffolds have been processed by both freeze-drying and electrospinning to introduce controlled microstructures in the resulting materials.[173] While GAGs are also commonly employed in coating methods, such as layer-by-layer deposition, we will focus in this chapter only on formation of 3D scaffolds, so the reader is referred elsewhere for more details on GAG-based coatings.[175-177] Specifically, in this section, advantages and disadvantages of two types of macroporous GAG-containing scaffolds are summarized.

7.4.2.1 Interpenetrating Scaffolds. Collagen is often combined with GAGs, such as CS, HA, and heparin, to form interpenetrating scaffolds for tissue engineering purposes due to its abundance in connective tissues.[178] The system was first introduced in 1989[179] and has been subsequently modified and adapted for primarily musculoskeletal tissues such as skin, cartilage, tendon, and bone.[157,165,172,180] The interpenetrating scaffold is unique in that it consists of two distinct but physically intertwined polymer networks[181] and has the advantage of high pore volume fraction (average of ~90% porous) and a large range of possible pore sizes (5–500 μm).[174] However, because the collagen–GAG scaffolds are prepared through physical crosslinking with the freeze-drying process, they exhibit low elastic moduli (~0.75 kPa).[182] Therefore, collagen–GAG scaffolds are often reinforced with chemical crosslinking with 1-ethyl-3-(3-dimethylaminopropyl)-carbodiimide (EDC) crosslinking,[164,165,183] dehydrothermal treatment at high temperature under vacuum[166,184] or a combination of both,[185] to promote additional intermolecular crosslinks in collagen.

For dermal applications, a collagen–HA–CS construct (average pore size ~110 μm) has been used as a carrier for rat dermal fibroblasts for grafting on the dorsum of rats.[165] A 9 : 1 : 1 v/v/v ratio collagen–HA–CS scaffold experienced minimal enzymatic degradation in collagenase after 5 hours with ~24% weight loss compared to collagen–HA scaffolds (~70% weight loss).[165] After 2 weeks, the tri-copolymer construct demonstrated the fastest rate of wound closure.[165] To restore other soft tissues, a EDC-crosslinked collagen–HA hydrogel scaffold (7.5% or 15% HA) with pores of 100–200 μm diameter was seeded with mouse preadipocytes for 8 days, which resulted in increased gene expression of adipsin, a marker for mature adipocytes, compared to a collagen-only control.[164] Some hydrolytic degradation of the collagen–HA scaffold occurred after 7 days (~11–13%). However, the presence of HA and its intermolecular bonding with collagen seemed to slow the degradation rate compared to that of pure collagen scaffolds (~19%).[164]

For load-bearing tissues such as cartilage, a collagen–HA scaffold (average pore size ~94 μm) was processed with dehydrothermal treatment, which resulted in a two-fold increase in compressive modulus (up to 0.55 kPa) relative to the un-crosslinked scaffolds.[166] Enzymatic degradation of a similarly crosslinked collagen–GAG scaffold occurred within 2 hours compared to non-crosslinked controls that solubilized in 90 minutes in collagenase.[186] After seeding scaffolds with rat MSCs, collagen II and SOX9 gene expression were significantly higher in the crosslinked collagen–HA scaffold by day 14 compared to a crosslinked collagen-only control scaffold.[166] In another study, a collagen–GAG scaffold was prepared with interpenetrating network of self-assembled collagen and both CS and HA methacrylate that were radically crosslinked.[157] Increasing levels of CS methacrylation slowed the rate of enzymatic degradation in collagenase type I with a loss of 40% after 60 days compared to the scaffolds containing non-methacrylated CS and HA, which degraded with 24 hours.[157] The increase in the amount of methacrylate conjugated to the CS also improved the compressive modulus of the scaffold to 45 kPa, which is almost three-fold higher than that of collagen and unmodified CS and HA.[157] It was found that the scaffold containing moderately methacrylated CS (17%) resulted in highest maintenance of chondrogenic phenotype *in vitro* with increased gene expression of aggrecan and SOX9 up to day 14 in rabbit chondrocytes.[157]

For tendon applications, a collagen–CS scaffold was crosslinked with dehydrothermal treatment and EDC and underwent directional solidification at different freezing temperatures to achieve a range of pore sizes (55–243 μm) that were aligned in the longitudinal plane of the scaffold.[185] Enzymatic degradation in either collagenase or chondroitinase of a similar collagen-GAG scaffold crosslinked by both dehydrothermal treatment and EDC showed minimal change after 120 minutes.[186] Equine tendon cells were seeded in the anisotropic scaffold for 14 days, showing higher cell metabolic activity in aligned scaffolds compared to isotropic controls. Supplementation with insulin-like growth factor 1 (IGF-1) during culture increased metabolic activity of the tendon cells over 7 days in scaffolds of all pore sizes as compared to non-supplemented controls. Culture in the presence of platelet-derived growth factor yielded similar results except for the 55 μm pore diameter scaffold at day 4.[185]

In further studies, three variations of collagen–GAG (HA, CS, or heparin) scaffolds were formed *via* freeze-drying and crosslinked with EDC to determine sulfation effects on equine tenocytes.[183] Because CS and heparin are highly sulfated, the scaffolds containing these two GAGs were able to pull down more IGF than the collagen–HA scaffold.[183] In a 14-day study with IGF supplementation, the equine tenocytes in the collagen–heparin construct exhibited higher metabolic activity and collagen II gene expression compared to other collagen–GAG scaffolds, indicating improved proliferation but a loss of tenocytic phenotype.[183] Further supporting these conclusions, gene expression of tenocytic markers, scleraxis homologue B (SCXB) and tenascin C (TNC), also decreased with the incorporation of increasing sulfated GAG content in the scaffold.[183]

In another CS-based porous scaffold containing gelatin and chitosan, osteogenic differentiation of rat MSCs was evaluated.[156] The CS–chitosan–gelatin mixture was freeze-dried and subsequently chemically crosslinked with EDC with a resulting mixture of pores ranging from 100 to 500 μm.[156] Hydrolytic degradation of the scaffold was measured and ~50% of the dry weight was retained after 21 days with ~40% of the weight lost occurring within the first 10 days. The decrease in rate of degradation was attributed to the presence of chitosan in the scaffold.[156] After 21 days of culture, rat MSCs seeded on the 3D scaffold demonstrated higher viability in the presence of osteogenic stimuli (dexamethasone, ascorbic acid, and β-glycerophosphate) relative to the unstimulated controls.[156]

The fact that at least two polymers are required for interpenetrating network scaffolds provides a great deal of versatility to this type of porous scaffold, both in terms of biological activity as well as mechanical properties. Incorporation of GAGs (usually CS or HA) as one of the scaffold components does provide some additional mechanical stability and the potential for biochemical signaling, as discussed in previous sections. However, it is interesting to note, particularly since these scaffolds have been explored with a variety of orthopedic cell types, that, without additional crosslinking, the overall mechanical properties of these scaffolds are very low (modulus well under 1 kPa), which may limit the applications for which these scaffolds are most appropriate unless sufficient post-fabrication crosslinking can be ensured.

7.4.2.2 Electrospun Scaffolds. Electrospinning with biopolymers, such as GAGs, has been used to create scaffolds with interconnected microstructure with control over porosity and fiber diameter.[173] Electrospinning is a technique that produces fibers based on electrostatic forces between a polymer solution charged by a high voltage source and a collector.[187] The size of fibers can be easily changed through the adjustment of the electrospinning parameters. Fibers of nanometer diameter (100–500 nm) are possible[188] and tissue engineering scaffolds composed of nanofibers have been shown to better mimic the structure of native ECM, thereby improving cell attachment and proliferation.[187] However, nanometer-scale pore sizes formed by the nanofibers limit cell migration and angiogenesis throughout the scaffold.[188] As a result, studies have incorporated techniques to introduce macropores to the scaffold.[189]

Scaffolds composed of electrospun nanofibers have been prepared using a PCL–heparin conjugate and photopolymerized for vascular tissue engineering with encapsulated human endothelial cells and VEGF.[190] The nonwoven PCL–heparin tubular scaffold exhibited a burst pressure of ~200 kPa, which indicated its ability to withstand physiological vascular conditions.[190] Hydrolytic degradation over 3 weeks was minimal, as determined by measuring the loss of surface heparin content over time.[191] The tubular PCL–heparin scaffold was determined to have 83% porosity and its loading efficiency and affinity for VEGF was higher compared to that of a PCL scaffold.[190] The scaffold was subsequently implanted in a ligated femoral artery of a canine model and was able to maintain patency for 4 weeks with coverage of the lumen by endothelial cell as shown by H&E staining.[190]

In other studies, electrospun nanofibers of CS methacrylate and PVA methacrylate have been fabricated and subsequently photopolymerized into a scaffold for chondrogenic differentiation of goat MSCs *in vitro*, which promoted increased collagen II deposition compared to PVA-only scaffolds.[192] The tissue repair potential of this scaffold was also investigated in a rat osteochondral defect model for 6 weeks. The

CS–PVA scaffold demonstrated more proteoglycan deposition than the untreated defects, as shown by safranin-*O* staining, and higher collagen II production than the PVA-only scaffold, as visualized by immunohistochemistry.[192]

Although the aforementioned scaffolds contained high porosity, the nanometer-scale mesh size may prevent adequate cell infiltration. To introduce micrometer-scale pores, electrospinning and salt leaching techniques were combined to form a macroporous and nanofibrous HA–collagen I scaffold for culture of bovine chondrocytes.[189] Salt particulates were deposited during the electrospinning process, followed by EDC crosslinking and salt leaching to form macropores of 50–100 µm in diameter. Hydrolytic degradation of the 80:20 weight ratio HA–collagen scaffold was the slowest, with ~60% weight remaining by day 12 relative to HA-only controls, which degraded completely after 5 days in cell culture media.[189] The addition of collagen increased the tensile strength of the nanofiber to ~430 kPa compared to ~270 kPa in HA only nanofibers. After 3 days of culture, there was greater chondrocyte proliferation in the 80:20 weight ratio HA–collagen scaffold relative to the 95:5 scaffold.[189]

Controlled photopolymerization can also be employed to increase porosity in electrospun scaffolds. Methacrylated HA and poly(ethylene oxide) were co-spun and subsequently crosslinked with a photomask to form pores close to 165 µm and 333 µm in diameter.[193] The moduli of the patterned scaffolds were ~400 kPa and did not differ significantly from the non-patterned controls. Degradation of the scaffold could be tuned by the methacrylation degree of the HA component as shown by previous work from the group.[193] Subcutaneous implantation of the porous construct in rats for 1 week showed significant improvement in cell infiltration through the thickness of construct compared to non-patterned controls.[193]

GAGs can be processed into porous scaffolds based on interpenetrating networks or electrospun into nanofiber meshes to achieve greater porosity than hydrogels, which can be very advantageous to promote cell infiltration as required in certain applications. While secondary crosslinking may be required in interpenetrating scaffolds to achieve mechanical stability, mechanical properties are less of a concern with electrospun scaffolds, depending on what material is co-spun with the GAG, but extremely small pore sizes remain an issue. In electrospinning, GAGs are primarily included due to their inherent bioactivity or as a means to achieve controlled release of growth factors. In the future, clever processing techniques, such as the addition of larger secondary porogens, may be further explored to achieve both the improved mechanical properties found in electrospinning as well as the greater cell infiltration capabilities seen in more macroporous scaffolds.

7.5 CONCLUSIONS

To harness the beneficial effects of GAGs for tissue engineering applications, they are often incorporated into various types of scaffolds, including highly hydrophilic hydrogels, as well as macroporous constructs made through either a freezing process or electrospinning. However, GAG incorporation often requires chemical modifications to the polymers that can change the innate properties of GAGs, which affect protein affinities as well as degradation rate. Therefore, as demonstrated throughout this chapter, more detailed knowledge of how the chemistry of native and modified GAGs affects protein binding and *in-vivo* degradation will allow better engineering of GAG-based biomaterials for tissue regeneration.

Apart from engineering considerations associated with fabrication of a particular type of construct (*e.g.* maintaining sufficient porosity over time) or designing optimal scaffold delivery mechanisms, current challenges with use of GAGs in tissue engineering scaffolds center on two primary issues: (1) *in-vivo* biocompatibility of modification chemistries; and (2) assurance of excretion of modified GAGs. Particularly if crosslinking into delivery systems occurs *in situ*, catalysts and other by-products of polymerization processes may introduce toxicity concerns, so additional modifications of these reactions may be necessary. Furthermore, maintaining *in-vivo* degradability is an important component. In the future, further studies on the macromolecular properties of GAGs are needed to investigate if GAG-based degradation products of scaffolds that were chemically modified prior to encapsulation can be metabolized and excreted from living organisms in a similar manner to their non-modified counterparts.

To date, GAG-based materials have proven safe in humans, as demonstrated by their incorporation in medical devices approved for clinical use (*e.g.* heparin-coated stents[194]). However, more precise spatial control of degradation as well as temporal control of biomolecule delivery may be required for tissue regeneration applications in the future, and involves increasingly sophisticated modifications to the GAG molecules to further tune affinity and degradation. With additional research further elucidating how to control growth factor binding and material degradation, GAG-based scaffolds have all potential to be an ideal carrier system for a wide range of regenerative medicine applications because of their biodegradability, non-toxicity and versatility for many different cellular and molecular cargos.

ACKNOWLEDGEMENTS

The authors are supported by funding from the NIH (R01 AR062006) and NSF (DMR 1207045).

REFERENCES

1. R. V. Stick and S. Williams, *Carbohydrates: The Essential Molecules of Life*, Elsevier Science, 2nd edn, 2008, p. 343.
2. E. D. Hay, *J. Cell Biol.*, 1981, **91**, 205s.
3. L. Robert, A. M. Robert and G. Renard, *Pathol. Biol.*, 2010, **58**, 187.
4. J. E. Scott, *Pathol. Biol.*, 2001, **49**, 284.
5. J. P. Urban, A. Maroudas, M. T. Bayliss and J. Dillon, *Biorheology*, 1979, **16**, 447.
6. J. D. Esko, K. Kimata and U. Lindahl, in *Essentials of Glycobiology*, ed. A. Varki, R. D. Cummings, J. D. Esko, H. H. Freeze, P. Stanley, C. R. Bertozzi, G. W. Hart and M. E. Etzler, Cold Spring Harbor, NY, 2nd edn, 2009.
7. A. Rapraeger, M. Jalkanen, E. Endo, J. Koda and M. Bernfield, *J. Biol. Chem.*, 1985, **260**, 11046.
8. J. P. Bali, H. Cousse and E. Neuzil, *Semin. Arthritis Rheum.*, 2001, **31**, 58.
9. C. Lallam-Laroye, Q. Escartin, A. S. Zlowodzki, D. Barritault, J. P. Caruelle, B. Baroukh, J. L. Saffar and M. L. Colombier, *J. Biomed. Mater. Res., Part A*, 2006, **79A**, 675.
10. L. Kock, C. C. van Donkelaar and K. Ito, *Cell Tissue Res.*, 2012, **347**, 613.
11. E. C. Huskisson, *J. Int. Med. Res.*, 2008, **36**, 1161.
12. T. Ehrenfreud-Kleinman, J. Golenser and A. J. Domb, in *Scaffolding in Tissue Engineering*, ed. P. X. Ma and J. Elisseeff, CRC Press, 2005, p. 27.
13. B. Casu, A. Naggi and G. Torri, *Semin. Thromb. Hemostasis*, 2002, **28**, 335.
14. G. Kogan, L. Soltes, R. Stern and P. Gemeiner, *Biotechnol. Lett.*, 2007, **29**, 17.
15. N. S. Gandhi and R. L. Mancera, *Chem. Biol. Drug Des.*, 2008, **72**, 455.

16. N. S. Gandhi and R. L. Mancera, *Biochim. Biophys. Acta, Proteins Proteomics*, 2012, **1824**, 1374.
17. B. M. Sattelle, J. Shakeri and A. Almond, *Biomacromolecules*, 2013, **14**, 1149.
18. S. Van Vlierberghe, P. Dubruel and E. Schacht, *Biomacromolecules*, 2011, **12**, 1387.
19. S. Guan, X. L. Zhang, X. M. Lin, T. Q. Liu, X. H. Ma and Z. F. Cui, *J. Biomater. Sci., Polym. Ed.*, 2013, **24**, 999.
20. J. J. Lim and J. S. Temenoff, *Biomaterials*, 2013, **34**, 5007.
21. G. Tae, Y. J. Kim, W. I. Choi, M. Kim, P. S. Stayton and A. S. Hoffman, *Biomacromolecules*, 2007, **8**, 1979.
22. T. Segura, B. C. Anderson, P. H. Chung, R. E. Webber, K. R. Shull and L. D. Shea, *Biomaterials*, 2005, **26**, 359.
23. O. Jeon, S. J. Song, K. J. Lee, M. H. Park, S. H. Lee, S. K. Hahn, S. Kim and B. S. Kim, *Carbohydr. Polym.*, 2007, **70**, 251.
24. A. D. McNaught, in *Glycoscience*, ed. K. T. B. O. Fraser-Reid, J. Thiem, G. L. Coté, S. Flitsch, Y. Ito, H. Kondo, S.-I. Nishimura and B. Yu, Springer-Verlag, Berlin Heidelberg, 2nd edn, 2008, vol. 1, p. 2727.
25. J. C. Boeyens, *J. Chem. Crystallogr.*, 1978, **8**, 317.
26. A. Fryer, Y. C. Huang, G. Rao, D. Jacoby, E. Mancilla, R. Whorton, C. A. Piantadosi, T. Kennedy and J. Hoidal, *J. Pharmacol. Exp. Ther.*, 1997, **282**, 208.
27. J. Hirsh, S. S. Anand, J. L. Halperin and V. Fuster, *Arterioscler., Thromb., Vasc. Biol.*, 2001, **21**, 1094.
28. P. Vongchan, M. Warda, H. Toyoda, T. Toida, R. M. Marks and R. J. Linhardt, *Biochim. Biophys. Acta, Proteins Proteomics*, 2005, **1721**, 1.
29. S. M. Knox and J. M. Whitelock, *Cell. Mol. Life Sci.*, 2006, **63**, 2435.
30. A. J. A. Groffen, C. A. F. Buskens, T. H. van Kuppevelt, J. H. Veerkamp, L. A. H. Monnens and L. P. W. J. van den Heuvel, *Eur. J. Biochem.*, 1998, **254**, 123.
31. T. Pap and J. Bertrand, *Nat. Rev. Rheumatol.*, 2013, **9**, 43.
32. H. B. Nader, S. F. Chavante, E. A. dos-Santos, F. W. Oliveira, J. F. de-Paiva, S. M. B. Jeronimo, G. F. Medeiros, L. R. D. de-Abreu, E. L. Leite, J. F. de-Sousa, R. A. B. Castro, T. Toma, I. L. S. Tersariol, M. A. Porcionatto and C. P. Dietrich, *Braz. J. Med. Biol. Res.*, 1999, **32**, 529.
33. L. C. J. Yong, *Exp. Toxicol. Pathol.*, 1997, **49**, 409.
34. C. L. R. Merry, M. Lyon, J. A. Deakin, J. J. Hopwood and J. T. Gallagher, *J. Biol. Chem.*, 1999, **274**, 18455.
35. M. Mobli, M. Nilsson and A. Almond, *Glycoconjugate J.*, 2008, **25**, 401.

36. R. A. Smith, K. Meade, C. E. Pickford, R. J. Holley and C. L. Merry, *Biochem. Soc. Trans.*, 2011, **39**, 383.

37. R. J. Baldwin, G. B. ten Dam, T. H. van Kuppevelt, G. Lacaud, J. T. Gallagher, V. Kouskoff and C. L. Merry, *Stem Cells*, 2008, **26**, 3108.

38. J. T. Gallagher, *J. Clin. Invest.*, 2001, **108**, 357.

39. B. Mulloy, S. Khan and S. J. Perkins, *Pure Appl. Chem.*, 2012, **84**, 65.

40. J. T. Gallagher, J. E. Turnbull and M. Lyon, *Int. J. Biochem.*, 1992, **24**, 553.

41. T. Barzu, J. C. Lormeau, M. Petitou, S. Michelson and J. Choay, *J. Cell. Physiol.*, 1989, **140**, 538.

42. R. Xu, A. Ori, T. R. Rudd, K. A. Uniewicz, Y. A. Ahmed, S. E. Guimond, M. A. Skidmore, G. Siligardi, E. A. Yates and D. G. Fernig, *J. Biol. Chem.*, 2012, **287**, 40061.

43. B. M. Loo, J. Kreuger, M. Jalkanen, U. Lindahl and M. Salmivirta, *J. Biol. Chem.*, 2001, **276**, 16868.

44. M. Maccarana, B. Casu and U. Lindahl, *J. Biol. Chem.*, 1993, **268**, 23898.

45. S. M. Rickard, R. S. Mummery, B. Mulloy and C. C. Rider, *Glycobiology*, 2003, **13**, 419.

46. S. Ricard-Blum, O. Feraud, H. Lortat-Jacob, A. Rencurosi, N. Fukai, F. Dkhissi, D. Vittet, A. Imberty, B. R. Olsen and M. van der Rest, *J. Biol. Chem.*, 2004, **279**, 2927.

47. M. Ishihara, *Glycobiology*, 1994, **4**, 817.

48. T. Spivak-Kroizman, M. A. Lemmon, I. Dikic, J. E. Ladbury, D. Pinchasi, J. Huang, M. Jaye, G. Crumley, J. Schlessinger and I. Lax, *Cell*, 1994, **79**, 1015.

49. H. Mori, Y. Kanemura, J. Onaya, M. Hara, J. Miyake, M. Yamasaki and Y. Kariya, *J. Biosci. Bioeng.*, 2005, **100**, 54.

50. F. J. Moy, M. Safran, A. P. Seddon, D. Kitchen, P. Bohlen, D. Aviezer, A. Yayon and R. Powers, *Biochemistry*, 1997, **36**, 4782.

51. D. B. Volkin, P. K. Tsai, J. M. Dabora, J. O. Gress, C. J. Burke, R. J. Linhardt and C. R. Middaugh, *Arch. Biochem. Biophys.*, 1993, **300**, 30.

52. B. M. Sattelle and A. Almond, *J. Comput. Chem.*, 2010, **31**, 2932.

53. B. M. Sattelle, J. Shakeri, I. S. Roberts and A. Almond, *Carbohydr. Res.*, 2010, **345**, 291.

54. F. Yu, J. J. Wolff, I. J. Amster and J. H. Prestegard, *J. Am. Chem. Soc.*, 2007, **129**, 13288.

55. A. J. Herrera, M. T. Beneitez, L. Amorim, F. J. Canada, J. Jimenez-Barbero, P. Sinay and Y. Bleriot, *Carbohydr. Res.*, 2007, **342**, 1876.

56. M. Hricovini, M. Guerrini, A. Bisio, G. Torri, A. Naggi and B. Casu, *Semin. Thromb. Hemostasis*, 2002, **28**, 325.

57. F. Y. Avci, N. A. Karst and R. J. Linhardt, *Curr. Pharm. Des.*, 2003, **9**, 2323.

58. S. K. Das, J. M. Mallet, J. Esnault, P. A. Driguez, P. Duchaussoy, P. Sizun, J. P. Herault, J. M. Herbert, M. Petitou and P. Sinay, *Chem.–Eur. J.*, 2001, **7**, 4821.

59. S. K. Das, J. M. Mallet, J. Esnault, P. A. Driguez, P. Duchaussoy, P. Sizun, J. P. Herault, J. M. Herbert, M. Petitou and P. Sinay, *Angew. Chem., Int. Ed.*, 2001, **40**, 1670.

60. S. Guglier, M. Hricovini, R. Raman, L. Polito, G. Torri, B. Casu, R. Sasisekharan and M. Guerrini, *Biochemistry*, 2008, **47**, 13862.

61. E. Seyrek and P. Dubin, *Adv. Colloid Interface Sci.*, 2010, **158**, 119.

62. J. E. Scott, F. Heatley and B. Wood, *Biochemistry*, 1995, **34**, 15467.

63. A. Malmstrom, B. Bartolini, M. A. Thelin, B. Pacheco and M. Maccarana, *J. Histochem. Cytochem.*, 2012, **60**, 916.

64. A. Malmstrom, *Biochem. J.*, 1981, **198**, 669.

65. L. Å. Fransson, F. Cheng, K. Yoshida, D. Heinegård, A. Malmström and A. Schmidtchen, in *Dermatan Sulfate Proteoglycans: Chemistry, Biology, Chemical Pathology*, ed. J. E. Scott, 1993, p. 11.

66. C. Tanford, E. Marler, E. A. Davidson and E. Jury, *J. Biol. Chem.*, 1964, **239**, 4034.

67. F. Dol, M. Petitou, J. C. Lormeau, J. Choay, C. Caranobe, P. Sie, S. Saivin, G. Houin and B. Boneu, *J. Lab. Clin. Med.*, 1990, **115**, 43.

68. M. Kobayashi, G. Sugumaran, J. A. Liu, N. W. Shworak, J. E. Silbert and R. D. Rosenberg, *J. Biol. Chem.*, 1999, **274**, 10474.

69. C. Malavaki, S. Mizumoto, N. Karamanos and K. Sugahara, *Connect. Tissue Res.*, 2008, **49**, 133.

70. S. Yung and T. M. Chan, *Peritoneal Dial. Int.*, 2007, **27**(suppl 2), S104.

71. S. Mizumoto and K. Sugahara, *FEBS J.*, 2013, 2462–2470.

72. F. Fernandez, J. Van Ryn, F. A. Ofosu, J. Hirsh and M. R. Buchanan, *Br. J. Haematol.*, 1986, **64**, 309.

73. P. Bendayan, H. Boccalon, D. Dupouy and B. Boneu, *Thromb. Haemostasis*, 1994, **71**, 576.

74. F. A. Ofosu, G. J. Modi, M. A. Blajchman, M. R. Buchanan and E. A. Johnson, *Biochem. J.*, 1987, **248**, 889.

75. R. M. Maaroufi, P. Giordano, P. Triadou, J. Tapon-Bretaudiere, M. D. Dautzenberg and A. M. Fischer, *Thromb. Res.*, 2007, **120**, 615.

76. T. T. Hong, C. L. Van Gorp, A. D. Cardin and B. R. Lucchesi, *Thromb. Res.*, 2006, **117**, 333.

77. J. K. Hennan, T. T. Hong, A. K. Shergill, E. M. Driscoll, A. D. Cardin and B. R. Lucchesi, *J. Pharmacol. Exp. Ther.*, 2002, **301**, 1151.

78. D. O'Keeffe, S. T. Olson, N. Gasiunas, J. Gallagher, T. P. Baglin and J. A. Huntington, *J. Biol. Chem.*, 2004, **279**, 50267.

79. T. Maruyama, T. Toida, T. Imanari, G. Yu and R. J. Linhardt, *Carbohydr. Res.*, 1998, **306**, 35.

80. C. I. Gama, S. E. Tully, N. Sotogaku, P. M. Clark, M. Rawat, N. Vaidehi, W. A. Goddard, 3rd, A. Nishi and L. C. Hsieh-Wilson, *Nat. Chem. Biol.*, 2006, **2**, 467.

81. A. Purushothaman, K. Sugahara and A. Faissner, *J. Biol. Chem.*, 2012, **287**, 2935.

82. K. Sugahara and T. Mikami, *Curr. Opin. Struct. Biol.*, 2007, **17**, 536.

83. J. M. Brown, J. Xia, B. Zhuang, K. S. Cho, C. J. Rogers, C. I. Gama, M. Rawat, S. E. Tully, N. Uetani, D. E. Mason, M. L. Tremblay, E. C. Peters, O. Habuchi, D. F. Chen and L. C. Hsieh-Wilson, *Proc. Natl. Acad. Sci. U. S. A.*, 2012, **109**, 4768.

84. S. S. Deepa, Y. Umehara, S. Higashiyama, N. Itoh and K. Sugahara, *J. Biol. Chem.*, 2002, **277**, 43707.

85. V. Hintze, A. Miron, S. Moeller, M. Schnabelrauch, H. P. Wiesmann, H. Worch and D. Scharnweber, *Acta. Biomater.*, 2012, **8**, 2144.

86. V. Hintze, S. Moeller, M. Schnabelrauch, S. Bierbaum, M. Viola, H. Worch and D. Scharnweber, *Biomacromolecules*, 2009, **10**, 3290.

87. D. Chen, M. Zhao and G. R. Mundy, *Growth Factors*, 2004, **22**, 233.

88. N. D. Miljkovic, G. M. Cooper and K. G. Marra, *Osteoarthr. Cartil.*, 2008, **16**, 1121.

89. T. Miyazaki, S. Miyauchi, A. Tawada, T. Anada, S. Matsuzaka and O. Suzuki, *J. Cell. Physiol.*, 2008, **217**, 769.

90. S. F. Penc, B. Pomahac, T. Winkler, R. A. Dorschner, E. Eriksson, M. Herndon and R. L. Gallo, *J. Biol. Chem.*, 1998, **273**, 28116.

91. A. Zamfir, D. G. Seidler, H. Kresse and J. Peter-Katalinic, *Glycobiology*, 2003, **13**, 733.

92. K. R. Taylor, J. A. Rudisill and R. L. Gallo, *J. Biol. Chem.*, 2005, **280**, 5300.

93. K. A. Radek, K. R. Taylor and R. L. Gallo, *Wound Repair Regen.*, 2009, **17**, 118.

94. T. N. Huckerby, *Prog. Nucl. Magn. Reson. Spectrosc.*, 2002, **40**, 35.

95. K. Meyer, A. Linker, E. A. Davidson and B. Weissmann, *J. Biol. Chem.*, 1953, **205**, 611.

96. J. L. Funderburgh, *Glycobiology*, 2000, **10**, 951.
97. J. J. Hopwood and H. C. Robinson, *Biochem. J.*, 1973, **135**, 631.
98. A. J. Quantock, R. D. Young and T. O. Akama, *Cell. Mol. Life Sci.*, 2010, **67**, 891.
99. E. Rodriguez, S. K. Roland, A. Plaas and P. J. Roughley, *J. Biol. Chem.*, 2006, **281**, 18444.
100. N. B. Schwartz, M. Domowicz, R. C. Krueger, H. Li and D. Mangoura, *Perspect. Dev. Neurobiol.*, 1996, **3**, 291.
101. P. J. Roughley, L. I. Melching, T. F. Heathfield, R. H. Pearce and J. S. Mort, *Eur. Spine J.*, 2006, **15**, S326.
102. T. N. Huckerby, G. H. Tai and I. A. Nieduszynski, *Eur. J. Biochem.*, 1998, **253**, 499.
103. S. Arnott, J. M. Guss, D. W. L. Hukins, I. C. M. Dea and D. A. Rees, *J. Mol. Biol.*, 1974, **88**, 175.
104. J. Iwaki, T. Minamisawa, H. Tateno, J. Kominami, K. Suzuki, N. Nishi, T. Nakamura and J. Hirabayashi, *Biochem. Biophys. Res. Commun.*, 2008, **373**, 206.
105. R. C. Hughes, *Biochimie*, 2001, **83**, 667.
106. B. P. Toole, *Nat. Rev. Cancer*, 2004, **4**, 528.
107. J. Dudhia, *Cell. Mol. Life Sci.*, 2005, **62**, 2241.
108. A. Almond, *Cell. Mol. Life Sci.*, 2007, **64**, 1591.
109. P. Pogany and A. Kovacs, *Carbohydr. Res.*, 2009, **344**, 1745.
110. J. Y. Lee and A. P. Spicer, *Curr. Opin. Cell Biol.*, 2000, **12**, 581.
111. A. Aruffo, I. Stamenkovic, M. Melnick, C. B. Underhill and B. Seed, *Cell*, 1990, **61**, 1303.
112. E. A. Turley, P. W. Noble and L. Y. W. Bourguignon, *J. Biol. Chem.*, 2002, **277**, 4589.
113. J. Entwistle, C. L. Hall and E. A. Turley, *J. Cell. Biochem.*, 1996, **61**, 569.
114. R. J. Peach, D. Hollenbaugh, I. Stamenkovic and A. Aruffo, *J. Cell Biol.*, 1993, **122**, 257.
115. S. Banerji, A. J. Wright, M. Noble, D. J. Mahoney, I. D. Campbell, A. J. Day and D. G. Jackson, *Nat. Struct. Mol. Biol.*, 2007, **14**, 234.
116. U. Hempel, V. Hintze, S. Moller, M. Schnabelrauch, D. Scharnweber and P. Dieter, *Acta. Biomater.*, 2012, **8**, 659.
117. H. Lyons and J. A. Singer, *J. Biol. Chem.*, 1971, **246**, 277.
118. O. Jeon, C. Powell, L. D. Solorio, M. D. Krebs and E. Alsberg, *J. Controlled Release*, 2011, **154**, 258.
119. S. Y. Choh, D. Cross and C. Wang, *Biomacromolecules*, 2011, **12**, 1126.
120. S. Ernst, R. Langer, C. L. Cooney and R. Sasisekharan, *Crit. Rev. Biochem. Mol. Biol.*, 1995, **30**, 387.

121. B. Casu and U. Lindahl, *Adv. Carbohydr. Chem. Biochem.*, 2001, **57**, 159.
122. R. J. Linhardt, P. M. Galliher and C. L. Cooney, *Appl. Biochem. Biotech.*, 1986, **12**, 135.
123. I. Vlodavsky, N. Ilan, A. Naggi and B. Casu, *Curr. Pharm. Des.*, 2007, **13**, 2057.
124. J. R. Baker and D. G. Pritchard, *Biochem. J.*, 2000, **348**, 465.
125. D. S. Pikas, J. P. Li, I. Vlodavsky and U. Lindahl, *J. Biol. Chem.*, 1998, **273**, 18770.
126. A. Naggi, B. Casu, M. Perez, G. Torri, G. Cassinelli, S. Penco, C. Pisano, G. Giannini, R. Ishai-Michaeli and I. Vlodavsky, *J. Biol. Chem.*, 2005, **280**, 12103.
127. Z. P. Xiao, B. R. Tappen, M. Ly, W. J. Zhao, L. P. Canova, H. S. Guan and R. J. Linhardt, *J. Med. Chem.*, 2011, **54**, 603.
128. A. Nilasaroya, P. J. Martens and J. M. Whitelock, *Biomaterials*, 2012, **33**, 5534.
129. M. V. Tsurkan, K. Chwalek, K. R. Levental, U. Freudenberg and C. Werner, *Macromol. Rapid Commun.*, 2010, **31**, 1529.
130. C. C. Lin and A. T. Metters, *Adv. Drug Delivery Rev.*, 2006, **58**, 1379.
131. H. H. Chu, J. Gao, C. W. Chen, J. Huard and Y. D. Wang, *Proc. Natl. Acad. Sci. U. S. A.*, 2011, **108**, 13444.
132. S. J. Bryant, J. A. Arthur and K. S. Anseth, *Acta. Biomater.*, 2005, **1**, 243.
133. E. L. Ferguson, J. L. Roberts, R. Moseley, P. C. Griffiths and D. W. Thomas, *Int. J. Pharm.*, 2011, **420**, 84.
134. N. Banu and T. Tsuchiya, *J. Biomed. Mater. Res., Part A*, 2007, **80A**, 257.
135. M. Kim, J. Y. Lee, C. N. Jones, A. Revzin and G. Tae, *Biomaterials*, 2010, **31**, 3596.
136. U. Freudenberg, A. Hermann, P. B. Welzel, K. Stirl, S. C. Schwarz, M. Grimmer, A. Zieris, W. Panyanuwat, S. Zschoche, D. Meinhold, A. Storch and C. Werner, *Biomaterials*, 2009, **30**, 5049.
137. A. D. Baldwin and K. L. Kiick, *Bioconjugate Chem.*, 2011, **22**, 1946.
138. L. H. Nguyen, A. K. Kudva, N. L. Guckert, K. D. Linse and K. Roy, *Biomaterials*, 2011, **32**, 1327.
139. N. A. Peppas, J. Z. Hilt, A. Khademhosseini and R. Langer, *Adv. Mater.*, 2006, **18**, 1345.
140. R. A. Scott and A. Panitch, *Wiley Interdiscip. Rev.: Nanomed. Nanobiotechnol.*, 2013, **5**, 388.
141. S. Prokoph, E. Chavakis, K. R. Levental, A. Zieris, U. Freudenberg, S. Dimmeler and C. Werner, *Biomaterials*, 2012, **33**, 4792.

142. T. Nie, R. E. Akins, Jr. and K. L. Kiick, *Acta. Biomater.*, 2009, **5**, 865.
143. R. Jin, L. S. M. Teixeira, P. J. Dijkstra, C. A. van Blitterswijk, M. Karperien and J. Feijen, *J. Controlled Release*, 2011, **152**, 186.
144. Y. Liang and K. L. Kiick, *Acta. Biomater.*, 2013, 1588–1600.
145. I. Levental, P. C. Georges and P. A. Janmey, *Soft Matter*, 2007, **3**, 299.
146. N. D. Leipzig and M. S. Shoichet, *Biomaterials*, 2009, **30**, 6867.
147. M. V. Tsurkan, K. R. Levental, U. Freudenberg and C. Werner, *Chem. Commun.*, 2010, **46**, 1141.
148. S. P. Seto, M. E. Casas and J. S. Temenoff, *Cell Tissue Res.*, 2012, **347**, 589.
149. J. L. Drury and D. J. Mooney, *Biomaterials*, 2003, **24**, 4337.
150. R. Jin, C. Hiemstra, Z. Zhong and J. Feijen, *Biomaterials*, 2007, **28**, 2791.
151. D. S. Benoit, A. R. Durney and K. S. Anseth, *Biomaterials*, 2007, **28**, 66.
152. D. S. Benoit and K. S. Anseth, *Acta. Biomater.*, 2005, **1**, 461.
153. D. S. W. Benoit, S. D. Collins and K. S. Anseth, *Adv. Funct. Mater.*, 2007, **17**, 2085.
154. L. Kjellen and U. Lindahl, *Annu. Rev. Biochem.*, 1991, **60**, 443.
155. S. Varghese, N. S. Hwang, A. C. Canver, P. Theprungsirikul, D. W. Lin and J. Elisseeff, *Matrix Biol.*, 2008, **27**, 12.
156. C. B. Machado, J. M. Ventura, A. F. Lemos, J. M. Ferreira, M. F. Leite and A. M. Goes, *Biomed. Mater.*, 2007, **2**, 124.
157. Y. Guo, T. Yuan, Z. Xiao, P. Tang, Y. Xiao, Y. Fan and X. Zhang, *J. Mater. Sci.: Mater. Med.*, 2012, **23**, 2267.
158. J. A. Simson, I. A. Strehin, Q. Lu, M. O. Uy and J. H. Elisseeff, *Biomacromolecules*, 2013, **14**, 637.
159. J. A. Simson, I. A. Strehin, B. W. Allen and J. H. Elisseeff, *Tissue Eng., Part A*, 2013, **19**, 1843.
160. D. A. Wang, S. Varghese, B. Sharma, I. Strehin, S. Fermanian, J. Gorham, D. H. Fairbrother, B. Cascio and J. H. Elisseeff, *Nat. Mater.*, 2007, **6**, 385.
161. I. Strehin, Z. Nahas, K. Arora, T. Nguyen and J. Elisseeff, *Biomaterials*, 2010, **31**, 2788.
162. I. Strehin, W. M. Ambrose, O. Schein, A. Salahuddin and J. Elisseeff, *J. Cataract Refractive Surg.*, 2009, **35**, 567.
163. A. Fakhari and C. Berkland, *Acta. Biomater.*, 2013, **9**, 7081.
164. N. Davidenko, J. J. Campbell, E. S. Thian, C. J. Watson and R. E. Cameron, *Acta. Biomater.*, 2010, **6**, 3957.
165. W. H. Wang, M. Zhang, W. Lu, X. J. Zhang, D. D. Ma, X. M. Rong, C. Y. Yu and Y. Jin, *Tissue Eng., Part C*, 2010, **16**, 269.

166. A. Matsiko, T. J. Levingstone, F. J. O'Brien and J. P. Gleeson, *J. Mech. Behav. Biomed. Mater.*, 2012, **11**, 41.

167. D. N. Shah, S. M. Recktenwall-Work and K. S. Anseth, *Biomaterials*, 2008, **29**, 2060.

168. J. L. Ifkovits, E. Tous, M. Minakawa, M. Morita, J. D. Robb, K. J. Koomalsingh, J. H. Gorman, 3rd, R. C. Gorman and J. A. Burdick, *Proc. Natl. Acad. Sci. U. S. A.*, 2010, **107**, 11507.

169. S. K. Seidlits, Z. Z. Khaing, R. R. Petersen, J. D. Nickels, J. E. Vanscoy, J. B. Shear and C. E. Schmidt, *Biomaterials*, 2010, **31**, 3930.

170. Q. Z. Chen, A. Bismarck, U. Hansen, S. Junaid, M. Q. Tran, S. E. Harding, N. N. Ali and A. R. Boccaccini, *Biomaterials*, 2008, **29**, 47.

171. N. S. Hwang, S. Varghese, H. Li and J. Elisseeff, *Cell Tissue Res.*, 2011, **344**, 499.

172. U. Hempel, S. Moller, C. Noack, V. Hintze, D. Scharnweber, M. Schnabelrauch and P. Dieter, *Acta. Biomater.*, 2012, **8**, 4064.

173. N. Annabi, J. W. Nichol, X. Zhong, C. Ji, S. Koshy, A. Khademhosseini and F. Deghani, *Tissue Eng., Part B*, 2010, **16**, 371.

174. I. V. Yannas, in *Scaffolding in Tissue Engineering*, ed. P. X. Ma and J. Elisseeff, CRC Press, 2005, p. 3.

175. J. M. Silva, N. Georgi, R. Costa, P. Sher, R. L. Reis, C. A. Van Blitterswijk, M. Karperien and J. F. Mano, *PloS One*, 2013, **8**, e55451.

176. H. Xu, Y. Yan and S. Li, *Biomaterials*, 2011, **32**, 4506.

177. Y. Gong, Y. Zhu, Y. Liu, Z. Ma, C. Gao and J. Shen, *Acta. Biomater.*, 2007, **3**, 677.

178. P. Bornstein, *Annu. Rev. Biochem.*, 1980, **49**, 957.

179. I. V. Yannas, E. Lee, D. P. Orgill, E. M. Skrabut and G. F. Murphy, *Proc. Natl. Acad. Sci. U. S. A.*, 1989, **86**, 933.

180. B. Huang, C. Q. Li, Y. Zhou, G. Luo and C. Z. Zhang, *J. Biomed. Mater. Res., Part B*, 2010, **92B**, 322.

181. G. C. Ingavle, A. W. Frei, S. H. Gehrke and M. S. Detamore, *Tissue Eng., Part A*, 2013, **19**, 1349.

182. B. A. Harley, J. H. Leung, E. C. Silva and L. J. Gibson, *Acta. Biomater.*, 2007, **3**, 463.

183. R. A. Hortensius and B. A. Harley, *Biomaterials*, 2013, **34**, 7645.

184. M. G. Haugh, M. J. Jaasma and F. J. O'Brien, *J. Biomed. Mater. Res., Part A*, 2009, **89**, 363.

185. S. R. Caliari and B. A. Harley, *Biomaterials*, 2011, **32**, 5330.

186. Y. S. Pek, M. Spector, I. V. Yannas and L. J. Gibson, *Biomaterials*, 2004, **25**, 473.

187. T. J. Sill and H. A. von Recum, *Biomaterials*, 2008, **29**, 1989.

188. Q. P. Pham, U. Sharma and A. G. Mikos, *Tissue Eng.*, 2006, **12**, 1197.
189. T. G. Kim, H. J. Chung and T. G. Park, *Acta. Biomater.*, 2008, **4**, 1611.
190. L. Ye, X. Wu, H. Y. Duan, X. Geng, B. Chen, Y. Q. Gu, A. Y. Zhang, J. Zhang and Z. G. Feng, *J. Biomed. Mater. Res., Part A*, 2012, **100**, 3251.
191. L. Ye, X. Wu, Q. Mu, B. Chen, Y. Duan, X. Geng, Y. Gu, A. Zhang, J. Zhang and Z. G. Feng, *J. Biomater. Sci., Polym. Ed.*, 2010.
192. J. M. Coburn, M. Gibson, S. Monagle, Z. Patterson and J. H. Elisseeff, *Proc. Natl. Acad. Sci. U. S. A.*, 2012, **109**, 10012.
193. H. G. Sundararaghavan, R. B. Metter and J. A. Burdick, *Macromol. Biosci.*, 2010, **10**, 265.
194. R. J. Parkinson, C. P. Demers, J. G. Adel, E. I. Levy, E. Sauvageau, R. A. Hanel, A. Shaibani, L. R. Guterman, L. N. Hopkins, H. H. Batjer and B. R. Bendok, *Neurosurgery*, 2006, **59**, 812.

Biomaterials: Spatial Patterning of Biomolecule Presentation Using Biomaterial Culture Methods

KYLE A. KYBURZ[a], NAVAKANTH R. GANDAVARAPU[a], MALAR A. AZAGARSAMY[a,b], AND KRISTI S. ANSETH*[a,b]

[a]Department of Chemical and Biological Engineering and the BioFrontiers Institute, University of Colorado, Boulder, USA; [b]Howard Hughes Medical Institute, University of Colorado, Boulder, USA
*E-mail: kristi.anseth@colorado.edu

8.1 INTRODUCTION

Signaling molecules that affect cellular functions can be broadly classified as insoluble matrix cues, such as integrin-binding adhesion proteins, or soluble and diffusible cues (*e.g.*, cytokines and chemokines). Spatial presentation of these signaling cues can be important in directing critical biological processes in all stages of life, often promoting diverse functional needs over multiple length and time scales, which was discussed in detail in Chapter 2 by Domowicz *et al.* Many physiological processes that necessitate recruitment of cells and secretion of extracellular matrix (ECM) are largely driven by the formation of gradients in signaling cues that sensitize cells to respond and elicit specific responses, such as migration, differentiation or

Mimicking the Extracellular Matrix: The Intersection of Matrix Biology and Biomaterials
Edited by Gregory A. Hudalla and William L. Murphy
© The Royal Society of Chemistry 2020
Published by the Royal Society of Chemistry, www.rsc.org

proliferation. For example, sonic hedgehog (Shh)[1] and bone morphogenetic proteins (BMPs)[2] instruct the spatial segregation and migration of cells important for development of the neural tube[3] (Figure 8.1A); angiogenic factors, such as vascular endothelial growth factor (VEGF) and stromal derived factor-1 (SDF-1), direct the recruitment of several different cell types (*e.g.*, leukocytes, endothelial cells, stem cells) during wound healing[4–6] (Figure 8.1B); and spatial patterns of different types of ECM proteins, such as laminins, are important in directing intestinal epithelial tissue development.[7–9]

Cells can sense these matrix signals either by simply binding to them or by internalization through receptors present on their surface. These events then trigger the initiation of a complex cascade of intracellular signaling pathways that drive cellular responses, depending on the strength of the signal, and thus give rise to important positional information to cells. For example, gradients of multiple cytokines [*e.g.*, tumor necrosis factor alpha (TNF-α), CXCL-1] induce neutrophils to become polarized and form large lamellopodia in the direction of higher cytokine concentrations, and a thin bulbous uropod at the distal end, resulting in preferential migration towards

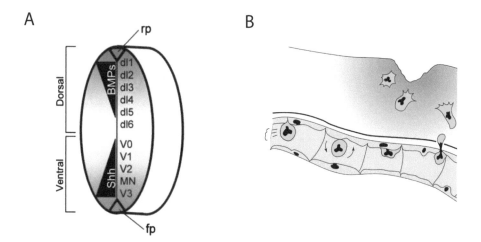

Figure 8.1 Physiological relevance of gradient presentation of bioactive cues in development and wound healing. (A) Gradient presentation of sonic hedgehog (Shh) and bone morphogenetic proteins (BMP) direct neuronal tube development and specify neural cell fate (dl1, dl2, V0, *etc.*... represent different classes of neurons, rp = roof plate, fp = floor plate). Image reprinted from Charron *et al.* with permission from Elsevier.[3] (B) Gradients of various growth factors released from wound sites direct migration of immune cells from the bloodstream and direct leukocytes to transmigrate to reach the site and participate in wound repair. Image reprinted from Martin *et al.* with permission from Elsevier.[6]

the site of infection.[4] With a growing interest in elucidation of these mechanisms and to satisfy the functional need for spatiotemporal patterning of signaling cues in various regenerative medicine and tissue-engineering applications, several biomaterial technologies have emerged in the past decade to control the microenvironment of cells.

Specifically, biomedical scientists and engineers have made great advancements in developing diverse materials and methods that enable spatial-gradient control over the presentation of signaling cues in bioscaffold systems for either 2D or 3D cell culture. Spatial control of presentation of gradients in biological signals using synthetic materials has the potential to bridge the gap between improving the field's understanding of the fundamentals of matricellular signaling and then using this knowledge to design intelligent and functional biomaterials for regenerating complex tissues and/or treating pathophysiological conditions. The increasing importance of patterning bimolecular signals in synthetic materials has resulted in numerous methods to achieve gradients *in vitro*. This chapter describes: (1) the general design principles that are often exploited to achieve spatial/gradient control over bimolecular presentation in synthetic scaffolds; and (2) current approaches and technologies that have been developed to allow such complex spatial presentation of signaling cues in synthetic biomaterial scaffolds.

8.2 DESIGN PRINCIPLES FOR CONTROLLING GRADIENTS IN SCAFFOLDS

Strategies to establish gradients of signaling molecules in synthetic biomaterials depend on the desired mode of presentation, *i.e.* either a diffusible/soluble form or an immobilized form. The choice of presentation typically depends on the final targeted application, as well as the mechanism of action of the biomolecules of interest on resident cells. This section provides a basic framework and discussion of various design principles that can used to establish gradients of signaling cues, while the subsequent section (Section 8.3) reviews the details of several experimental methods that have been developed for generating gradients in bioscaffold material systems.

8.2.1 Controlling Diffusive Transport of Signaling Cues

Diffusion is defined as spontaneous movement of molecules from a region of high concentration to an area of lower concentration. Typically, diffusing molecules travel from a source towards a sink region and reach steady state when the flux of the diffusing molecules through a unit area is constant (Figure 8.2A). However, in the presence of cells, signaling molecules can be continuously consumed, and this depletion can significantly affect the local concentration. After sufficient time, which depends on the diffusion co-efficient of the signaling cue and kinetics of consumption by the cells, this reaction and diffusion process reaches a steady state, resulting in a constant concentration gradient. The concentration gradient profile can be modeled by solving the diffusion governing equation, along with the relevant cellular consumption model. The steady-state concentration profile is given by:

$$D\frac{\partial^2 C}{\partial x^2} = r \tag{8.1}$$

where D is the diffusion coefficient of the delivery molecule, C is its concentration, and r is the rate of consumption by cells.

Since the steady-state concentration profile is dependent on the source and sink concentrations (*i.e.*, the boundary conditions), methods using this design approach need to maintain constant source and sink concentrations throughout the experimental period. The

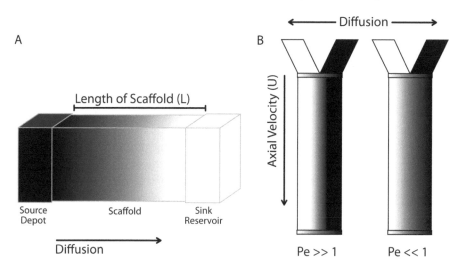

Figure 8.2 Design parameters to control diffusive and convective biomolecule presentation. (A) Diffusive biomolecule presentation involves a source and sink depot as biomolecules diffuse from volumes of high concentration to volumes of low concentrations and reach a steady state gradient across the scaffold. (B) Convective presentation is performed by flowing two laminar, miscible streams alongside one another. The gradient that develops between the streams is governed by the Peclet number (Pe) for the two streams.

time required to reach steady state is an important design parameter as cells are continuously exposed to the signaling cues during the transient period, which scales as L^2/D, where D is the molecular diffusivity of the diffusing species and L the diffusing distance. Also, the dimensionless Damkohler number (Da), defined as the ratio of the reaction rate to the diffusion rate, is desired to be $\ll 1$, in which case any change in the local concentration (due to consumption by cells) is quickly replenished by the diffusion of molecules from the higher concentration area, allowing a constant gradient to be maintained.

8.2.2 Controlling Convective Transport

Fluids flowing in the laminar flow regime [Reynolds number (Re) $\ll 1$] allow juxtaposing miscible streams to mix by diffusion in a direction perpendicular to the flow and without turbulent eddies (Figure 8.2B). This basic phenomenon is often exploited in microfluidic devices where it is used to form soluble gradient patterns by convective flow of multiple streams with varying concentrations of signaling cues side by side to present gradient profiles over cells. Gradient profiles generated with this mechanism depend on volumetric flow rates of the streams and the diffusion co-efficient of the signaling cues (eqn (8.2)). The governing equation for fluid flowing in the x-direction with diffusion in the y-direction, where U is the average axial velocity of the fluid:

$$U\frac{\partial C}{\partial x} = D\frac{\partial^2 C}{\partial y^2} \tag{8.2}$$

The Peclet number (Pe), a dimensionless number defined as the ratio of the advective transport rate, which depends on the velocity of the stream, to the diffusive transport rate, given by the diffusion co-efficient, describes the mixing of the flowing streams at the interface. Non-dimensionalizing eqn (8.2) yields a useful design equation (eqn (8.3)) involving the Pe and the concentration of the signaling cues:

$$u\frac{\partial C'}{\partial \xi} = \frac{1}{Pe}\frac{\partial^2 C'}{\partial \eta^2} \tag{8.3}$$

In such a system, a Pe $\gg 1$ results in low diffusive flux perpendicular to flow, while a Pe $\ll 1$ results in good diffusive mixing of the streams (Figure 8.2B). Hence, simply by controlling the Pe of the flows, a wide variety of gradient profiles with various slopes in these profiles can be generated. Note that the gradient is established at the onset of the flow, and cells can be continuously exposed to a constant gradient without any transition period, in contrast to the case of pure diffusive processes. This limits the exposure of cells to signaling cues at undesired concentrations or for prolonged times during the development of the desired steady-state gradient. Since this design phenomenon requires exquisite control of the flow of the fluids, methods to establish the gradients are generally based on microfluidic technologies that allow excellent control of the dynamics of the fluid flow, as well as the geometry of the flows.

8.2.3 Immobilization

Presentation of signaling factors in an immobilized form represents an orthogonal, complementary approach to delivery of soluble factors, where immobilization can allow persistent signaling and can minimize the quantity of signaling molecules needed (*e.g.*, no reservoir is required). In general, immobilization of signaling cues is achieved either through physical or covalent interactions with material matrices used for cell culture. Physical immobilization can be achieved by exploiting either hydrogen bonding or hydrophobic interactions, affinity interactions, physical conformation, or by charge-based interactions between the biomolecule and the scaffold. Further, depending on the strength of the interaction and the timescale of the cellular response, this immobilization can be sensed as reversible or irreversible. Alternatively, covalent immobilization can be achieved by various conjugation methods that react the molecules directly, and often times irreversibly, to the biomaterial scaffold, as discussed in Chapter 5 by Kilian. Spatial control over the presentation of the cues with this approach is typically achieved by controlling where the conjugation reactions occur (*e.g.*, with light) or by the spatially controlling the delivery of the biomacromolecule (*e.g.*, microinjection).

Physical immobilization methods can be achieved by mimicking those found in the native ECM to present signaling cues efficiently. For example, heparin sulfate (HS) is known to sequester stromal derived factor 1α (SDF-1), a known chemoattractant for directing stem cell migration, through positively charged lysine moieties on SDF-1 and the sulfated groups on HS. When SDF-1 is bound in this manner, it results in reduced enzymatic cleavage of SDF-1, increased localized concentrations, and the effective presentation of the SDF-1 active site to cells. Physical immobilization of biomolecules can occur either non-specifically, using molecules such as heparin that bind many proteins, or specifically, using a binding pair such as avidin–biotin. The non-specific interactions allow for a versatile platform, as many different proteins can be bound to the substrate; however, this introduces less control when multiple proteins are present. Other methods to physically immobilize biomolecules exploit conformational entrapment, such as molecular imprinting, where a synthetic receptor pocket is engineered within the material.[10,11]

Chemical immobilization necessitates mild reaction conditions to maintain bioactivity of the macromolecule and allow immobilization to be performed in the presence of cells and/or tissues. It is important when designing chemical immobilization techniques to ensure the target molecule retains its bioactivity and is not affected by either the necessary chemical modification or the conjugation to the scaffold. Additionally, minimizing steric hindrance is important, as cells must be able to appropriately bind and interact with the conjugated molecule to elicit the desired cell signaling. Various successful conjugation techniques exist in the literature and have been reviewed extensively by Azagarsamy *et al.* and Jabbari.[12,13] The development of bio-orthogonal "click" reactions, highly efficient reactions with a specificity that minimizes cross-reactions within a complex biological milieu, has greatly expanded the library of viable conjugation methods in recent years. These diverse methods act as handles to covalently bind bioactive molecules to biomaterial scaffolds at biologically relevant concentrations, and in some instances, with spatial and temporal control of the loading.

8.3 METHODS TO CONTROL SPATIAL DELIVERY OF SIGNALING CUES

Methods to achieve spatial presentation of signaling cues in synthetic biomaterials depend on numerous factors, including the mechanism of action, relevant timescales of signaling (*e.g.*, sustained, transient), spatial length scales over which the signaling occurs (*e.g.*, over a cell body *vs.* a group of cells), dosage, potency of the signaling molecule, and the choice of the biomaterial. The following section describes some of the common methods that have been developed and published in the literature to generate patterns of biomolecules to study and manipulate cellular functions *in vitro*.

8.3.1 Diffusion-Based Methods

8.3.1.1 Source and Sink Method. This approach represents one of the simplest experimental methods to establish gradients of signaling cues at the micrometer scale across a region of interest. Typically, the region across which a gradient is desired is sandwiched between a source and sink of the biological signal, relying on simple diffusion of the cue, and resulting in a steady-state linear profile across the region (Figure 8.2A). Liquid reservoirs containing dissolved signaling cues at known concentrations act as the "source" and "sink" compartments that are physically separated by the biomaterial. The signaling cue diffuses across the biomaterial or over the cell culture area, depending on the experimental set-up.[14–20] The need to maintain a steady-state gradient profile requires source and sink compartments to be replenished frequently, the use of very large volumes compared to the gradient volume within the biomaterial matrix, or the controlled release of the biomolecules from the source compartment.

This method has been used in many applications to observe the effect of biomolecules on cell function. Abhyankar *et al.* developed membrane-covered source and sink chambers allowing formation of a linear gradient of formyl-Met-Leu-Phe (fMLP) across a 3 mm-long straight channel separating the source and sink. This device was used to study fMLP dependent chemotaxis of neutrophils.[17] Further, by controlling the spatial position of source and sink chambers, Atencia *et al.* demonstrated creation of 2D gradient profiles of up to three different cues and studied the chemotaxis of bacteria to glucose gradients.[18] Using simple source and sink chambers, Cao and Shoichet *et al.* created gradients of neurotrophic factors [nerve growth factor (NGF), neurotrophin-3 (NT-3), brain-derived neurotrophic factor (BDNF)] and studied their role on neurite extension of embryonic lumbar dorsal root ganglion cells (DRGs).[20] A linear gradient of neurotrophic factors over 5 mm was achieved by maintaining large volumes of source and sink reservoirs, and the studies demonstrated that neurite extension towards the gradient was more dependent on the absolute concentration than the fractional concentration slope.[21]

While providing a relatively straightforward experimental system, the gradient profiles generated by the above methods are static and cannot be controlled dynamically *in situ* over the experimental period. To overcome this issue, experimental set-ups have evolved using static cell culture chambers sandwiched between source and sink manifolds with continuously flowing fluid (Figure 8.3B).[22] By controlling the concentration of signaling cues in the flowing fluid, the fluid pressure in the manifold, and the influx/outflux area (*i.e.*, area of contact between the gradient generation region and the manifolds), linear and non-linear gradients can be achieved over millimeter-wide cell culture regions in both two and

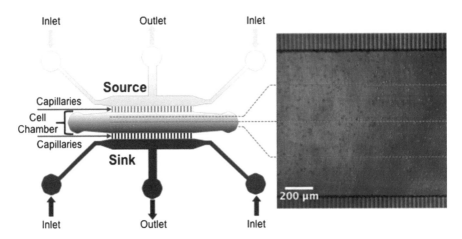

Figure 8.3   Shamloo *et al.* utilized a manifold design to present a linear gradient of Kit ligand to migrating murine bone-marrow-derived mast cells. Streams containing Kit ligand in media are introduced through the two source stream inputs and subsequently introduced to the cell chamber by flowing through the capillaries. Similarly, media lacking Kit ligand is introduced in a similar manner on the sink side of the cell chamber. The source and sink chambers are replenished by flow leaving from their respective outlets. Shown on the right is a phase-contrast image of the mast cells migrating in the cell chamber, where the capillaries are present on the top and bottom of the image. Reproduced from ref. 22 with permission from The Royal Society of Chemistry.

three dimensions.[14,15,23–25] Fluid flow in the manifolds is controlled using microfluidic techniques to allow *in situ* control over signaling cue concentration and fluid velocity. This method has been used to direct bone marrow mast cell migration through stable presentation of Kit ligand (Figure 8.3). The ability to directly visualize cell migration allowed for the elucidation of two concentration regimes that resulted in chemoattraction or chemorepulsion, as well as the effect of Kit ligand gradient exposure over time.[22] As another example, this method was used to study the role of VEGF gradients, known to be important during healing of vascular structures, on polarization and chemotaxis of human umbilical vein endothelial cells (HUVEC).[25] The results demonstrated that while overall VEGF concentration influenced the number of filopodia extensions per cell, gradients across the cell surface were necessary for HUVEC polarization and subsequent directed chemotaxis.

8.3.1.2   Controlled Release.   An alternative approach to establish gradients based on diffusion is to mimic natural morphogen signaling, in which a localized mass of cells secretes morphogens that diffuse to surrounding cells. In this method gradients are established by designing biomaterials with localized "depots" that can release a signaling cue in a sustained manner. Gradient profiles are tuned by controlling the spatial organization of the depots or by varying their release kinetics through the biomaterial properties. For example, Peret *et al.* encapsulated protein-loaded microspheres inside a poly(ethylene glycol) diacrylate (PEGDA) hydrogel that was laminated to a second hydrogel in which the gradient was desired.[26] The protein hydrodynamic radius—which affects the diffusion co-efficient—hydrogel microsphere concentration—which determines the total protein loading—and hydrogel mesh size were exploited to tune the slope, magnitude, persistence length and persistence time of the formed gradients. This method can be easily extended to achieve gradients of multiple signaling cues by using additional depots or through the introduction of multiple cues. Using a similar technique, Chen *et al.* encapsulated platelet-derived growth factor (PDGF)- and VEGF-loaded poly(lactide-*co*-glycolide) (PLG) microspheres inside a PLG scaffold and achieved local and spatially defined gradient profiles of these two growth factors.[27] Implantation of the microparticle-laden scaffolds in rat hind limbs showed significant blood vessel maturation in the presence of both gradients compared to VEGF alone. To increase spatial control of the soluble gradient and decrease long-range effects, Yuen *et al.* co-delivered VEGF and anti-VEGF antibodies in opposing gradients.[28] This approach shows how one can mimic natural *in vivo* processes to control activator (in this study VEGF) concentration by using opposing factors (in this study anti-VEGF antibodies), and demonstrated increased angiogenesis in mouse ischemic hind limb models through restriction of VEGF activity only to tissue regions in close proximity to the gel.

As a final example, a piston-based gradient generator was engineered by Wang *et al.* to suspend varying concentrations of growth factor loaded microspheres in silk fibroin scaffolds. They used this process to create continuous gradients of BMP-2 and IGF-1 inside silk scaffolds due to diffusion from the microspheres.[29] The total number of microspheres present at each location determines the local growth factor concentration. The efficacy of this delivery method was demonstrated by following the cellular response of human mesenchymal stem cells (hMSCs). In particular, hMSCs exhibited osteogenic or chondrogenic differentiation markers as a result of the gradient presentation of BMP-2 and opposing gradients of BMP-2 and IGF-1, but not in IGF-1 gradients alone.

8.3.2 Convective Flow-Based Approaches

Microfluidic technologies that combine the ability to fabricate devices with micrometer scale resolution of fluid flow have been used to generate gradients of various soluble signaling cues.[23,30-32] Flow properties of fluids operating in the laminar flow regime are dominated by viscous properties of the fluid as opposed to the inertial forces of the flow.[33] The laminar flow regime of fluids is well characterized and understood mathematically and allows prediction of mass flow characteristics of chemical species with great accuracy. In this regime, miscible fluid streams of varying concentrations of signaling cues are flowed side by side without any turbulent eddies. This property allows reliable approaches to form various gradient profiles simply by defining the concentration of the species in each of the flowing streams.

As an early example of this technology, Jeon *et al.* first reported the classic "Christmas tree" like microchannel network that allows diffusive mixing of streams to generate output streams with gradient concentrations of chemical species (Figure 8.4A).[34] In this method, a small number of input streams with specified concentrations are repeatedly split, mixed and positioned side by side to generate streams having different proportions of the specific input streams. The concentration of signaling cues in each of the output streams then allows generation of desired gradients. Since efficient diffusive mixing is desired in the microfluidic network, flow conditions are selected such that moderate Pe number and proper residence times are achieved.

Dertinger *et al.* extended this method by combining output streams from multiple "tree-like" networks to form a single stream and demonstrated generation of periodic sawtooth and parabolic gradient profiles within a single channel (Figure 8.4B).[35] The output streams were flowed through or over a biomaterial, resulting in gradient presentation of the stream signals to cells. The mixing of the fluid streams in the network can be modeled mathematically, allowing prediction of the gradient profiles in the output streams precisely, and hence appropriately program the design of the device to generate various concentration ranges and profile shapes.[35-38] This type of advanced network architecture has been used to generate gradients of cytokines to gain a fundamental understanding of biomolecule effects on a variety of biological mechanisms, including migration of metastatic tumor cells, polarization and chemotaxis of immune cells, proliferation and differentiation of neural stem cells and axon extension of neurons.[5,24,39-47] Using a pyramidal microchannel network, Chung *et al.* generated epidermal growth factor (EGF), fibroblast growth

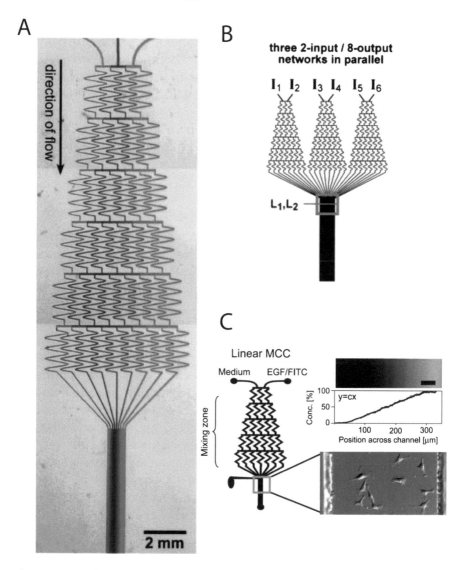

Figure 8.4 (A) "Christmas tree" like microfluidic mixer for generation of gradient streams. Three separate input streams of specific concentrations are increasingly mixed as they flow through the device to generate nine streams juxtaposed, resulting in a gradient across the desired region. Image reprinted from Dertinger *et al.*[35] Copyright 2001 American Chemical Society. (B) Multiple input/output configurations can be used to generate a variety of gradient profiles. Shown here, three 2-input and an 8-output microfluidic channel network is used to form gradients. Image reprinted from Dertinger *et al.*[35] Copyright 2001 American Chemical Society. (C) Schematic showing a linear microfluidic chemotactic chamber (MCC) to form epithelial growth factor (EGF) gradients across a channel where migrating breast cancer tumor cells are tracked. A corresponding image characterizes the percentage of EGF as a function of position across the channel. Fluorescein isothiocyanate (FITC) was introduced into the device with EGF for imaging and characterization purposes. Image reprinted from Wang *et al.*[5] with permission from Elsevier.

factor 2 (FGF-2) and PGDF gradients that were collectively presented to human neural stem cells. They subsequently studied the impact of their local presentation on cellular proliferation and astrocyte differentiation, which showed a clear linear dependence on growth factor concentration. In another study by Wang *et al.*, the effect of an EGF growth factor gradient on migration of breast cancer tumor cells, MDA-MB-231, was investigated.[5] Results demonstrated that linear gradients were not effective in inducing cell migration, while non-linear polynomial gradient had a strong influence.

8.3.3   Immobilization

Immobilization of signaling cues allows for generation of persistent gradients of biomolecules (*e.g.*, growth factors, cytokines and matrix adhesion molecules) in a spatiotemporally defined manner, often without the need for any special conditions/requirements after functionalization. Such strategies are particularly relevant to situations where there is a desire to present signaling cues that are bound to the extracellular environment and can interact with cells in a longer-term and continuous manner. For example, immobilization can mimic some physiological conditions, such as the role of various adhesive cues on haptotaxis, which is defined as migration of cells up a gradient of adhesive cues. Such processes have been studied in relation to metastatic tumor cells and immune cells. Immobilization has the practical benefit of using small amounts of physiologically relevant chemical cues, as the strategy does not require large source reservoirs often required for purely diffusion-based strategies, and additionally does not require continuous flow as with the convective methods (discussed in above sections). In general, growth factor immobilization is typically achieved either by covalently tethering (chemically) proteins to the matrix or by affinity binding to matrix functionalities through physical interactions.

8.3.3.1   Chemical Immobilization.   Several bioconjugation reactions, including thiol–ene, thiol–acrylate, acrylate–acrylate, thiol–maleimide, hydroxysuccinimide–amine, Michael addition, and azide–alkyne, have been used for modification of proteins and can be adopted to achieve chemical immobilization of numerous biochemical cues to bioscaffolds, as discussed in Chapter 5 by Kilian. By controlling the stoichiometry of the reactive functional groups, the extent of reaction, or by dictating the spatial position of the reactions, signaling cues can be conjugated to biomaterial scaffolds at user-defined regions to generate patterns and gradients. Several methods exploit the above chemical strategies through the use of gradient mixing, microfluidics (which were discussed in detail in Section 8.3.2), and photolithographic processes to covalently immobilize biological signals in a spatially defined manner that recapitulates aspects of the complex *in vivo* milieu.

Methods based on diffusion and convection principles have been adopted to establish soluble gradients of signaling cues and subsequently conjugate them into biomaterials, resulting in gradients dictated by their diffusion/convection currents. For example, Vepari and Kaplan formed linear gradients of horseradish peroxidase (HRP) by diffusion or forced convection along the length of silk scaffolds; the HRP was then covalently bound to the scaffold by 1-ethyl-3-(3-dimethylaminopropyl)carbodiimide (EDC) coupling between carboxylic acid and alcohol functionalities.[48] Similarly, gradient generators based on "tree-like" microfluidic network,[49,50] evaporation-induced back-mixing[51] and hydrodynamic stretching[52] have all been used to generate soluble gradients of signaling cues, which were reacted *in situ* into hydrogel networks, resulting in stable gradients of functionalized biomolecules.

Light-driven polymerization chemistries have been readily adopted to integrate signaling cues into biomaterial scaffolds. Such photochemical reactions allow control over the extent of conjugation and concentration of the signal by simply controlling the dose of photons delivered to a region of interest through regulation of the light intensity and/or exposure time. Using these methods, spatial resolution and fidelity is achieved using standard photolithographic techniques, such as photomasks and collimated light (Figure 8.5A), or *via* stereolithographic techniques, using confocal microscopy to raster a pulsed laser light at user defined regions in 3D space (Figure 8.5B). Collectively, these techniques confine the conjugation reactions to regions where the photons are delivered, rather than spatially controlling the delivery of the biological signal. Using multiphoton light provides excellent *z*-dimension spatial resolution by confining photo-conjugation reactions to highly defined focal volumes, because the absorption of two-photon light falls off to the sixth power away from focal volume as opposed to single-photon absorption, which scales to the third power. Hence, multiphoton photopatterning is often adopted where 3D control over the immobilization is required. Finally, since the reaction is triggered only in the presence of light, immobilization of signaling cues can be controlled temporally and in a user-defined manner to achieve conjugation at any time during cell culture.

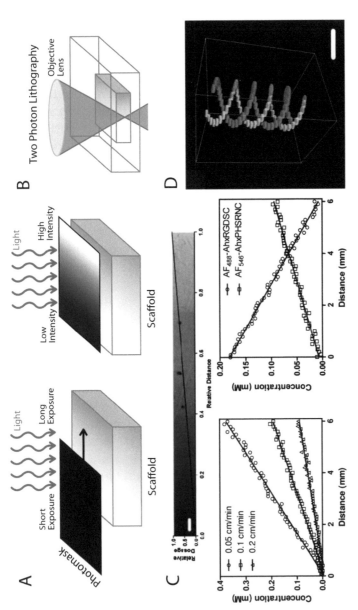

Figure 8.5 Spatial patterning of biomolecules. (A) Photomasks control the intensity and duration of light presented to the biomaterial volume, which spatially controls the conjugation reaction kinetics. The left diagram shows a moving photomask to control light exposure, and the right diagram shows a graded photomask that allows different intensities of light to pass through to the biomaterial volume. (B) Two-photon lithography controls light exposure to a specific three-dimensional volume in the biomaterial scaffold. This method allows for high spatial resolution in the z-axis that is lacking when using photomasks. (C) DeForest *et al.* took advantage of a moving photomask to control the intensity of light to conjugate of thiol-containing peptides to the excess enes found in the pre-formed hydrogel. It was observed that increasing the duration of light exposure increased the concentration of conjugated peptide. Further, this method was used to form two opposing gradients of thiol-containing, cell adhesive peptides over the length of the material. Images reprinted with permission from DeForest *et al.*,[60] copyright 2010 American Chemical Society. (D) Complimentary, DeForest *et al.* used two-photon photolithography and thiol-ene chemistry to spatially tether fluorescently labeled thiol-containing peptides in a helical structure showing high pattern fidelity. Scale bar = 200 μm. Image reprinted with permission from DeForest *et al.*[58]

Therefore, photopatterning approaches provide the opportunity to test intricate hypotheses of spatial and temporal biomolecule presentation on cell function.

A light-based technique used by Hahn *et al.* covalently tethered an acrylated arginine-glycine-aspartic acid-serine (RGDS) peptide, a fibronectin-derived cell adhesion motif, at user-defined regions *via* a radical-mediated reaction between acrylated peptides and excess acrylates present on partially crosslinked PEGDA hydrogels.[53] Similarly, Lee *et al.* synthesized PEGDA hydrogels with spatially patterned RGD by two-photon laser scanning photolithography and demonstrated control of HUVEC migration in response to spatial variations in adhesivity.[54] Hoffman *et al.* extended this method to show sequential immobilization of two distinct fibronectin peptides, RGDS and CS-1, at spatially defined regions in 3D.[55] While simple and versatile, the acrylate photopolymerization techniques can lead to final structures and microenvironments that are difficult to define (*e.g.*, complex final network structures and control of the extent of unreacted acrylate groups in the initial step). These methods also necessitate post-synthetic modification of the bioactive ligands (*e.g.*, acrylation of the peptide) to enable conjugation and photopatterning.

Recognizing the value of photochemical reactions, thiol–ene photopolymerization has recently emerged as a complementary click chemistry tool to synthesize hydrogel biomaterials in the presence of cells, as well as to functionalize materials with biochemical cues.[56] Fairbanks *et al.* synthesized step-growth PEG hydrogels functionalized with peptides, and demonstrated that by leaving excess enes as pendant functionalities (*e.g.*, using an off-stoichiometry reaction), a sequential patterning of thiolated biological signals could be achieved (*e.g.*, CRDS, thiolated proteins).[57] Since the method uses standard solid phase peptide synthesis to create cysteine-containing peptide sequences, the requirement of any post-synthetic modification of the bioactive moiety is eliminated.

Extending the concept, DeForest *et al.* demonstrated photopatterning of thiol-containing peptides and proteins using either photomasks (Figure 8.4C) or two-photon lithography (Figure 8.4D).[58–60] In this example, two orthogonal click chemistries were utilized in sequence: a strain-promoted azide–alkyne cycloaddition (SPAAC) to form the hydrogel and the thiol–ene conjugation to immobilize the biochemical functionalities with spatial control. The speed of the photochemical conjugation allowed tethering signals at physiologically relevant concentrations in seconds to minutes (*i.e.*, typically scanning speeds used for imaging cells). In general, the concentration of the signaling cue is linearly dependent on intensity of light, initiator concentration and time of exposure, allowing experimenters a high degree of control. This linear dependence was exploited to generate continuous gradients of a cell adhesive ligand with varying slopes by simply regulating the light dosage.

As one final example, DeForest *et al.* demonstrated covalent tethering and removal of signaling cues using two distinct wavelengths of light.[61] The thiol–ene reaction was used to conjugate biological signals at 405 nm and a photocleavable, nitrobenzyl ether group was used to remove the signal at 365 nm. Although not specifically demonstrated, one might envision how multiphoton control provides benefits for both introducing and removing signaling cues.

While *in situ* addition and removal of biochemical signals is important, other techniques have also emerged that borrow from principles used to cage molecules to study intracellular signaling. Such caged chemical moieties can be "uncaged" to expose important functional groups, rendering dormant signals active. As an indirect example of uncaging, Shoichet and co-workers used such methods to create RGD peptide channels at specific locations in agarose hydrogels to direct neurite outgrowth and migration of rat DRG cells.[20,21,62] Specifically, thiols were uncaged with light to create spatial patterns that were subsequently reacted with maleimide functionalized RGDS peptides.[21] Such two-step approaches avoid exposing sensitive biomolecules to light. More recently, Aizawa *et al.* used a photo-labile coumarin (which has different spectral characteristics from nitrobenzyl) to cage sulfides that were subsequently exposed using two-photo lithography.[62] The sulfides were exposed in a spatially defined manner and reacted with maleimide functionalized VEGF to study how endothelial cells can maintain retinal stem and progenitor cells (RSPCs) in a quiescent phenotype by inhibiting their proliferation and differentiation.

8.3.3.2 Physical Immobilization. While chemical immobilization serves as a powerful tool to immobilize bioactive signals at specific locations within a biomaterial scaffold, most of these methods involve chemical modification of the immobilizing unit and oftentimes radical initiation or light exposure (*e.g.*, during photo-patterning). While these methods can be quite versatile for immobilizing peptide signals, they can lead to aggregation, denaturation, or loss of activity of full-length proteins. Because of these complications, careful control experiments must be performed to test the activity of the ligand, its accessibility for binding, and ability to signal cells when immobilized on biomaterial scaffolds. Thus, to provide alternative methods for presentation of these signals, methods based on principles of affinity binding (*i.e.*, non-covalent interactions between proteins and pendant scaffold functionalities) have been pursued. From a basic perspective, these non-covalent interactions can be either non-specific, such as electrostatics and/or van der Waals interactions, or specific, such as ligand–receptor (*e.g.*, antigen–antibody) interactions.

8.3.3.2.1 Non-Specific: Electrostatic Interactions. While non-specific, electrostatic interactions of proteins to charged surfaces is highly ubiquitous in biological systems, regulating such interactions in a spatially defined or gradient manner requires that the cell–material interface be intricately controlled. While non-specific interactions can be effective in sequestering a myriad of cues, the ability to sequester specific biomolecules is lacking. For example, Lee *et al.* gradiently absorbed heparin, a sulfated glycosaminoglycan (GAG), to a poly-caprolactone/pluronic F127 (PCL/PL-F127) scaffold and studied sequestration of three different heparin-binding growth factors, BMP-7, transforming growth factor beta 2 (TGF-β_2), and VEGF.[63]

Using microfluidics to create charged surfaces, Dertinger *et al.* exploited "tree-like" gradient generators to create fluid streams with varying concentrations of laminin that physically adsorbed onto negatively charged poly(lysine) substrates.[64] The influence of the laminin gradient on rat hippocampal neurons was studied, especially how the protein affected outgrowth and axonal specification. Beyond a critical threshold in the laminin signal, axon specification was dictated towards increasing laminin concentrations.

8.3.3.2.2 Specific: Receptor–Ligand-Mediated Interactions. In addition to non-specific interactions of proteins with bioscaffolds and surfaces used for cell culture, *specific* ligand–receptor interactions have also been developed to immobilize and present signaling molecules. Since most ligand–receptor binding interactions are orthogonal to each other, Lutolf *et al.* demonstrated the immobilization of two different protein molecules site-specifically (Figure 8.6A).[65,66] Avidin–biotin and proteinA–Fc were chosen as the two orthogonal receptor–ligand binding pairs, and BSA–FITC functionalized with biotin and Alexa546 functionalized with Fc were chosen as model proteins. Two receptors: avidin and protein A were first covalently incorporated throughout a hydrogel matrix randomly. The protein gradients were then generated by using microfluidic channels to control the local delivery of BSA–FITC–biotin and Alexa546–Fc. For example, to create overlapping opposing gradients, BSA–FITC–biotin and AF–546–Fc were passed sequentially through opposing inlets into a gel, but to achieve overlapping parallel gradients, the two protein

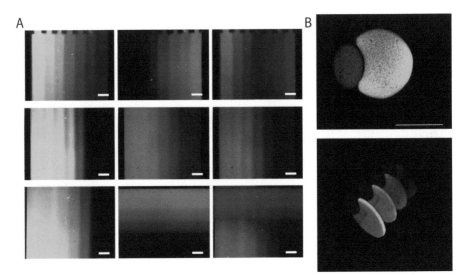

Figure 8.6 Physical immobilization to spatially pattern biomolecules. (A) Cosson *et al.* took advantage of physical immobilization and microfluidic techniques to form overlapping molecule gradients with directional control. Shown here, fluorescence microscopy images of immobilized BSA–FITC–biotin and AF–546–Fc on PEG hydrogel in opposing gradients (top three images), overlapping gradients (middle three images) and in orthogonal gradients (bottom three images). Scale bar = 100 µm. Image reprinted with permission from Cosson *et al.*[66] (B) Wylie *et al.* exploited specific physical interactions and two-photon photolithography to spatially pattern sonic hedgehog–barstar and ciliary neurotrophic factor–biotin at increasing depths within the hydrogel. The top image is a confocal micrograph showing the patterning at 400 µm, and the lower image shows patterning at the 400, 500, 600, and 700 µm depths. Scale bar = 100 µm. Adapted by permission from Macmillan Publishers Ltd: Wylie *et al.*[67]

solutions were passed sequentially through same inlet. Imaging of the fluorophore-tagged proteins was used to estimate the fidelity of the resultant gradients.

To avoid the use of microfluidic channels, Wylie *et al.* used photochemical reactions to create multiple functional groups in spatially defined patterns.[67] The patterning then dictated conjugation of growth factor partners from a complex solution, allowing one-step protein conjugation method with cell-laden scaffolds (Figure 8.6B). In their approach, the avidin–biotin binding pair and "barnase–barstar" pair (Kd) were exploited to sequester ciliary neurotrophic factor (CNTF) and sonic hedgehog (SHH), respectively. The representative spatially well-defined 3D images for the immobilized fluorescently labeled SHH–barstar and CNTF–biotin are shown in Figure 8.6B.

8.4 CONCLUDING REMARKS

Cells receive a complex milieu of important signals from their extracellular microenvironment, which can dynamically change in both space and time. Biomaterial scaffolds that allow one to study or manipulate aspects of these extracellular signals are providing opportunities to answer basic biological questions about cell-matrix signaling and strategies to manipulate cell function for applications in drug delivery and tissue regeneration. By combining the fields of microfluidics, polymer chemistry and engineering, and drug delivery, innovative and impactful systems are being engineered that allow for dynamic user-mediated control over biomolecule presentation in cell culture systems.

This chapter outlines many of the current methods in patterning biomaterials and gives examples of their application to study and direct cell function. Clearly, opportunities abound to use these techniques and materials to answer more advanced cell biology questions or provide improved *in vitro* models of *in vivo* processes, such as wound healing and development. Advanced imaging methods combined with real-time cell tracking are now beginning to merge with the bioscaffold strategies presented in this chapter. Collectively, integration of these methodologies will provide unprecedented information about how cells receive, process, and exchange signals with their microenvironment, which should ultimately improve strategies for the design of biomaterial carriers for cell-based therapies or delivery of biomolecules to treat disease.

REFERENCES

1. I. Patten and M. Placzek, *Cell. Mol. Life Sci.*, 2000, **57**, 1695–1708.
2. D. S. Eom, S. Amarnath, J. L. Fogel and S. Agarwala, *Development*, 2011, **138**, 3179–3188.
3. F. Charron, E. Stein, J. Jeong, A. P. McMahon and M. Tessier-Lavigne, *Cell*, 2003, **113**, 11–23.
4. P. Friedl and B. Weigelin, *Nat. Immunol.*, 2008, **9**, 960–969.
5. S. J. Wang, W. Saadi, F. Lin, C. Minh-Canh Nguyen and N. Li Jeon, *Exp. Cell Res.*, 2004, **300**, 180–189.
6. P. Martin and S. J. Leibovich, *Trends Cell Biol.*, 2005, **15**, 599–607.
7. A. L. Bolcato-Bellemin, O. Lefebvre, C. Arnold, L. Sorokin, J. H. Miner, M. Kedinger and P. Simon-Assmann, *Dev. Biol.*, 2003, **260**, 376–390.
8. I. C. Teller and J. F. Beaulieu, *Expert Rev. Mol. Med.*, 2001, **3**, 1–16.

9. P. Simo, P. Simon-Assmann, F. Bouziges, C. Leberquier, M. Kedinger, P. Ekblom and L. Sorokin, *Development*, 1991, **112**, 477–487.

10. M. Watanabe, T. Akahoshi, Y. Tabata and D. Nakayama, *J. Am. Chem. Soc.*, 1998, **120**, 5577–5578.

11. M. K. Nguyen and E. Alsberg, *Prog. Polym. Sci.*, 2014, **39**, 1235–1265.

12. M. A. Azagarsamy and K. S. Anseth, *ACS Macro Lett.*, 2012, **2**, 5–9.

13. E. Jabbari, *Curr. Opin. Biotechnol.*, 2011, **22**, 655–660.

14. W. Saadi, S. W. Rhee, F. Lin, B. Vahidi, B. G. Chung and N. L. Jeon, *Biomed. Microdevices*, 2007, **9**, 627–635.

15. B. Mosadegh, C. Huang, J. W. Park, H. S. Shin, B. G. Chung, S. K. Hwang, K. H. Lee, H. J. Kim, J. Brody and N. L. Jeon, *Langmuir*, 2007, **23**, 10910–10912.

16. U. Haessler, Y. Kalinin, M. A. Swartz and M. Wu, *Biomed. Microdevices*, 2009, **11**, 827–835.

17. V. V. Abhyankar, M. A. Lokuta, A. Huttenlocher and D. J. Beebe, *Lab Chip*, 2006, **6**, 389–393.

18. J. Atencia, J. Morrow and L. E. Locascio, *Lab Chip*, 2009, **9**, 2707–2714.

19. H. Wu, B. Huang and R. N. Zare, *J. Am. Chem. Soc.*, 2006, **128**, 4194–4195.

20. X. Cao and M. S. Shoichet, *Neuroscience*, 2003, **122**, 381–389.

21. T. A. Kapur and M. S. Shoichet, *J. Biomed. Mater. Res., Part A*, 2004, **68**, 235–243.

22. A. Shamloo, M. Manchandia, M. Ferreira, M. Mani, C. Nguyen, T. Jahn, K. Weinberg and S. Heilshorn, *Integr. Biol.*, 2013, **5**, 1076–1085.

23. T. M. Keenan and A. Folch, *Lab Chip*, 2008, **8**, 34–57.

24. B. G. Chung, F. Lin and N. L. Jeon, *Lab Chip*, 2006, **6**, 764–768.

25. I. Barkefors, S. Le Jan, L. Jakobsson, E. Hejll, G. Carlson, H. Johansson, J. Jarvius, J. W. Park, N. Li Jeon and J. Kreuger, *J. Biol. Chem.*, 2008, **283**, 13905–13912.

26. B. J. Peret and W. L. Murphy, *Adv. Funct. Mater.*, 2008, **18**, 3410–3417.

27. R. R. Chen, E. A. Silva, W. W. Yuen and D. J. Mooney, *Pharm. Res.*, 2007, **24**, 258–264.

28. W. W. Yuen, N. R. Du, C. H. Chan, E. A. Silva and D. J. Mooney, *Proc. Natl. Acad. Sci. U. S. A.*, 2010, **107**, 17933–17938.

29. X. Wang, E. Wenk, X. Zhang, L. Meinel, G. Vunjak-Novakovic and D. L. Kaplan, *J. Controlled Release*, 2009, **134**, 81–90.

30. P. Gravesen, J. Branebjerg and O. S. Jensen, *J. Micromech. Microeng.*, 1993, **3**, 168.

31. I. Meyvantsson and D. J. Beebe, *Annu. Rev. Anal. Chem.*, 2008, **1**, 423–449.
32. N. K. Inamdar and J. T. Borenstein, *Curr. Opin. Biotechnol.*, 2011, **22**, 681–689.
33. T. M. Squires and S. R. Quake, *Rev. Mod. Phys.*, 2005, **77**, 977–1026.
34. N. L. Jeon, S. K. W. Dertinger, D. T. Chiu, I. S. Choi, A. D. Stroock and G. M. Whitesides, *Langmuir*, 2000, **16**, 8311–8316.
35. S. K. W. Dertinger, D. T. Chiu, N. L. Jeon and G. M. Whitesides, *Anal. Chem.*, 2001, **73**, 1240–1246.
36. J. Sager, M. Young and D. Stefanovic, *Langmuir*, 2006, **22**, 4452–4455.
37. D. Irimia, D. A. Geba and M. Toner, *Anal. Chem.*, 2006, **78**, 3472–3477.
38. W. Yi, M. Tamal and L. Qiao, *J. Micromech. Microeng.*, 2006, **16**, 2128.
39. B. G. Chung, L. A. Flanagan, S. W. Rhee, P. H. Schwartz, A. P. Lee, E. S. Monuki and N. L. Jeon, *Lab Chip*, 2005, **5**, 401–406.
40. J. Y. Park, S. K. Kim, D. H. Woo, E. J. Lee, J. H. Kim and S. H. Lee, *Stem Cells*, 2009, **27**, 2646–2654.
41. J. Y. Park, C. M. Hwang, S. H. Lee and S. H. Lee, *Lab Chip*, 2007, **7**, 1673–1680.
42. O. C. Amadi, M. L. Steinhauser, Y. Nishi, S. Chung, R. D. Kamm, A. P. McMahon and R. T. Lee, *Biomed. Microdevices*, 2010, **12**, 1027–1041.
43. J. Mai, L. Fok, H. Gao, X. Zhang and M. M. Poo, *J. Neurosci.*, 2009, **29**, 7450–7458.
44. J. Lii, W. J. Hsu, H. Parsa, A. Das, R. Rouse and S. K. Sia, *Anal. Chem.*, 2008, **80**, 3640–3647.
45. N. Li Jeon, H. Baskaran, S. K. Dertinger, G. M. Whitesides, L. Van de Water and M. Toner, *Nat. Biotechnol.*, 2002, **20**, 826–830.
46. W. Saadi, S. J. Wang, F. Lin and N. L. Jeon, *Biomed. Microdevices*, 2006, **8**, 109–118.
47. A. M. Taylor, M. Blurton-Jones, S. W. Rhee, D. H. Cribbs, C. W. Cotman and N. L. Jeon, *Nat. Methods*, 2005, **2**, 599–605.
48. C. P. Vepari and D. L. Kaplan, *Biotechnol. Bioeng.*, 2006, **93**, 1130–1137.
49. J. A. Burdick, A. Khademhosseini and R. Langer, *Langmuir*, 2004, **20**, 5153–5156.
50. Z. Liu, L. Xiao, B. Xu, Y. Zhang, A. F. Mak, Y. Li, W.-Y. Man and M. Yang, *Biomicrofluidics*, 2012, **6**, 024111.
51. J. He, Y. Du, J. L. Villa-Uribe, C. Hwang, D. Li and A. Khademhosseini, *Adv. Funct. Mater.*, 2010, **20**, 131–137.

52. Y. Du, M. J. Hancock, J. He, J. L. Villa-Uribe, B. Wang, D. M. Cropek and A. Khademhosseini, *Biomaterials*, 2010, **31**, 2686–2694.

53. M. S. Hahn, L. Taite, J. J. Moon, M. C. Rowland, K. A. Ruffino and J. L. West, *Biomaterials*, 2006, **27**, 2519–2524.

54. S. H. Lee, J. J. Moon and J. L. West, *Biomaterials*, 2008, **29**, 2962–2968.

55. J. C. Hoffmann and J. L. West, *Soft Matter*, 2010, **6**, 5056–5063.

56. C. E. Hoyle and C. N. Bowman, *Angew. Chem.*, 2010, **49**, 1540–1573.

57. B. D. Fairbanks, M. P. Schwartz, A. E. Halevi, C. R. Nuttelman, C. N. Bowman and K. S. Anseth, *Adv. Mater.*, 2009, **21**, 5005–5010.

58. C. A. DeForest and K. S. Anseth, *Angew. Chem., Int. Ed.*, 2012, **51**, 1816–1819.

59. C. A. DeForest, B. D. Polizzotti and K. S. Anseth, *Nat. Mater.*, 2009, **8**, 659–664.

60. C. A. DeForest, E. A. Sims and K. S. Anseth, *Chem. Mater.*, 2010, **22**, 4783–4790.

61. C. A. DeForest and K. S. Anseth, *Nat. Chem.*, 2011, **3**, 925–931.

62. Y. Aizawa and M. S. Shoichet, *Biomaterials*, 2012, **33**, 5198–5205.

63. S. H. Oh, T. H. Kim and J. H. Lee, *Biomaterials*, 2011, **32**, 8254–8260.

64. S. K. Dertinger, X. Jiang, Z. Li, V. N. Murthy and G. M. Whitesides, *Proc. Natl. Acad. Sci. U. S. A.*, 2002, **99**, 12542–12547.

65. S. Allazetta, S. Cosson and M. P. Lutolf, *Chem. Commun.*, 2011, **47**, 191–193.

66. S. Cosson, S. A. Kobel and M. P. Lutolf, *Adv. Funct. Mater.*, 2009, **19**, 3411–3419.

67. R. G. Wylie, S. Ahsan, Y. Aizawa, K. L. Maxwell, C. M. Morshead and M. S. Shoichet, *Nat. Mater.*, 2011, **10**, 799–806.

C. ECM Assembly, Organization, and Dynamics

CHAPTER 9

Biomaterials: Controlling Properties Over Time to Mimic the Dynamic Extracellular Matrix

LISA SAWICK[a] AND APRIL KLOXIN*[a]

[a]University of Delaware, Newark, USA
*E-mail: akloxin@udel.edu

9.1 INTRODUCTION

Materials whose properties can be controlled in time have been utilized in numerous biomedical applications for the past several decades, from controlled drug delivery[1,2] to tissue engineering and regenerative medicine.[3] For example, seminal works have demonstrated that hydrolytically degradable moieties, such as anhydrides[4] or esters,[5] can be incorporated within the backbone of biomaterials and, upon cleavage and subsequent material erosion, release embedded or encapsulated drugs at a pre-programmed rate.[6,7] These functionalities subsequently were integrated within polymers and polymer networks for the support or delivery of mammalian cells in the replacement or regeneration of tissues.[8,9] More recently, additional classes of degradable units have been utilized, including those responsive to enzymes[10,11] and light,[12] to tailor the

Mimicking the Extracellular Matrix: The Intersection of Matrix Biology and Biomaterials
Edited by Gregory A. Hudalla and William L. Murphy
© The Royal Society of Chemistry 2020
Published by the Royal Society of Chemistry, www.rsc.org

temporal properties of these biomaterials *in situ* for specific applications. Upon this backdrop, our knowledge has grown rapidly of how the biophysical and biochemical properties of a cell's microenvironment, particularly the extracellular matrix (ECM), influence cellular functions and fate. Degradable biomaterials have emerged as scaffolds to mimic the dynamic ECM not only for tissue regeneration but also for asking biological questions within an environment closer to that of the native ECM (*e.g.*, three-dimensional, soft, or responsive).[13–16]

The growing area of research related to designing and utilizing soft biomaterials with temporally controlled properties as ECM mimics is the focus of this chapter and lies at the interface between biomaterials science and ECM biology. The native ECM is complex and dynamic, presenting biochemical and biophysical cues at specific locations and times that drive cellular functions and subsequently is remodeled by these resident cells (*i.e.*, dynamic reciprocity).[17,18] Understanding this complex interplay between cells and their microenvironment, the ECM, is essential in designing new therapeutic approaches to mitigate disease progression or direct tissue regeneration. Biomaterials whose properties can be controlled in time are being utilized as tools to probe the role of microenvironment signals in cell fate and to direct cell behavior. These materials need to mimic key aspects of the ECM, and engineered soft biomaterials with programmed or tunable temporal property control are well suited for this.

In this chapter, we will provide an overview of how biomaterial properties can be controlled over time to mimic changes in the ECM and subsequently summarize the material chemistries and methods that can be utilized to enable and measure these temporal changes in biophysical or biochemical material properties. Additionally, select examples will be presented of how biomaterials with temporally controlled properties are being applied as synthetic ECM mimics to probe and direct cell function and fate in numerous applications.

We primarily focus on approaches to control biomaterial properties *in vitro* for their use as ECM mimics. In particular, we will concentrate our discussions on hydrogel-based biomaterials as soft tissue mimics owing to their appropriate moduli and the ease of engineering their properties for control over time. However, the chemistries and methods presented can easily be applied to other biomaterial systems, such as electrospun scaffolds.

9.2 TEMPORAL CHANGES IN THE ECM AND BIOMATERIAL-BASED APPROACHES TO MIMICRY

In this section, we present a description and comparison of key properties that change over time in native tissues and in synthetic biomaterials towards mimicking the native ECM. Specifically, changes in biophysical properties, including structure and stiffening, and biochemical properties, including soluble and insoluble cues, are described. We will use the example of wound healing to give context to the changes in these properties that occur in the native ECM and how they can be modulated within synthetic mimics.

9.2.1 Biophysical Properties

9.2.1.1 Structure. ECM structure plays a critical role in providing support for cells within their native tissue. Proteins create the nano- to micro-structure that surrounds cells and their organization directs the macroscopic properties of the ECM. The ability to mimic temporal changes in the nano- to micro-structure of biomaterials to alter local or macroscopic properties is an important task in engineering controlled environments to understand cell response to structural changes.

9.2.1.1.1 Native. The structure of the native ECM is tissue specific and differs based on the type and concentration of structural proteins that define the tissue of interest. Collagen and elastin are the most prevalent structural proteins in the native ECM, providing support or flexibility. Collagen has a triple-helical conformation between three polypeptide strands that are held together by hydrogen bonds. These triple-helices self-assemble to form larger fibers that make up much of the strength and structure of the natural ECM.[19] Elastin is composed of two segments along its polypeptide chain—a hydrophobic region and an alpha-helical region—rich in lysine and alanine that allow covalent cross-linking between elastin polypeptides. This structure overall is coil-like and gives elastin its ability to stretch and flex to a high degree.[20]

A number of other structural proteins are present in native ECMs, including fibronectin, laminin, and vitronectin. Fibronectin, a glycoprotein commonly found within connective tissue (*e.g.*, bone marrow), is associated with cell adhesion to the ECM. It interacts with other proteins, cells, and with itself, assembling into fibrils that form linear and branched mesh structures around cells.[21] Laminin, found in the basal lamina, a structure within the basement membrane of tissues, promotes cellular functions including adhesion, migration, and differentiation. The basement membrane is a thin layer of the ECM that underlies epithelial and endothelial cells, and its structure is partly attributed to the assembly of laminins and their interactions with other ECM proteins.[22] Vitronectin, another fibrillar cell adhesion protein, also promotes spreading of cells within the ECM. This characteristic is partly responsible for vitronectin's role in promoting tumor cell survival and invasion in tissue as cancer progresses.[23]

Native ECM structure is not static with time as cells naturally secrete enzymes and new structural proteins that degrade and rebuild tissues, respectively. Wound healing and disease result in ECM remodeling as cells respond to disruptions in their surrounding environment. For example, in the case of healing a skin wound (Figure 9.1B), a series of steps occur that cause remodeling of the native ECM. After wounding, the formation of a blood clot provides a provisional or temporary matrix that stimulates the migration and proliferation of fibroblasts in the wound site.[24] Enzymes secreted by cells in response to the wound degrade existing structural proteins and fibroblasts deposit new collagen to rebuild the ECM. The new collagen at the wound site is aligned randomly and is denser than the original tissue, resulting in the different structural properties of scar tissue.[25,26] While this is a simplified example of wound healing, understanding the mechanism through which structural remodeling occurs over time is crucial to designing new materials that mimic biological processes, such as wound healing for directed tissue regeneration.

9.2.1.1.2 Biomaterial-Based Mimicry. The ability to mimic the structure of the native ECM with synthetic biomaterials has been an area of growing interest among researchers. Hydrogels, widely studied soft biomimetic materials, have been used in a range of applications from tissue engineering to disease models. Hydrogels to mimic the ECM specifically have been formed by covalently crosslinking bioinert synthetic polymers, such as poly(ethylene glycol) (PEG)[27–30] and poly(vinyl alcohol) (PVA),[8,31] or functionalized polysaccharides, such as hyaluronic acid (HA)[32,33] and alginate,[34] and incorporating proteins or related peptides to induce bioactivity.[2] Careful selection of chemical functionality that allows the dynamic tuning of biophysical properties in these covalently crosslinked hydrogels will be described in Section 9.3.1. Self-assembly mechanisms also have allowed the formation of structures found within the native ECM in synthetic hydrogels. For example, collagen-mimetic peptide strands

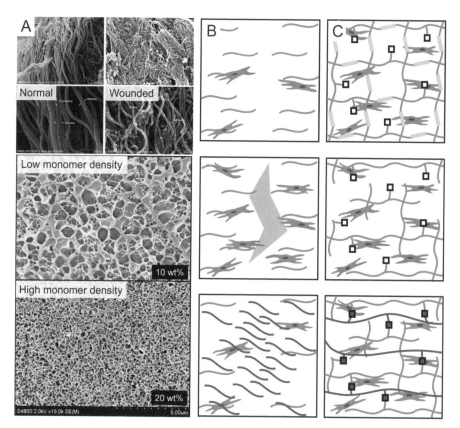

Figure 9.1 Biophysical properties of the ECM. (A) Scanning electron microscope images of collagen fibril size and alignment in normal and wounded skin samples. Collagen fibers align in healthy skin, whereas the collagen fibrils in wounded and repaired skin are randomly aligned and more crosslinked, resulting in an increased modulus. Reproduced with permission from ref. 26. © 2011 Elsevier. Biomaterials can mimic these properties with increasing monomer density for an increased modulus more characteristic of scar tissue. Reproduced with permission from ref. 39. © 2012 Elsevier. (B) Schematic of the native ECM during wound healing. After injury, cells are recruited to the wound site. Existing ECM proteins are degraded by cell-secreted factors and new proteins are deposited to rebuild the tissue. (C) Schematic of a dynamic ECM mimic. Cells cultured inside the material degrade crosslinks that mimic native ECM proteins. At a user-controlled time point, new structural monomer is added to the matrix to mimic rebuilding of the wound site.

that self-assemble through physical crosslinks form the triple-helical fibers seen in the native ECM are being used to make novel collagen-based hydrogels.[35–38] Furthermore, densities of the matrix components within these synthetic materials may be altered to affect other ECM properties related to structure,[39] such as matrix modulus or 'stiffness'. In this chapter we thus focus on hydrogels as soft biomaterials whose properties can be temporally controlled to mimic the dynamic native ECM.

While this chapter will focus on hydrogel-based biomaterials as soft, synthetic ECMs, we would like briefly to mention electrospun polymer scaffolds, as these biomaterials are widely used for tissue engineering applications as structural mimics of the ECM and offer some temporal property control depending on base polymer selection [*e.g.*, non-degradable poly(ethylene oxide) (PEO) *vs.* degradable poly(D,L-lactide-*co*-glycolide) (PLGA) or poly(caprolactone) (PCL)]. The electrospinning process creates a nanofibrous structure with fiber sizes that mimic native ECM protein size and porosity. Alignment of the fibers can maintain phenotype and direct cell fate, and the incorporation of chemical functionality and degradable polymers can allow their degradation as cells enter the synthetic material.[40] In principle, many of the chemistries discussed in this chapter are also applicable to these structural ECM mimics.

Capturing the components and properties of natural structure as they change in time with a synthetic biomaterial can provide an even more accurate model that mimics the dynamic nature of the ECM. For example, to mimic wound healing (Figure 9.1C), peptides cleaved by cell-secreted enzymes can be incorporated into the biomaterial matrix to mimic the degradation of the initial protein composition in the native ECM. Secondary proteins or peptides can be added at a later time that bind to or bond with reactive chemistries present in the biomaterial to mimic cellular remodeling of the matrix. Pre-programmed matrix degradation can be achieved with the incorporation of hydrolytic chemistries that degrade under cell culture conditions; however, greater control over matrix degradation in response to cells can be achieved through the enzymatically degradable peptides described. The concentrations of enzymatically degradable peptides present in the initial matrix and secondary peptides added to the material can be varied so that the final structure of the biomaterial mimics the remodeled ECM. For spatiotemporal control of these processes, the incorporation of photolabile chemistries has allowed *in situ* topographical patterning *via* surface erosion and bulk degradation for biophysical cue tuning of ECM mimics.

9.2.1.2 Stiffness. Stiffness describes how rigid a material is in response to an external force and varies for tissues in the body depending on their function. For example, bone is required to be stiff so that it may support body weight, while skin tissue is more flexible so that it can stretch as the body moves. While stiffness is a defined a material property (k), the term 'stiffness' is often used as a descriptor of how stiff or soft a material is and modulus (E, Young's modulus; G, shear modulus) is used as an indirect measure of this, where:

$$G = \rho_x RTQ^{-1/3} \tag{9.1}$$

and ρ_x is the crosslink density, R is the ideal gas constant, T is the temperature, and Q is the volumetric swelling ratio of the material. Furthermore, the structure of the ECM defines tissue stiffness and is based on the protein composition, arrangement, and density.

9.2.1.2.1 Native. Native tissues have a range of stiffnesses depending on their function in the body. Modulus, which can easily be measured experimentally, has been used to describe a wide range of native tissue mechanical properties. Table 9.1 provides the elastic moduli of various native tissues.[41–45] Tissues such as bone exhibit much higher moduli (E ~ 5.4–22 GPa), which denote a much stiffer material, while tissues such as bone marrow have lower moduli (~600 Pa), indicating softer, less rigid mechanical properties.

Studies on the moduli of tissues over time have also been conducted to understand processes such as aging, healing, and disease. The ECM naturally stiffens with age, which is attributed to an increase in random crosslinking between collagen fibers in the tissue.[46] This causes tissues such as bone to become less flexible and thus more susceptible to fracture, while tissue such as skin becomes tough and more difficult to stretch. In the case of wound healing, the deposition of new proteins at a wound site leads to an increasingly fibrotic environment. The increase in density of protein fibers leads to a more densely packed matrix with higher modulus than uninjured tissue. Similar behavior is also observed in diseases such as pulmonary fibrosis, where the lungs stiffen as they are in a continuous wound healing state.[47]

Table 9.1 Elastic moduli of various native tissues. The techniques used to measure the modulus of each tissue are listed.

Tissue type	Measurement	Elastic modulus
Bone	Tensile, micro-bending, indentation	5.4–22 GPa [41]
Bone marrow	Shear	~600 Pa [42]
Brain	Indentation	260–490 Pa [43]
Cartilage	Compression	450–800 kPa [44]
Liver	Compression	640 Pa [43]
Lungs	Tension	5–6 kPa [43]
Skin	Shear	420–850 kPa [45]

9.2.1.2.2 Biomaterial-Based Mimicry. The ability to mimic tissue stiffness and dynamic changes in it is an important topic in designing novel biomaterials. Cells respond to substrate stiffness and thus the ability to capture changes in stiffness with synthetic materials is important in studying disease and processes in the body such as wound healing.[48] Initial material modulus can be tailored by controlling crosslink density, with an increased crosslink density corresponding to an increased modulus or stiffer material (eqn 9.1). From this initial material's modulus, there are several ways in which biomaterial stiffness can be altered over time and different size scales on which these properties can be controlled.

To mimic a decrease in stiffness, various degradable chemistries that respond to conditions such as light, pH, or cell-secreted enzymes can be incorporated within the backbone or crosslinks of the material. Nonnatural chemistries such as the *o*-nitrobenzyl photolabile group can be incorporated into crosslinks of the synthetic material and degrade in response to the application of light.[49] Hydrolytic chemistries can also allow for matrix degradation within aqueous environments.[50,51] The incorporation of peptides that mimic structural proteins such as collagen and degrade in response to cell-secreted enzymes can be used to mimic a decrease in matrix stiffness in the presence of cells. The use of the photolabile chemistry allows for highly controlled matrix degradation (*e.g.*, user directed in space and time) that only occurs when exposed to light. Enzyme-triggered degradation can occur in bulk in response to the external application of enzymes or locally through the secretion of enzymes by cells within the matrix, affording varying degrees of *in situ* control. Hydrolytic degradation occurs in bulk within hydrogels, owing to their high water content, and the rate of degradation is pre-programmed synthetically through degradable group selection, where increasing the number of degradable bonds present (*e.g.* esters) increases the probability of degradation and thus the degradation rate.[52]

The versatility of soft synthetic biomaterials also permits control of matrix stiffening over time. In forming synthetic polymer matrices, end groups on backbone and crosslinking monomers react to form a polymer network. If the monomer end groups are reacted off stoichiometry (backbone end groups > crosslinker end groups), there will be a number of free reactive end groups available within the matrix. New crosslinking monomer can be diffused into the material at a later time and reacted to mimic stiffening. Another method to mimic matrix stiffening is to create multiple reactive end groups on a single type of monomer as shown in Figure 9.1C. One of the end group types will react to initially form the hydrogel matrix while the other remains unreacted. A second monomer that has selective end groups for the unreacted groups present in the matrix can be diffused into the material and reacted to stiffen the matrix.

9.2.2 Biochemical Properties

Cellular response to the dynamic ECM is not limited to changes in its biophysical properties, as changes in biochemical properties of the ECM also play a major role in directing cell behavior. Biochemical cues include proteins and small molecules that can signal cells to attach and spread within their environment, align, or even differentiate. Understanding the temporal changes in biochemical cues present in the ECM will further aid in designing novel biomimetic materials.

9.2.2.1   Soluble Cues

9.2.2.1.1   Native.   A number of soluble cues reside in the native ECM, including growth factors or cytokines and smaller molecules such as chemokines and hormones. Growth factors can stimulate cell growth, proliferation, and differentiation, thus playing a major role in directing cell behavior within tissues. For example, vascular endothelial growth factor (VEGF) contributes to processes such as wound healing by promoting angiogenesis, the formation of new blood vessels from existing vessels. VEGF stimulates the migration and proliferation of endothelial cells that degrade and re-deposit basement membrane at wound sites while aligning to form new capillaries.[53] Cytokines were originally defined as soluble factors associated with hematopoietic and immune cells. Like growth factors, they function to promote growth, proliferation, and differentiation, and thus the term cytokine is often used interchangeably with growth factor.[54] For instance, interleukin-17 (IL-17) is a cytokine associated with the immune system that is upregulated upon injury to epithelial cells (wounding) and induces secretion of factors that control bacterial and fungal pathogens to prevent infection at the site of injury.[55] Chemokines are a class of small cytokines that direct the migration of cells within the ECM where they are expressed. They are often associated with leukocytes and immune response to wounding.[56] During chronic infections, many different chemokines are responsible for recruiting immune cells to affected tissue sites. For example, the chemokine CCL5 (RANTES) has been shown to play a role in sustaining immune response to viral infections by supporting the function of CD8 T immune cells.[57] Hormones are small chemicals secreted by cells and affect the behavior of cells in other tissues of the body. They vary widely in function, including small peptides such as insulin, which regulate metabolism of carbohydrates and fats, lipid-derived hormones such as aldosterone, which helps regulate blood pressure, and monoamines such as serotonin, which plays a role in regulating feelings of well-being.

Soluble cues in the native ECM are in a constant state of flux as the body responds to changes in the environment. Returning to the example of wound healing, soluble factors secreted by cells at the wound site can stimulate the production and deposition of new proteins to rebuild the ECM (Figure 9.2B). For example, in response to transforming growth factor beta 1 (TGF-β1) secretion, fibroblasts proliferate and deposit collagen-I within a wound site.[58] Cells sense these cues as gradients on the micrometer scale where different portions of the cell are in contact with different concentrations of cues or as larger gradients on the millimeter scale with cues that appear uniform locally

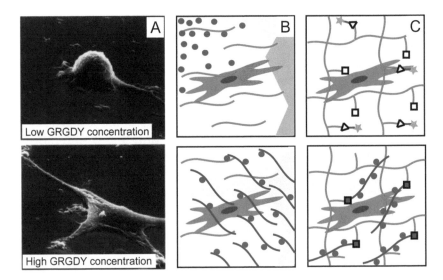

Figure 9.2   Biochemical properties of the ECM. (A) Scanning electron microscope images of cells on surfaces presenting different concentrations of the GRGDY adhesion peptide. Higher concentrations of the cue resulted in increased cell spreading over the surface. Reproduced with permission from ref. 65. © Rockefeller University Press. Originally Published in S. P. Massia, *et al.*, *J. Cell Biol.*, 1991, **114**, 1089–1100. (B) Schematic of biochemical cues presented in the native ECM during wound healing. After injury, small soluble molecules diffuse into the environment. These cues can be sequestered by proteins present in the environment as the wound heals. (C) Schematic of a dynamic ECM mimic presenting various biochemical cues. Initial cues presented within the matrix are removed and replaced with secondary cues to mimic changes in the native ECM at a wound site.

but change spatially as cells move along the gradient.[59] The behavior of cells in response to soluble factors and their gradients in the body is complex, and understanding the response of cells as these cues change in time is important to creating new treatments for injury and disease.

9.2.2.1.2 Biomaterial-Based Mimicry. The static and dynamic presentation of soluble cues in synthetic biomaterial mimics of the ECM has been the focus of extensive study, owing to their importance in biomedical applications as overviewed above. Dynamic presentation of cues is inherently challenging as factors must be either added or removed at specified times without significantly affecting biomolecule efficacy to direct cellular response. A variety of methods thus have been devised to present cues in synthetic ECMs by controlling biomaterial properties over time, including: (1) the external addition and diffusion of the cue into the material; (2) the chemical attachment of the cue to the material through covalent bonds; (3) the degradation of linkers between the cue and the material using cleavable chemistries; and (4) the binding of factors to the synthetic matrix through noncovalent interactions.

The simplest of these methods is adding or removing the soluble factor externally at a specified point in time to allow its diffusion into or out of the matrix. In this method, there is little independent control over how fast the cue enters the material as diffusion is controlled by the pore size of the material, which often is directly linked to crosslink density and modulus.[60] Another method to mimic the removal of soluble factors in a synthetic biomaterial is through the incorporation of cleavable tethers to which soluble cues are attached (Figure 9.2C).[61] These tethers can be designed for removal by various stimuli, including cell-secreted enzymes and light, allowing for a more controlled release of the soluble cues. Designing the matrix with multiple types of reactive end groups can allow covalent bonding of soluble cues to the material at desired time points. Similarly these reactive end groups can link peptides and proteins that sequester soluble cues on their surface,[62] a method that will be discussed in greater detail in the next section.

9.2.2.2 Insoluble Cues

9.2.2.2.1 Native. Insoluble biochemical cues within the native ECM are considered to be a part of its structure, but function to interact with cell surface receptors promoting behaviors such as adhesion and proliferation. One major type of insoluble biochemical cue in the native ECM is short peptide sequences that are part of whole ECM proteins. For example, arginine-glycine-aspartic acid-serine (RGDS), a peptide found within the protein fibronectin, as well many others, has been shown to bind cells through a variety of cell-surface integrins ($\alpha V\beta 3$ (strongest), $\alpha 5\beta 1$, $\alpha 8\beta 1$, $\alpha V\beta 1$, $\alpha V\beta 5$, $\alpha V\beta 6$, $\alpha V\beta 8$, $\alpha IIb\beta 3$)[63,64] and promote adhesion.[65] There are many types of proteins within the native ECM, including glycosaminoglycans (GAGs) and proteoglycans, that bind soluble factors, intermittently immobilizing them within the ECM for cell binding and controlled release. For example, insulin-binding growth factors (IGFs) must be bound to insulin-like growth factor binding proteins (IGFBPs) to access cell surface receptors and stimulate the growth of cells in various tissues. Cells synthesize and deposit IGFBPs in the native ECM, which determines the amount of IGF that may bind and become available to surrounding cells.[66] This type of sequestration, depicted in Figure 9.2B, protects the factors from degradation and can also increase their activity.

9.2.2.2.2 Biomaterial-Based Mimicry.

Dynamic changes in insoluble cues present in the native ECM can be captured with biomaterial mimics. As shown in Figure 9.2C, insoluble cues can be removed and added to the matrix, which can elicit cellular responses such as adhesion or differentiation. These cues can be short peptide sequences mimicking whole ECM proteins that specifically target integrin receptors on the cell surface, or peptides and GAGs that sequester soluble cues on their surface. The fibronectin-based adhesion peptide RGDS has been incorporated into dynamic synthetic mimics of the ECM through covalent addition (Section 9.3.1.1) and irreversible cleavage mechanisms (Section 9.3.1.2). RGDS has been diffused into pre-formed synthetic materials at select time points and reacted to immobilize the cue within the material. It also has to be removed at select time points through externally (*e.g.*, photolytic) or cell-triggered (*e.g.*, enzymatic) degradation or over time through pre-programmed (*e.g.*, hydrolytic) degradation. Combining these techniques, such as sequestering plus integrin-binding sequences, have led to highly tunable dynamic mimics of the ECM.[67] For example, the proteoglycan heparan sulfate binds the soluble cue basic fibroblast growth factor (FGF-2) and induces fibroblast growth during wound healing. In addition to playing a role in the activity of FGF-2, it stabilizes the factor so that it is able to induce the growth of fibroblasts.[68] Incorporating insoluble cues such as RGDS and heparan sulfate within biomaterials mimics can provide greater understanding of how changes in the ECM affect processes such as wound healing.

9.3 CHEMICAL HANDLES TO MODULATE BIOMATERIAL PROPERTIES AND MIMIC THE DYNAMIC ECM

In designing biomaterial mimics of the ECM, a variety of methods have been developed to allow temporal control of their properties. These approaches include covalent coupling and cleavage chemistries that allow attachment and removal of soluble and insoluble cues within the synthetic matrix at different time points. There are also a variety of noncovalent chemistries including self-assembly and binding that have been used to modulate biomaterial properties in time and impart dynamic character. In this section, we will give an overview of a range of covalent and noncovalent chemistries for changing properties in the presence of cells for modification of synthetic ECMs *in situ*.

9.3.1 Covalent Chemistries

9.3.1.1 Coupling. Addition of biophysical and biochemical cues to synthetic matrices plays a large role in the temporal modulation of synthetic ECM mimics. Covalent chemistries allow coupling of cues within synthetic matrices, providing a high degree of tunability to the material properties. Herein, several cytocompatible mechanisms used to initiate coupling of cues within synthetic matrices are described. Various irreversible and reversible covalent chemistries that have been used in models of the ECM are also described.

9.3.1.1.1 Triggering Mechanisms. A key to controlling biomaterial properties in time is dictating when and where modification reactions happen. In the case of adding cues to synthetic ECMs, chemical reactions that can be triggered with initiators or catalysts are utilized to begin material formation or modification. Further, for designing dynamic ECM mimics, these chemistries must be cytocompatible so that changes to the biophysical and biochemical properties of the material can be made in the presence of cells. Several cytocompatible coupling methods exist, including photoinitiation, redox initiation, and base-catalyzed mechanisms, producing radicals or anions that drive polymerization or addition reactions to add crosslinks or pendant groups within biomaterials. These coupling methods are used to trigger the reaction of a wide range of chemistries that will be described in detail in Sections 9.3.1.1.2 and 9.3.1.1.3.

Photoinitiation. Photoinitiation mechanisms to form and modify biomaterials containing various functional groups (Table 9.2) must

Table 9.2 Initiation mechanisms for various cytocompatible coupling chemistries.

Chemistry	Photoinitiation	Redox initiation	Base-catalyzed addition
Acrylate[79,88]	♦	♦	
Diels–Alder[a,101]			
Disulfide[b,97,99]			
SPAAC[a,90]			
Tetrazine–norbornene[a,92,93]			
Thiol–ene:			
1. Thiol–acrylate[50,80,84]	♦	♦	♦
2. Thiol–allyl, norbornene[75,85,137]	♦	♦	
3. Thiol–maleimide[83]			♦

[a]These reactions proceed in aqueous conditions without initiator or catalyst.
[b]Formed *via* oxidation using base.

use both cytocompatible initiators and doses of light so that they can be used in the presence of cells to dynamically tune the environment. Ultraviolet (UV; long wave, centered at 365 nm), visible (400–600 nm), and two-photon infrared (IR) irradiation have been used to initiate polymerization and dynamic property changes in many biomaterial applications. Irradiating cells with these wavelengths and exposure to initiator-generated radicals for longer periods of time can result in cell death or DNA damage;[69,70] however, reducing exposure times or light intensity can lessen the negative effects.

The combination of wavelength, intensity, and exposure time determines the appropriate dose of light and can be selected to minimize cell death and DNA damage while triggering initiation (or cleavage in the case of degradation). To control biomaterial properties in the presence of cells, several cytocompatible wavelengths and irradiation doses (time * intensity) have been reported. UV light exposure of 6–10 mW cm^{-2} at 365 nm (long wavelength UV) has limited to no adverse effects on cell survival if applied for less than 10 minutes, as measured by cytotoxicity, metabolism, p53 expression; however, longer exposure times to 10 mW cm^{-2} (30 minutes) does result in significant DNA damage (p53 activation).[70,71] Visible light is much less damaging to cells: for example, exposure to 80 mW cm^{-2} at 470–490 nm results in minimal adverse effects after periods of exposure longer than 5 minutes.[70] Moving into the infrared, cells remain viable under two-photon irradiation when exposed to pulse energies at or below 4 nJ and do not have significant intracellular ablation after exposure to 1.5 nJ irradiation.[72] Thus, selecting two-photon irradiation pulse energies less than 1.5 nJ will maintain cell viability while preserving the intracellular structure.

Depending on the light source, initiators can be selected from amongst the two classes of radical photoinitiators, which differ in their mechanism of radical generation. Type I photoinitiators cleave into two radicals upon application of light, whereas type II initiators enter an excited state after application of light and abstract a hydrogen from a coinitiator species.[73] Several known water-soluble, cytocompatible initiators include Irgacure 2959 (I2959, Type I), lithium phenyl-2,4,6-trimethylbenzoylphosphinate (LAP, Type I), and eosin Y (Type II). I2959 works best with UV irradiation and has been shown to promote rapid polymerization rates with ~10 mW cm^{-2} at 365 nm.[74] LAP can be used with UV and limited visible wavelengths of light and has been shown to reduce polymerization times when compared to I2959.[73] Eosin Y has been used in visible light polymerizations, which may reduce any negative effects observed with UV light (*e.g.*, DNA damage); however, as a Type II initiator it often requires a coinitiator and a catalyst to match the polymerization rates achieved with Type I photoinitiators. Despite its slower rate, a thiol–norbornene crosslinking system was recently reported to occur within minutes with only eosin Y as the photoinitiator, no longer requiring a coinitiator or catalyst for the reaction to occur at a reasonable rate.[75] In two-photon polymerization, the light can be directed within specific cross-sections of the material to allow photopatterning of both biochemical and biophysical cues. A number of water-soluble photoinitiators can be used in two-photon photoinitiation, including I2959, rose bengal, eosin Y, erythrosin, flavin adenine dinucleotide, methylene blue, WSPI, and G2CK, and have been used in synthetic ECM applications, including forming hydrogels and crosslinking proteins.[76] While traditionally photoinitiators have been used to initiate free radical chain or step growth polymerization reactions for biomaterial modification, they also can be used to drive the copper-catalyzed azide–alkyne click reaction by reducing Cu(II) to Cu(I).[77]

Redox Initiation. Reduction–oxidation (redox) initiation mechanisms similarly can be used to initiate polymerization reactions over a range of temperatures, including cytocompatible temperatures (37 °C), at reasonable reaction rates for the formation or modification of biomaterials.[78] Traditional redox initiation mechanisms using metal ions for reduction are numerous; however, many of these are not considered cytocompatible. Over the years, new redox mechanisms using enzymes such as glucose oxidase (GOx)[79,80] and initiators such as ammonium persulfate (APS) with water-soluble catalysts, such as ascorbic acid (AA)[81] or tetraethylmethylene diamine (TEMED),[82] have been used in the presence of cells as an alternative to light initiation mechanisms.

Base-Catalyzed Addition. Of the various *in situ* coupling mechanisms, base-catalyzed addition is less commonly seen in biomaterial applications, as there are fewer options for its use. One of the most prevalent applications of Michael-type addition is to form hydrogels with chemistries and tune their properties in the presence of cells. The use of a water-soluble base, such as triethanolamine (TEA), may be required for rapid *in situ* formation or modification reactions, such as the reaction of the Michael-type reaction of acrylates and thiols[50] as will be described in the next section. Further, biomaterial modification can be controlled in both time and space by using this mechanism in combination with photodeprotection (*e.g.*, photodeprotection of thiols and subsequent coupling to a maleimide).[83]

9.3.1.1.2 Irreversible Coupling Chemistries. Many irreversible coupling chemistries have been used in biomaterial-based synthetic mimics of the ECM, as depicted in Figure 9.3. These chemistries provide stable bonds over time in the presence of cells and under cell culture conditions. Furthermore, their reactions are considered cytocompatible with limited to no toxic byproducts and can be used to control biomaterial temporal properties to mimic changes in the ECM.

Thiol–ene. Thiol–ene chemistries, which have been widely used to create dynamic biomaterials-based ECM mimics, belong to a type of highly selective and efficient orthogonal reactions that have been termed 'click' chemistries.[84,85] The most common mechanism for temporal control of the thiol–ene reaction is free-radical polymerization. Upon free radical initiation, a thiyl radical is created which attacks the carbon=carbon double bond (-ene). The radical propagates along the -ene and then abstracts a hydrogen from a thiol to form a new thiyl radical in a chain transfer step. The process continues to alternate between radical propagation and chain transfer as the material polymerizes for formation or modification. Thiol–ene reactions can also occur as Michael-type additions in the presence of a base such as TEA. Michael-type addition follows a similar scheme to free-radical polymerization, in which the propagation of an anion replaces radical propagation. In thiol–ene Michael-type addition, the -ene must be electron deficient for the process to occur, where 'enes' that have been used in the presence of cells include acrylates,[50] vinyl sulfones,[86] and maleimides.[87] Reacting the

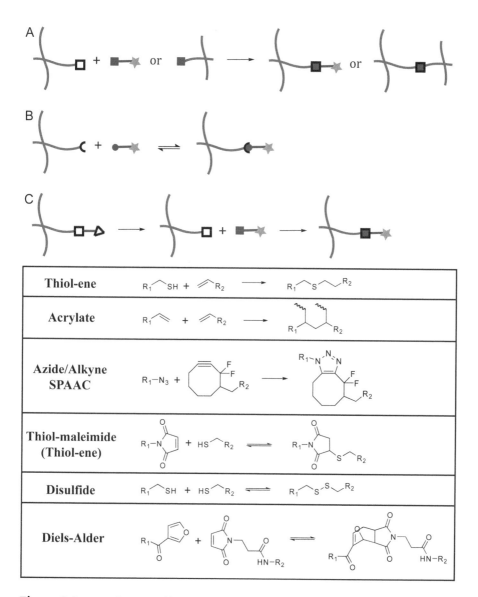

Figure 9.3 Covalent coupling chemistries. Three general methods of dynamically controlling ECM properties through covalent coupling chemistries are: (A) the addition of cues to unreacted functional groups; (B) the reversible coupling of cues; and (C) the removal of a protecting group to expose a functional group for cue attachment. Generalized reaction schematics for the coupling chemistries are shown.

thiols and -enes off-stoichiometry can provide sites to which bio-chemical and biomechanical cues can be added at a later time in the matrix.

Acrylate.   The use of acrylate chemistry to create and modify bioma-terials can be considered a simplified technique compared to 'click' chemistries such as the thiol–ene mechanism, since only one type of reactive group is present. It forms a less homogeneous network as it relies on entanglement of polymer strands to create its network. Acry-late end groups react through a homopolymerization process and can be initiated through the various methods as described above. Addi-tionally, acrylates can be coupled to thiols by step growth mechanism using Michael-type reaction[50] or by step growth polymerization using free radicals, which results in a mixed mode polymerization (*i.e.*, thiol–acrylate addition reactions and acrylate homopolymerization by chain polymerization).[88] The presence of extra acrylate end groups after polymerization can allow its dynamic modification and property control over time.[71,89]

Orthogonal Reactions for Decoupling Biomaterial Formation and Modification.   Several high efficiency, orthogonal reactions recently have been utilized to form biomaterials in the presence of multi-ple functional groups for their later modification. Copper-catalyzed azide–alkyne cycloaddition (CuAAC), another click chemistry, was originally investigated in various polymer applications; however, the presence of the copper catalyst limited its biological applications due to cytotoxicity. To overcome this, a strain-promoted azide–alkyne cycloaddition (SPAAC) mechanism was developed for use in the presence of cells and to allow dynamic modulation of biochemical and biomechanical cues in an ECM mimic.[90] This mechanism takes advantage of the ring strain on a cyclooctyne combined with a diflu-oromethylene electron-withdrawing moiety to promote its reactivity with azides.[91] The difluorinated cyclooctyne (DIFO) chemistry is more readily reactive than monofluorinated cyclooctyne (MOFO) and has been used to modify various monomers for rapidly forming bioma-terials. To create new biomaterials with the benefits of SPAAC while using more synthetically tractable monomers, a new inverse electron demand Diels–Alder click reaction between tetrazine and norbornene or *trans*-cyclooctene was designed.[92,93] These materials can form read-ily at near physiological conditions in less than 5 minutes, faster than that observed for the SPAAC polymerization, demonstrating their potential as bioorthogonal reactions for the design of novel dynamic biomaterials.

9.3.1.1.3 Reversible Coupling Chemistries. Bonds between monomers formed through reversible coupling chemistries can degrade over time in the presence of cell culture conditions or various stimuli, including pH changes. These chemistries can be tuned so that material degradation rates can mimic biomechanical and biochemical changes of the native ECM. Like the irreversible coupling chemistries, new biomechanical and biochemical cues can be added at any time point to further mimic ECM remodeling.

Thiol–Maleimide. Thiol–maleimide click chemistry is another form of thiol–ene chemistry formed through the Michael addition of a pendant thiol to an ene-containing maleimide ring. Their reaction is rapid under aqueous conditions, and biomaterials formed through this chemistry have been used in a variety of applications, including the crosslinking of soft hydrogels.[94] While there have not been many cases reported using thiol–maleimide chemistry for cell-based applications, viable murine C2C12 myoblasts have been encapsulated in thiol–maleimide hydrogels, showing promise in applications beyond seeding cells on the surface of materials formed with this chemistry.[95] Generally considered stable, reverse Michael-type reaction and controlled thiol–maleimide degradation has been shown in the presence of glutathione,[96] a peptide secreted by cells, which has implications in drug delivery and degradable ECM mimics.

Disulfides. The covalent bonding of two free thiol groups through oxidation results in the formation of a disulfide bond, which is considered to be weak when compared to other covalent bonds. This 'weak' bond is relatively easy to cleave to re-generate thiols and so the reaction is considered reversible. Various methods to dissociate disulfides to alter biomechanical properties have been studied, including exposure to glutathione,[96] LAP,[97] and dithiothreitol (DTT).[98] Gels and films that have been formed by disulfide bonds between thiol-modified monomers show a high degree of cytocompatibility;[99] however, disulfide bonds generally form on a slower timescale than the other chemistries that have been described (~hours),[98] so their *in situ* modification is not as facile.

Diels–Alder. The Diels–Alder reaction occurs between a diene and dienophile to form a cyclohexene and can occur under cytocompatible, aqueous conditions.[100] The reaction is reversible primarily at high temperatures; however, a recent development has led to a Diels–Alder reaction mechanism that is reversible under physiological conditions.[101] Bonds between terminal maleimides and furans allow a reversible Diels–Alder cycloaddition at cytocompatible temperatures (37 °C), although more rapid release was observed with increasing temperatures. These studies demonstrated the temporal release of biochemical cues, with future implications in designing an ECM mimic where both biochemical and biomechanical cues are dynamically altered.

9.3.1.2 Cleavage. Development of various cleavage chemistries has led to the ability to further modulate changes in the biomechanical and biochemical properties of biomaterial-based ECMs. While reversible chemistries are considered a type of cleavage mechanism, for clarity and conciseness these reactions were covered in the prior section. Here, we focus on the description of irreversible cleavage chemistries that provide additional methods to temporally control the properties of synthetic ECM mimics.

9.3.1.2.1 Cleavage Mechanisms. A variety of methods to cleave biophysical and biochemical cues from biomaterials homogeneously (*i.e.*, bulk) or spatially specific (*i.e.*, targeted or surface erosion) have been designed for different applications. Bulk degradation maintains an even distribution of cues within the synthetic matrix, whereas surface erosion or targeted removal can be used to create gradients or patterns of cues on or within the material. These techniques include functional group cleavage and material degradation by enzymes, light, and hydrolysis (Figure 9.4).

Degradation by enzymes requires the incorporation of a specific chemistry that responds to enzymes that are present or may be introduced into the material environment. This type of cleavage is typically used for bulk degradation, as it is difficult to target specific areas within the material with enzymes, which readily diffuse throughout synthetic hydrogel-based matrices. A common example of this type of bulk degradation is the inclusion of enzymatically degradable peptides that are cleaved in the presence of cell-secreted matrix metalloproteinases (MMPs). However, MMPs can be bound to cell membranes for cell-driven local degradation.[20,102] Hydrolysis, occurring when the bonds in a material cleave in the presence of water, is often utilized for bulk degradation of hydrogel-based ECM mimics, as exposure to water is difficult to target to specific portions of these materials; however, hydrophobic biomaterials, such as electrospun polyester-based matrices do exhibit surface erosion.[103] Hydrolysis has wide use in the degradation of biomaterial cues due to its simplicity and cytocompatibility as water is a major component in the ECM environment.

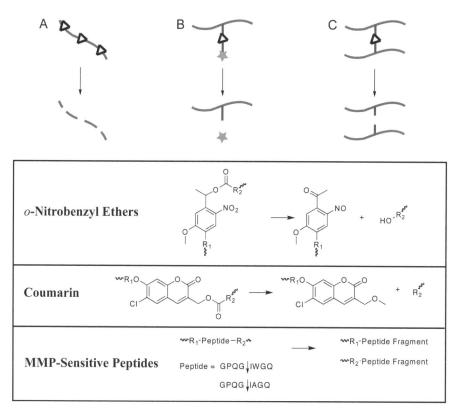

Figure 9.4 Irreversible cleavage chemistries. There are various methods of utilizing cleavage chemistries to control material properties in time, including: (A) the cleavage of groups present within the backbone of a monomer; (B) the removal of a group attached to a cleavable chemistry; and (C) the cleavage of a crosslink connecting monomers together. Examples of *o*-nitrobenzyl ethers,[49,82] coumarin,[114] and MMP-sensitive peptide[86,104] cleavage chemistries are shown.

Cleavage by light, or photocleavage, has application in both bulk and targeted degradation as light may be applied to all or part of the material. The incorporation of various photolabile moieties allows control over degradation rates depending on their response to irradiation. Similar to photoinitiation mechanisms, the wavelengths of light used in photocleavage should be in the longwave UV, visible, or two-photon light ranges for cytocompatibility and be applied at low intensity to prevent damage to cells.

9.3.1.2.2 Cleavage Chemistries

Enzymatically Degradable Peptides. Peptides, or short protein-derived sequences, are used in a range of biomaterial applications to create better mimics of the native ECM. Enzymatically degradable peptide crosslinks are often incorporated to allow cell migration, proliferation, and protein secretion within the matrix.[104] These are responsive to cell-secreted enzymes, most commonly MMPs, and are derived from various structural ECM proteins such as collagen. Degradation rates vary depending on the sequence of amino acids and can be tuned depending on their application. Common sequences used within hydrogel-based matrices are listed in Table 9.3.[104] Synthetic matrix bulk degradation (*i.e.*, decreasing crosslink density) can be predicted based on tabulated or measured rate constants for each sequence as shown in eqn (9.2) and (9.3):

$$-\frac{\partial[S](z, t)}{\partial t} = \frac{\partial[P](z, t)}{\partial t} = [E](z, t)k_{cat}\frac{[S]}{K_m + [S]} \tag{9.2}$$

$$t_c = \frac{\left(1 - \frac{1}{\left[r(f_1 - 1)(f_2 - 1)\right]^{1/2}}\right)}{k_{cat}[E]}[S]_0 \tag{9.3}$$

The amount of substrate (peptide) cleaved [S] or cleavage product formed [P] over time can be predicted based on the tabulated k_{cat} and K_m values, and the enzyme concentration [E] using eqn 9.2.[105] Additionally, in eqn 9.3, the time at which gel dissolution occurs can be predicted based on enzyme concentration [E], initial substrate concentration [S]$_0$, tabulated k_{cat}, the functionalities of the monomers f, and the molar ratio of the functional groups r.[86]

Photolabile Moieties. Cleavage of biochemical and biomechanical cues by light allows for a high degree of spatial and temporal control over matrix properties. The *o*-nitrobenzyl photolabile moiety has been used in both bulk and surface patterning applications and its cleavage

Table 9.3 Common enzymatically degradable peptide sequences utilized for cell-triggered control of biomaterial properties. Cleavage enzymes and k_{cat}/K_m (M^{-1}s^{-1}) values are reported for each sequence.[104] Larger k_{cat}/K_m corresponds with a faster degradation rate.

Sequence	MMP-1	MMP-2	MMP-3	MMP-7	MMP-8	MMP-9	MTI-MMP
GPQG↓IAGQ	60.6	180	16.7	110	1570	93.9	—
GPQG↓IWGQ	434	555	56.0	—	11 100	214	—
PVG↓LIG	—	121 000	—	—	—	3600	—
IPVS↓LRSG	98	82 000	2300	9700	—	11 500	4300

can occur under cytocompatible doses of UV (365 nm, 6–10 mW cm^{-2}, <10 minutes), visible (470–490 nm, 80 mW cm^{-2}), or two-photon (1.5 nJ) irradiation with the formation of no cytotoxic byproducts when attached to a polymer network.[106] There are many types of *o*-nitrobenzyl photolabile groups that have different pendant substituents, affecting their light absorbance, as well as quantum yields, and thus cleavage rates (Table 9.4).[107–111] Irreversible cleavage of the ester or amide bond proximate to the nitro functional group leads to the release of biophysical (*e.g.*, crosslink density) and biochemical (*e.g.*, pendant group) cues attached by this photolabile moiety.

The development and use of photolabile coumarins have been described for various biomaterials applications. Coumarins have been used individually and in combination with other photolabile chemistries to enhance their sensitivity to degradation upon the application of light.[111–117] UV, visible, near infrared (NIR), and two-photo irradiation at cytocompatible doses can cause the irreversible cleavage of the ester (or amide) in this photolabile chemistry, releasing attached cues.

Synthetic matrix bulk degradation of optically thin gels can be predicted based on the number of degradable groups per crosslink (b), the moles of crosslinks (n_x), and tabulated or measured rate constants (k) or characteristic degradation time constant (τ) for each photolabile group as shown in eqn (9.4) and (9.5).[109,118] Biomaterial thickness and photolabile group concentration and absorbance at the selected irradiation wavelength will dictate whether the matrix bulk degrades or surface erodes.[82,118–120]

$$\frac{\mathrm{d}n_x}{\mathrm{d}t} = -kbn_x = -\frac{bn_x}{\tau} \tag{9.4}$$

where the degradation rate constant is directly proportional to quantum yield (ϕ), molar absorptivity (ε), wavelength (λ), and incident light

Table 9.4 Characteristic degradation time constants for *o*-nitrobenzyl photolabile groups with varying substituents. Varying the pendant substituents can tune degradation rate for different applications.[109]

Degradable group	τ (min)
4-(4-(Hydroxymethyl)-2-methoxy-5-nitrophenoxy) butanoic acid	65 ± 3.5
4-(4-(1-Hydroxyethyl)-3-methoxy-5-nitrophenoxy) butanoic acid	5.0 ± 0.03
4-(3-(Hydroxymethyl)-4-nitrophenoxy) butanoic acid	40 ± 3.3
4-(3-(1-Hydroxyethyl)-4-nitrophenoxy) butanoic acid	2.0 ± 0.03
1,3-Hydroxymethyl-2-nitrobenzene	6.5 ± 0.6

intensity (I_0), and inversely proportional to Avogadro's number (N_A), Planck's constant (h), and the speed of light (c):

$$k = \frac{\phi \varepsilon \lambda I_0 (2.303 \times 10^{-6})}{N_A hc} \tag{9.5}$$

Ester, Amide, and Thioester Hydrolysis. Hydrolysis of esters, amides, and thioesters is considered a cytocompatible cleavage mechanism since it naturally occurs in the presence of water; however, the acid generated upon cleavage can cause an inflammatory response *in vivo*. This mechanism typically is used for the bulk removal of cues (*e.g.*, cleavage of crosslinks or pendant groups) within hydrogels or the surface of erosion of hydrophobic polymer scaffolds. A water molecule breaks the ester, amide, or thioester and supplying an –OH to form a carboxylic acid and an –H to create an alcohol, amine, or thiol. In comparison to other chemistries described above, the degradation of esters, amides, and thioesters often is relatively slow and not as easily controlled *in situ*. Common hydrolytic groups and their respective rate constants or half-lives are listed in Table 9.5, with rate constants dependent on neighboring functional groups and covalently bonded molecules.[121–123] Instead, the degradation rates of these systems are preprogrammed with the number of cleavable repeat units within a crosslink or pendant group, where increasing the number of repeats increases the probability of cleavage and thus the rate of degradation. Additionally, towards increasing the rate of degradation, researchers have incorporated charged chemistries and molecules such as

Table 9.5 Half-lives of various hydrolytically degradable groups commonly used within hydrogel-based biomaterials. Esters exhibit the fastest rates of degradation amongst this group, approximately days as opposed to on the order of months for thioesters and years for amides, and are the most widely used of the hydrolytic chemistries in degradable synthetic mimics of the ECM. For both ester and amide groups, it is shown that degradation rates can be increased and decreased by including different substituent groups to which they are attached (*e.g.*, electron withdrawing groups).

Degradable group	$\tau_{1/2}$, half-life
Ester: *e.g.*, **PEG–ester–peptide hydrogels**[121]	
1. Peptide = GRCRGGRCRG	6.56 days
2. Peptide = GDCDGGDCDG	36.1 days
Amide:[122]	
1. Adjacent functional groups = CH_3, CH_3, and H	1.46 years
2. Adjacent functional groups = $ClCH_2$, H, and H	38 000 years
Thioester:[123]	
S-methyl thioacetate	155 days

amino acids so that hydrolysis rates are faster and controlled to better mimic dynamic changes in the native ECM.[121] Charged or nucleophilic groups added to materials containing hydrolytic chemistries act as 'catalysts' during hydrolysis by attacking the functional group and donating an electron pair during cleavage. The resulting cleavage rates are increased from weeks to a few days. Synthetic matrix bulk degradation can be predicted based on tabulated or measured rate constants for each hydrolytically degradable group, as shown in eqn (9.6):[124]

$$\frac{\mathrm{d}n_t}{\mathrm{d}t} = -k'n_t \tag{9.6}$$

These kinetics for hydrolysis can be applied to highly swollen gels (volumetric swelling ratio, $Q > 10$) using pseudo first-order kinetics where n_t is the moles of degradable blocks in the hydrogel and k' is the pseudo first-order degradation rate constant. Further, mass loss can be predicted using a statistical-kinetic model.[52,125]

9.3.2 Noncovalent Interactions

A large number of covalent chemistries are available for temporal changes in synthetic biomaterial-based ECMs, owing to the ease and predictability of controlling their cleavage and related matrix properties. However, a number of noncovalent chemistries have also been studied, as these physical interactions within biomaterials may better mimic the dynamic nature and structure of the native ECM. Interactions between atoms and molecules in these chemistries can control the structure and composition of the material in time. This section will focus on two of the most common types of noncovalent interactions that have been studied for biomaterials applications: self-assembly and binding.

9.3.2.1 Self-Assembly

9.3.2.1.1 Peptides. Select peptides and proteins naturally assemble into higher ordered structures owing to hydrogen bonding (noncovalent) between amino acids that is stabilized by disulfide bonds (covalent). These include collagen mimetic peptides (CMPs) based on the sequence $(GPO)_n$,[126] stimuli-responsive peptides such as the sequence $C(FKEF)_2C$,[127] and triblock proteins made through recombinant DNA techniques.[128] Hydrogen bonding in proteins and peptides occurs between the oxygen on a carbon and the hydrogen on the amine in the 'backbone' of two neighboring amino acids. This bonding leads to assembled structures, beta sheets or alpha helices (Figure 9.5A and B), that can influence how cells interact with proteins in the native

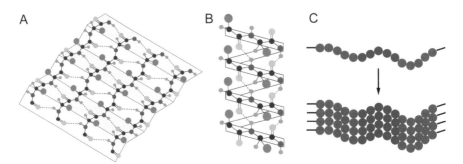

Figure 9.5   Self-assembling structures used within hydrogels. (A) β-Sheet secondary peptide structure. (B) α-Helical secondary peptide structure. (C)
Diblock co-polymers.

environment. Researchers are designing methods to mimic peptide
and protein structures within biomaterials, as these are key components of the native ECM, through noncovalent interactions that promote self-assembly.

Peptide–amphiphiles are one of the most widely studied type of
self-assembling peptides and have been used to create scaffolds that
mimic the native ECM structure and composition.[129] The assembly of
hydrophobic and hydrophilic regions of the peptides into beta sheets
or alpha helical structures can be controlled by pH and structural
changes can be made by altering the pH over time. For example, a
peptide designed with alternating valine (V) and lysine (K) amino acid
residues, which have a high propensity toward forming β-sheet structures, adopts a β-hairpin secondary structure promoted by the inclusion of the tetrapeptide V^DPPT. Changes in pH promote the folding
of the designed peptide across the tetrapeptide that is further stabilized by the alternating hydrophobic and hydrophilic V and K amino
acid sequence.[130] Thus, selective peptide design has allowed progress in controlling the formation of nanostructures within synthetic
matrices.

9.3.2.1.2   Block Copolymers.   Block copolymers contain multiple
regions or 'blocks' of different polymers (Figure 9.5C) within their
polymer chain. They have been used to make self-assembling biomaterials that respond to environmental cues such as pH and temperature, which can dynamically alter their structural properties.
These materials work in a similar manner to peptide–amphiphiles,
containing hydrophobic and hydrophilic polymer regions that assemble when exposed to various environmental cues like the presence of
water at varying pH. Some have also been designed to contain blocks

of peptides and blocks of polymer to provide a more accurate mimic of the native ECM, allowing cells to adhere and interact better within the synthetic material.[131]

There are a number of block copolymer hydrogels for applications in drug delivery and tissue engineering, including polystyrene-poly(ethylene oxide) (PS-PEO) diblock and PS-PEO-PS triblock copolymer blends that form hydrogels with spherical domains resulting from assembly of the hydrophobic and hydrophilic blocks of the copolymer.[132] Hydrogels formed from oppositely charged poly(allyl glycidyl ether-*b*-ethylene glycol-*b*-allyl glycidyl ether) triblock copolyelectrolytes were shown to soften with increasing salt concentration, which has implications for the design of future block copolymers with tunable mechanical properties.[133] The sizes of the structures formed by the different domains in self-assembled block copolymers exist on the nano- to micro-scale and can be controlled by changing the ratio and type of monomers, the lengths of the blocks within the monomers, and the pH or salt concentration in the environment.[134,135]

9.3.2.2 Binding. Binding allows the sequestration and release of various soluble cues within the ECM. Capturing this process with a synthetic material can help direct cell behavior over time and better mimic interactions occurring in the native ECM. Methods to mimic growth factor binding, protein adsorption on surfaces, and the design of responsive hydrogels to binding will be described in this section.

9.3.2.2.1 Growth Factor Binding. The incorporation of proteins, peptides, or other insoluble cues that sequester growth factors in synthetic matrices can amplify the activity of the factor and control its release over time. These insoluble cues can be covalently bound throughout or patterned into the material and then introduced to soluble factors at desired time points. The soluble factor will bind to the surface of the cue, where it is protected from degradation and can be released over time (Figure 9.6A).

Heparin, a GAG known to sequester various growth factors, is one insoluble cue that has been widely studied in growth factor binding biomaterials. Researchers have successfully linked heparin into polymer hydrogel scaffolds and studied the release rates of various growth factors.[136] The incorporation of affinity peptides to bind growth factors for protection and controlled release within materials has also been studied. For example, the peptide sequences WHSW and KRI-WFIPRSSWY were shown to bind TGFβ1 when incorporated in PEG hydrogels, preserving their bioactivity, and a number of other peptide

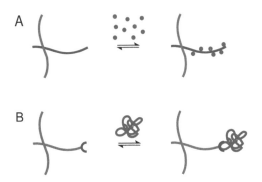

Figure 9.6 Noncovalent binding. (A) Growth factors are sequestered on the surface of proteins and peptides incorporated within synthetic matrices to maintain and enhance bioactivity. (B) Proteins adsorb on surfaces functionalized with groups such as phosphates or charged ions to direct cell adhesion and fate.

sequences also exist in native ECM proteins that can be used in other similar applications.[137] In the various growth factor binding biomaterials, the release kinetics of bound factors can be determined through the equilibrium binding expression:[1]

$$P + L \overset{k_f,\, k_r}{\Longleftrightarrow} P \cdot L \tag{9.7}$$

where P is the insoluble cue (protein, peptide, *etc.*), L is the ligand (growth factor), and k_f and k_r are the forward and reverse reaction rate constants. Assuming that this reaction is fast and reversible, the transport of proteins and factors through the material can be determined by:[1]

$$\frac{D}{K_b + 1} \nabla^2 C_p = \frac{\partial C_p}{\partial t} \tag{9.8}$$

where C_p is the concentration of free protein, D is the diffusion coefficient, and K_b is the ratio of ligand to the dissociation constant ($[L]/K_d$). The binding of ligand to insoluble cues is accounted for in the $K_b + 1$, which decreases the diffusivity of the ligand within the material.[1] Research in this area continues to broaden the range methods to create dynamic ECM mimics and demonstrate how changes in the native ECM directing cell behavior in response to a combination of soluble and insoluble cues.

9.3.2.2.2 Protein Adsorption on Surfaces. Protein adsorption on surfaces has applications in biomaterial-based mimics of the native ECM to direct cell adhesion, cell response, and material cytocompatibility. In the case of implanted materials, adsorption of proteins on the surface helps direct initial cell response to the material so that the body does not reject the implant and begins to incorporate and naturally alter the material.[138] In the context of synthetic biomaterial-based ECMs, cellular synthesis and deposition of different types of proteins onto synthetic surfaces can direct cell fate. For example, PEG scaffolds functionalized with phosphate groups directed osteogenesis of human mesenchymal stem cells (hMSCs) through increased production and deposition of osteopontin, whereas those functionalized with *tert*-butyl directed adipogenesis through peroxisome proliferating antigen receptor gamma expression.

Various methods to allow or promote protein adsorption to a surface have been designed, taking advantage of the charges on the amino acids that define hydrophilic and hydrophobic regions within the protein. Surfaces can be modified with charged ions for a net positive or negative charge to which proteins of the opposite charge can adhere. Hydrophobic and hydrophilic polymer surfaces can also adsorb proteins to their surface in a manner similar to that described in Section 9.3.2.1.1 for peptide self-assembly, and changes in pH can cause desorption of the protein from the surface.

9.3.2.2.3 Binding-Responsive Hydrogels. The use of natural environmental cues to dynamically alter biomaterials is an area of interest to many researchers as it provides natural mechanisms to modulate material behavior. Enzymatically degradable peptides that cleave in response to cell-secreted MMPs is one such mechanism; however, other developments have focused on binding of soluble cues in the ECM to direct changes in synthetic material structure. In the native ECM, proteins undergo conformational changes due to the binding of soluble cues. Taking advantage of these mechanisms, researchers are developing protein-based materials that bind soluble cues to change conformation to dynamically alter their biophysical properties.[139]

9.3.3 Methods to Characterize *In Situ* Property Changes

Methods to observe *in situ* property changes have allowed researchers to characterize changes in biomaterial properties as they occur over time. In this section, we describe a variety of methods commonly used to study biophysical and biochemical changes in synthetic biomaterial-based ECM mimics. These involve a variety of chemical techniques, such as fluorescent dyes and colorimetric assays, and direct or indirect measurements of structure using microscopy or rheometry, respectively.

9.3.3.1 Fluorescent Dyes. The use of fluorescent dyes to observe changes in the properties of biomaterials can provide both visual and quantitative confirmation that the desired dynamic changes are occurring within these synthetic ECM mimics. Fluorescence can be incorporated with different techniques to report the presence or lack of cues in the material environment. Methods include, but are not limited to, tagging biochemical cues with fluorophores and dyes, conjugation of enzyme-sensitive fluorophores into the matrix, and immunolabeling. By observing the intensity of fluorescence, it is possible to determine relative amounts cues present in the synthetic matrix.

One method is the addition of a fluorophore-tagged cue that fluoresces upon excitation by light when the cue is conjugated to the matrix, but does not fluoresce after the cue has been removed from the material. Biochemical cues such as peptides have been tagged with fluorophores prior to their addition to the synthetic matrix. Free amine and thiol reactive groups within these biochemical cues have been used for covalent modification with fluorophores so that they are stably attached.[83,140] This method is spatially specific or targeted and a practical option for monitoring dynamic changes in a synthetic ECM mimic since the dye can only appear when tagged cue is added or removed from the matrix. Alternatively, a Forster resonance energy transfer (FRET) fluorophore–quencher pair can be employed where the proximity of the acceptor and donor chromophores dictate fluorescence intensity. For example, in monitoring the binding of soluble factors with peptides conjugated to synthetic matrices, FRET has been employed to visualize and quantify binding events.[141] Similar to FRET, self-quenching Bodipy dyes have been incorporated into enzymatically degradable peptide crosslinks that fluoresce upon cleavage of the crosslink.[142] Such methods have implications in observing dynamic cell response to materials and might be applied to other cell–ECM interactions. Another technique to observe changes in synthetic ECM mimics is through immunolabeling. In this method, fluorophore-conjugated antibodies are used to label proteins that have been incorporated into the biomaterials, identifying the presence of the protein. This attachment cannot be dynamically controlled, so it can only be used to identify the presence or lack of cues after addition or cleavage.

9.3.3.2 Rheology. Rheology, or the study of the flow of matter, is a technique that has been widely used to characterize the biophysical properties of soft biomimetic materials. A rheometer measures a material's response to an applied stress, which can be used to determine its properties including elastic modulus and, indirectly, mesh size. These properties change as the biophysical composition of a synthetic matrix changes, and the ability to characterize mechanical properties *in situ* is key to tuning the material for biological applications.

The modulus of a material is expected to increase with the addition of crosslinks, correlating to a decrease in mesh size, while the cleavage or removal of these crosslinks or cues decreases modulus while increasing mesh size. Tests on a rheometer can be used to measure these changes in modulus over time so that the rate of increase or decrease in modulus as it relates to the addition or cleavage of cues can be determined. Additionally, microrheology techniques can be used to characterize mechanical properties as they change in time and in the presence of cells. These techniques have been used to track particle movement and thus temporally evolving modulus during the degradation of hydrolytically[143] and enzymatically[144] degradable hydrogels. Such techniques are very useful in understanding degradation at the gel–sol transition point and hold promise for tracking cells in materials over time. Using information gained from rheology, researchers can temporally tune biophysical properties of the material so that the rate of change in modulus mimics that of the native ECM.

9.3.3.3 Monitoring Dynamic Cell Behavior. In addition to monitoring changes in the biochemical and biophysical properties of synthetic ECM mimics, the ability to monitor the migration, differentiation, and proliferation of cells in real time allows for observation of dynamic cellular response to changes in a synthetic biomaterial-based ECM. Lentiviral vectors that encode green fluorescence protein (GFP) have been used to monitor cell activation and gene expression in real time.[145] Vectors are designed to infect cells and encode the production of GFP in response to environmental cues, allowing the visualization of cell response to the addition or cleavage of particular biochemical or biophysical cues of interest. Sophisticated imaging software has also been developed to track cell migration through materials in response to gradients of biochemical and biophysical cues that have been patterned into the environment. Cell speed has been determined for cancer migration using software to track cell position over time in microscopy images, even if cells disappear and reappear within the field of view.[146] Imaging software is not limited to studying migration and has also been used to determine changes in properties such as cell area, shape, and orientation over time.[72,147,148]

9.4   CONTROLLING BIOMATERIAL BIOPHYSICAL PROPERTIES *IN SITU*

Research on controlling synthetic matrix stiffness, microstructure, and topography has advanced our understanding of how changes in biophysical cues direct cell behavior. In this section, we will give an overview of some relevant recent works on temporally controlling the biophysical properties of synthetic biomaterials.

9.4.1   Controlling Synthetic Matrix Stiffness in Time

Two- and three-dimensional (2D and 3D) models have been created to study how changes in ECM biochemical and biophysical properties affect cell response. Response to material stiffness has been studied in 2D systems, where cells are grown on top of a softening or stiffening surface, and in 3D with cells encapsulated within the softening or stiffening material. The coupling and cleavage chemistries described in earlier sections have been used to modulate these material properties and tuned based on their application.

9.4.1.1   Photoresponsive Biomaterials.   Light responsive materials are widely studied for controlling biomaterials in time and space. Photoaddition and photocleavage mechanisms have allowed dynamic alteration of material biophysical properties. Many light-responsive chemistries rely on the application of long wavelength UV light, and the rates at which cues are added and cleaved are often rapid to reduce the dose of light applied to the material and cells on or within it. Using light, changes are easy to externally trigger, allowing spatial and temporal control over the removal and addition of cues.

   Matrix softening in the presence of cells has been achieved using a photodegradable PEG hydrogel containing an *o*-nitrobenzyl ether-derived moiety as a photolabile group. A photodegradable acrylate monomer with a pendant carboxylic acid was reacted with PEG-bis-amine to create a diacrylate macromer used to form the hydrogels. Polymerization of the diacrylate monomer occurred under radical redox initiation conditions in the absence of light to prevent the degradation of the photolabile monomer during gel formation. Subsequent exposure to cytocompatible doses of UV light led to bulk and surface degradation of the material. Rheometry was used to characterize bulk degradation by monitoring the dynamic decrease in modulus in response to light application. Temporal control over modulus for thin hydrogels (<100 μm) was demonstrated by periodic irradiation, with the modulus decreasing upon application of light and remaining constant in the absence of light (Figure 9.7A). Spatial control over gel degradation and surface erosion of thick hydrogels (500 μm) was demonstrated by the patterning of ridges on the hydrogel surface. Profilometry was used to characterize this patterned surface erosion of the gels, with longer exposure times resulting in deeper ridges. The high degree of spatial and temporal control over biophysical property changes demonstrate the material's versatility as a dynamic ECM mimic.[49] For example, these materials have been used to probe how matrix stiffness temporally regulates myofibroblast activation and deactivation.[149,150] More recently, photodegradation has been used in conjunction with photocoupling for temporal microenvironment regulations, as discussed in Section 5.1.1.[151] Additionally, *o*-nitrobenzyl ether groups with different substituents, dictating their degradation rates, have been incorporated within the same hydrogel for selective temporal release of specific cell populations with selection of the irradiation wavelength.[109]

A variety of photoinitiated chemistries have been utilized to control biomaterial matrix stiffening. In one such chemistry, HA was modified with methacrylates to make a long HA chain with multiple -enes along its backbone. This methacrylated hyaluronic acid (MeHA) was then reacted in various proportions with DTT in TEA (base catalyzed thiol–ene Michael-type addition) to initially form gels. At select time points, the initiator I2959 was diffused into the matrix and then exposed to cytocompatible doses of UV light, whereupon the remaining free methacrylates along the HA backbone reacted (radical diacrylate), resulting in the stiffening of the matrix (Figure 9.7A). It was found that not only does stiffening direct cell behavior, but that the time when the material was stiffened affected differentiation of hMSCs. If the hMSCs remained on the soft substrates for longer periods of time before stiffening occurred, adipogenic differentiation was favored, while osteogenic differentiation was favored if the material stiffening occurred earlier. This result further demonstrates the importance of designing dynamic biomaterials so that the effect of temporal changes in the native ECM on cell phenotype and function may be better understood.[71]

9.4.1.2 Enzymatically Responsive Biomaterials.

Enzymatically responsive biomaterials are an important category of synthetic ECM mimics because they can be designed to respond to specific cell-secreted enzymes and tuned for the material's application. A number of enzymatically degradable peptides derived from sequences in ECM proteins have been reported for use as crosslinks in responsive hydrogels. Designing biomaterials containing these enzyme-degradable

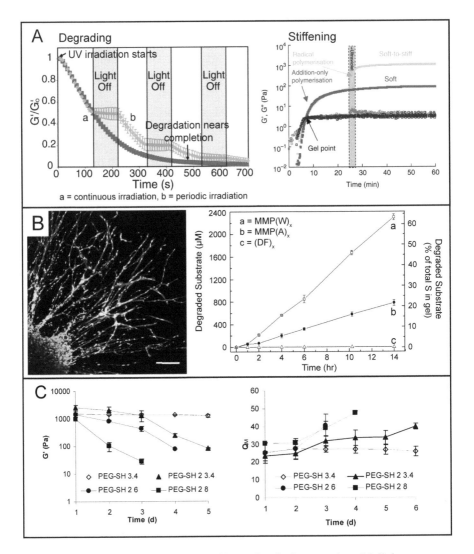

Figure 9.7 Examples of modulating biomechanical properties with light, enzymes, or hydrolysis. (A) Light: Rheological data of gels exposed to periodic doses of UV light to temporally degrade[49] or stiffen[71] these synthetic matrices. Reproduced with permission from ref. 49. © 2009 AAAS. Reproduced with permission from ref. 71. © 2012 Nature Publishing Group. (B) Enzymes: Fibroblasts invade a PEG hydrogel containing MMP-cleavable crosslinks (GPQG↓IWGQ [MMP(W)$_x$] or GPQG↓IAGQ [MMP(A)$_x$]) or non-cleavable crosslinks (GDQGIAGF [(DF)$_x$]). The highest rate of degradation is observed for MMP(W)$_x$ while (DF)$_x$ is not degraded. Reproduced with permission from ref. 86. © 2003 National Academy of Sciences, USA. (C) Hydrolysis: PEG-based hydrogels with varying PEG crosslinker molecular weights and numbers of methylene moieties between ester and thiol groups. The PEG-SH 3.4 kDa containing only one methylene between the ester and thiol does not degrade over the time course. The remaining PEG-SH 2 3.4 kDa, PEG-SH 2 6 kDa, and PEG-SH 2 8 kDa, containing two methylenes between the ester and thiol, hydrolytically degrade over the time course with higher degradation rates observed for higher molecular weight PEG. Reproduced with permission from ref. 152. © 2010 American Chemical Society.

peptides can allow cells to behave more naturally in synthetic biomaterials, enabling processes such as proliferation and migration.

Researchers who use enzymatically degradable materials often work to tune degradation rates of the matrix so that alterations in biophysical cues mimic that in the body and drive desired cell response. In the initial development of enzymatically degradable peptides, the cleavage of natural proteins has been studied to determine specific sites at which they cleave in response to secreted enzymes such as MMPs. From this, short site-specific sequences of amino acids have been identified and the rates at which they cleave ($\sim k_{cat}$) calculated. One highly reported collagen-I-based peptide sequence, GPQG↓IWGQ cleaves in the presence of various MMPs (MMP-2, -9, *etc.*), but more interestingly, its cleavage rate was shown to increase from the 'wild-type' (GPQG↓IAGQ) with a simple amino acid substitution (A→W).[86] With this in mind, further developments in tuning enzyme degradation rate to drive dynamic cell behavior have been made by varying k_{cat} values with peptide sequence design. When these sequences were used as crosslinks for enzyme-responsive hydrogels, degradation was observed upon application of various MMPs. Crosslinks with higher k_{cat} values degraded more rapidly than those with lower k_{cat} values. Taken a step further and used for the culture of fibroblasts, the hydrogels formed with high k_{cat} peptides were shown to support growth and proliferation and allow cell migration through the matrix in comparison to those with lower k_{cat} values (Figure 9.7B). Tuning degradation in this way can allow the design of ECM mimics to study processes such as cell migration and proliferation.

9.4.1.3 Hydrolytically Degradable Biomaterials. Hydrolytic degradation of biomaterials allows the pre-programmed change of biophysical properties *in vitro* and *in vivo* native-like environments since tissues are comprised of large amounts of water. Varying the number of ester, amide, and thioester linkages, the molecular weight of the material, and the position of the linkages within the material has allowed the tuning of degradation rates to suit various applications. Mimicking PEG-based hydrogel scaffolds have been designed to hydrolytically degrade with the incorporation of different ester bonds in the PEG crosslinking monomer.[152] Crosslinking monomers of varying molecular weights were incorporated in the hydrogels and degradation was monitored by rheology. In this system, it was shown that a higher monomer molecular weight resulted in an initially increased degradation rate. The effects of ester bond location on degradation rate was also studied, and decreasing the distance between the ester and thiol groups in the hydrogel system increased degradation rates

(Figure 9.7C). By using a combination of various factors, such as molecular weight and cleavable group location, degradation may be tuned for temporal control over biophysical properties. A photocross-linked alginate hydrogel with tunable degradation rate was reported for applications to support naturally occurring processes such as tissue regeneration. A 2-aminoethyl methacrylate modification of the alginate created both amide and ester bonds that could be hydrolytically cleaved. It was shown that increasing the percent methacrylation of alginate decreased the degradation rate of the hydrogel, allowing control over its degradation and thus properties in time.[153]

9.4.2   Controlling Topography in Time

Cells are in immediate contact with their microenvironment, attaching and responding to the physical structure of the ECM. The ability to control structure from the nano- to the micro-scale that surround cells in time can elucidate the role that biophysical properties, such as microstructure and topography, play in directing cell fate. Although studies on how changing biophysical properties affect cell behavior have been focused on more macroscopic properties such as modulus, several studies that investigate cell response to dynamic changes in microstructure and topography will be described here.

Cell fate can be directed partly by topographical changes in the ECM. This naturally occurs in the body, thus the ability to modulate changes in topography of biomaterial-based matrices can be used to understand how cells respond to such biophysical changes. Recently, a hydrogel system whose topographical features can be dynamically controlled *in situ* was described.[154] The incorporation of an *o*-nitrobenzyl acrylate within PEG monomers was used to create a photodegradable PEG-based hydrogel in which surface topography could be altered dynamically upon exposure to cytocompatible doses of UV light under photomasks with varying patterns. The features eroded into the gel surfaces were on the micrometer scale (~5 µm), which are small enough for cells to detect variations in topography. Alignment of hMSCs cultured on the surface of unpatterned gels and gels patterned with lines and rectangles were compared. Cells cultured on the line-patterned surface exhibited higher levels of alignment than those on unpatterned surfaces. In demonstrating dynamic modulation of topography, cells that were initially cultured on an unpatterned surface aligned after the erosion of a line-patterned surface. Subsequent erosion to create a square-patterned surface reversed this alignment, with future implications in directing cell fate in mimics of the native ECM by controlled dynamic topographical changes (Figure 9.8A).

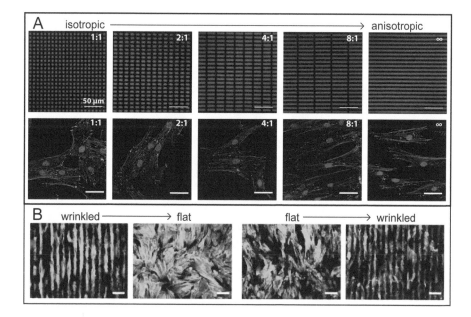

Figure 9.8 Examples of controlling matrix topography. (A) Photodegradation of wells into the surface of PEG-based hydrogels containing an *o*-nitrobenzyl photolabile moiety is used to create topographic features and direct hMSC alignment on the gel surface. hMSCs align with the direction of lines (anisotropic) patterned into the surface; however, patterning with squares (isotropic) resulted in a more rounded morphology. Sequential temporal patterning of the surface (anisotropic → isotropic) demonstrated reversible changes in cell morphology. Reproduced with permission from ref. 154. © 2012 John Wiley and Sons. (B) Substrates that wrinkle in response to applied strain are used to direct hMSC alignment. hMSCs seeded onto flat, stretched substrates align upon release of the strain and formation of micro-scale wrinkles, whereas those seeded on initially wrinkled substrates become randomly oriented after strain is applied to flatten the material surface. Reproduced with permission from ref. 155. © 2013 John Wiley and Sons.

Research to understand cell response to changes in topography also has been conducted in materials that wrinkle in response to changes in applied strain. Stretched sheets of poly(dimethylsiloxane) were exposed to ultraviolet/ozone to create a stiffer surface while the inner material remained soft, allowing the formation of wrinkles upon release from strain. hMSCs seeded on top of the wrinkled surfaces initially aligned with wrinkle direction, but upon stretching to remove the topography became randomly spread on the material surface. The reverse behavior was observed for the flat to wrinkled surface transition, further indicating that the ability to control topography over time is an important factor to consider in designing novel biomaterials (Figure 9.8).[155]

9.5 CONTROLLING BIOMATERIAL BIOCHEMICAL PROPERTIES *IN SITU*

9.5.1 Insoluble Cues

Insoluble cues in the native ECM change as tissues remodel. With the range of techniques we have described for addition and removal of these cues within biomaterial-based ECM mimics, researchers have been able to capture various cell responses to changes in them. Herein we describe several examples of the addition and removal of insoluble biochemical cues that have demonstrated a dynamic effect on cell fate.

9.5.1.1 Covalent Addition of Peptides and Proteins. One well-researched technique for the dynamic incorporation of biochemical cues in ECM mimics is through selective covalent modification. Various combinations of the chemistries that were described in Section 9.3.1.1 have been used to model the dynamic native ECM. These cytocompatible reactive chemistries have been designed to allow spatiotemporal control over the addition of cues depending on the application of the system that is being studied.

Systems using multiple orthogonal chemistries to allow the initial formation and later modification of the material have been studied to create new materials in which biomechanical and biochemical properties are independently controlled in time. There are numerous examples of the covalent addition of peptides to synthetic matrices with photopatterning and orthogonal chemical crosslinking strategies; however, the spatiotemporal addition of larger, complex proteins is limited due to difficulty in maintaining the bioactivity of the molecule. Recently, a PEG-based hydrogel system was described for the patterning of proteins with light-controlled enzymatic coupling.[156] Activated transglutamase factor XIII (FXIIIa), an ECM-crosslinking enzyme, catalyzes the reaction between an ε-amine on lysine (K) and a γ-carboxamide on glutamine (Q). To enable spatiotemporally controlled crosslinking with FXIIIa, the amines on lysine-containing peptides were caged with a photolabile group, and the peptides were crosslinked into hydrogels with a thiol–ene Michael-type addition between a cysteine-containing peptide and vinylsulphone-modified PEG. Upon exposure to UV light, the cage was released, allowing the free-amine to react with the carboxamide on glutamine-containing proteins in the presence of FXIIIa. Patterns on the micrometer scale were created with photomasks, and biological activity of the proteins was maintained, demonstrating the system's potential for studying the effects of patterning complex and physiologically relevant proteins on directing cell behavior (Figure 9.9A).[156]

9.5.1.2 Photorelease of Peptides and Proteins. In the native ECM, insoluble biochemical cues appear and disappear as cells remodel their environment. While there are a great number of chemistries for the addition of insoluble biochemical cues, advances in the dynamic control mimics of the ECM have led to the development of various photolabile chemistries that allow the removal of these cues. An RGDS tether using the photolabile *o*-nitrobenzyl acrylate group was described for the removal of biochemical cues. The tether was initially polymerized within the hydrogels and later removed upon application of UV

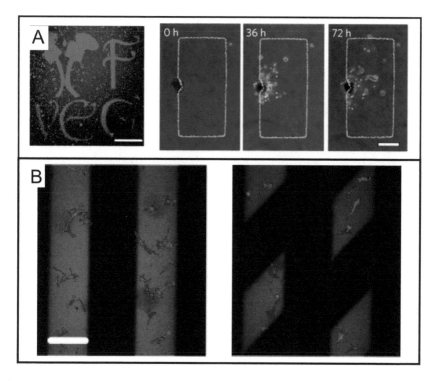

Figure 9.9 Examples of controlling biochemical properties in time. (A) PEG-based hydrogels are spatially patterned with fluorescently labeled VEGF with micron-scale resolution using an enzymatic photopatterning reaction. Cells invade into patterned sections of the gel network using this reaction scheme. Reproduced with permission from ref. 156. © 2013 Nature Publishing Group. (B) Cell response to photoaddition and photocleavage of chemical cues from matrices. Cells adhere to an *o*-nitrobenzyl-containing RGDS patterned sections of PEG-based hydrogel and detach upon photocleavage of the cue. Reproduced with permission from ref. 61. © 2012 John Wiley and Sons.

light. hMSC response was studied and viability of the cells was shown to decrease over time due to the lack of adhesion to the surrounding matrix. While this result describes the removal of the tether, one could use this described chemistry to add cues and then remove them at any points in time for a highly controlled dynamic mimic of the ECM.

Combining photoaddition and photocleavage mechanisms, an orthogonal PEG-based hydrogel modified with RGDS tethers was designed to mimic the addition and removal of FN from the environment surrounding cells.[61] Hydrogels were initially formed through SPAAC, leaving pendant -enes on the crosslinks within the matrix unreacted for later modification. RGDS was modified with a cysteine (thiol group) to allow attachment to the pendant -enes, an *o*-nitrobenzyl ether moiety to allow its cleavage from the matrix, and an AlexaFluor-488 fluorescent group to allow imaging of cue addition within the gel. Eosin Y photoinitiator and the modified RGDS were diffused into the pre-formed gels and exposed to visible wavelengths of light under a photomask to pattern in the cue. After removal of unreacted RGDS and eosin Y, the gels were imaged and it was shown that the cues had successfully attached to the regions of gel exposed to light. Removal of the cues was achieved by exposure to cytocompatible doses of UV light, and secondary gradients or patterns of the cue were created within the gels with photomasks during its cleavage. Cells initially attached to the patterned RGDS and detached after cleavage of the cue, indicating a high level of spatiotemporal control over material properties (Figure 9.9B). This spatiotemporal control is not limited to the addition and removal of biochemical cues but was also applied previously in creating a material in which the addition of biochemical cues and degradation of biophysical cues were controlled in time.[151]

9.5.2 Soluble Cues

The incorporation of soluble cues, such as growth factors, cytokines and chemokines, in biomaterial-based mimics of the native ECM can be done most simply by allowing their diffusion into the material. However, there is little control over the dynamics with this method unless they have been incorporated using more sophisticated devices (*e.g.*, microfluidics). To address this, methods have been designed to sequester soluble cues to allow their controlled release within synthetic microenvironments and better mimic their presentation in the native ECM. Examples of these methods to sequester and release soluble cues will be described in this section.

9.5.2.1 Growth Factor Binding. Binding of soluble cues to synthetic matrices through interactions with large biomolecules such as heparin or protein-mimicking peptides can provide insight into how the sequestration and release of factors over time within the native ECM influences cell function and can be used to direct cell function and ultimately fate. Most studies on heparin binding of growth factors have been focused on *in-vivo* drug delivery and therapeutic applications; however, its addition to mimics of the ECM to direct cell behavior would prove useful in providing control over the dynamic release of soluble cues. One study in using heparin for controlled growth factor release describes a synthetic mimic of the ECM for the *in-vitro* culture of hepatocytes.[157] Thiolated heparin was reacted with PEG diacrylate in a cytocompatible Michael-type addition mechanism to encapsulate cells within the heparin-based hydrogel. Human growth hormone (HGF) added to these heparin-based gels was sequestered along the strands of heparin and released at a slower rates than hydrogels made with only PEG, indicating a controlled release mechanism. Hepatocytes cultured in the heparin hydrogels with HGF maintained their hepatic phenotype for longer periods than those cultured in gels without heparin and the PEG-only gels (Figure 9.10). This type of approach to designing a material with controlled release of soluble cues could be applied to other culture systems to better mimic the behavior of soluble cues in the native ECM and demonstrate cellular response to its dynamic changes.

Biomimetic peptide sequences also have been used to sequester soluble factors to direct cellular function in synthetic scaffolds. Targeting cell populations with small proteins and biomolecules such as growth factors presents a challenge in designing ECM mimics due to their activity at low concentrations and short half-lives *in vivo*. One example of materials aimed at addressing this issue is hydrogels containing peptides that have been designed with enhanced binding affinity for FGF-2. The sequence KRTGQYKL, which exhibits binding affinity for FGF-2 was covalently linked into the gels to sequester the factor after its diffusion into the material. Enhanced FGF-2 binding was achieved by the conjugation of multiple peptides to poly(acrylic acid) polymer chains prior to hydrogel polymerization. The spacing created by the poly(acrylic acid) chain decreased steric hindrance, and the presence of multiple peptides increased the number of binding sites, creating greater affinity for the growth factor. This increase in affinity binding also resulted in extended growth factor release, with potential in applications such as controlled growth factor delivery and directing cell fate.[158] This work has motivated subsequent research in designing

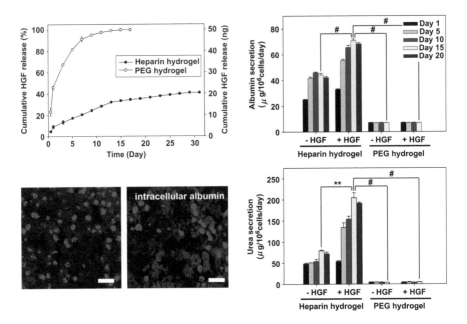

Figure 9.10   Example of growth factor binding. The release of HGF sequestered within heparin-based hydrogels is slower than the release observed in PEG-based hydrogels, in which all of the HGF is released in 15–20 days. Hepatocytes cultured within the heparin-based hydrogels secreted albumin and urea, indicating hepatic function, which was further upregulated by the addition of HGF. The PEG-based hydrogels without did not support hepatic function even after the addition of HGF. Reproduced with permission from ref. 157. © 2010 Elsevier.

other peptide-functionalized biomaterials that bind growth factors to maintain bioavailability and provide protection from exposure to UV polymerization conditions while allowing controlled release of the factor over time.[137]

9.5.2.2 Encapsulation of Polymer Microspheres within Synthetic Matrices.

The release of soluble cues from microspheres embedded in synthetic materials is a less studied method for controlled release of biochemical cues, nevertheless, it has led to the design of novel dynamic materials to mimic changes in the native ECM. Liming Bian, *et al.* reported the encapsulation of alginate microspheres in hyaluronic acid hydrogels for the controlled release of TGF-B3. Coated alginate microspheres demonstrated slower release rates and a lower initial burst than uncoated alginate microspheres, promoting the survival and chondrogenesis of hMSCs in the culture system. This controlled release mechanism provides yet another tool for researchers to mimic changes in the dynamic ECM.

9.6 CONCLUSION

The ECM changes in time and is key in influencing cellular functions. Towards understanding these complex interactions, synthetic biomaterials with temporal property control have been created to mimic the dynamic ECM. Numerous covalent and non-covalent chemistries have been used to form hydrogel-based biomaterials and modulate the biomechanical and biochemical properties of synthetic matrices in the presence of cells, providing a range of tools for the design of dynamic ECM mimics. *In situ* measurement techniques have been developed to observe temporal changes in both material and biological response. Additionally, property changes can be predicted based on the kinetics of the reaction chemistries and the connectivity of the underlying polymer network. Designing materials to mimic dynamic changes in the native ECM is an important area of research, and the information provided here demonstrates the progress made by scientists in creating methods to understand biological systems, especially tissue regeneration and disease, *in vitro* and to design new therapies.

REFERENCES

1. C.-C. Lin and A. T. Metters, *Adv. Drug Delivery Rev.*, 2006, **58**, 1379–1408.
2. N. A. Peppas, J. Z. Hilt, A. Khademhosseini and R. Langer, *Adv. Mater.*, 2006, **18**, 1345–1360.
3. B. V. Slaughter, S. S. Khurshid, O. Z. Fisher, A. Khademhosseini and N. A. Peppas, *Adv. Mater.*, 2009, **21**, 3307–3329.
4. N. Kumar, R. S. Langer and A. J. Domb, *Adv. Drug Delivery Rev.*, 2002, **54**, 889–910.
5. K. E. Uhrich, S. M. Cannizzaro, R. S. Langer and K. M. Shakesheff, *Chem. Rev.*, 1999, **99**, 3181–3198.
6. S. Sershen and J. West, *Adv. Drug Delivery Rev.*, 2002, **54**, 1225–1235.
7. M. Biondi, F. Ungaro, F. Quaglia and P. A. Netti, *Adv. Drug Delivery Rev.*, 2008, **60**, 229–242.
8. K. S. Anseth, A. T. Metters, S. J. Bryant, P. J. Martens, J. H. Elisseeff and C. N. Bowman, *J. Controlled Release*, 2002, **78**, 199–209.
9. P. A. Gunatillake and R. Adhikari, *Eur. Cells Mater.*, 2003, **5**, 1–16.
10. M. Lutolf and J. Hubbell, *Nat. Biotechnol.*, 2005, **23**, 47–55.
11. J. Patterson, M. M. Martino and J. A. Hubbell, *Mater. Today*, 2010, **13**, 14–22.
12. J. S. Katz and J. A. Burdick, *Macromol. Biosci.*, 2010, **10**, 339–348.

13. M. Mehta, K. Schmidt-Bleek, G. N. Duda and D. J. Mooney, *Adv. Drug Delivery Rev.*, 2012, **64**, 1257–1276.
14. M. W. Tibbitt and K. S. Anseth, *Biotechnol. Bioeng.*, 2009, **103**, 655–663.
15. M. S. Rehmann and A. M. Kloxin, *Soft Matter*, 2013, **9**, 6737–6746.
16. J. S. Mohammed and W. L. Murphy, *Adv. Mater.*, 2009, **21**, 2361–2374.
17. C. M. Nelson and M. J. Bissell, *Annu. Rev. Cell Dev. Biol.*, 2006, **22**, 287.
18. R. Xu, A. Boudreau and M. J. Bissell, *Cancer Metastasis Rev.*, 2009, **28**, 167–176.
19. M. D. Shoulders and R. T. Raines, *Annu. Rev. Biochem.*, 2009, **78**, 929.
20. A. J. Bruce Alberts, J. Lewis, M. Raff, K. Roberts and P. Walter, *Molecular Biology of the Cell*, Garland Science, New York, 4th edn, 2002.
21. P. Singh, C. Carraher and J. E. Schwarzbauer, *Annu. Rev. Cell Dev. Biol.*, 2010, **26**, 397.
22. M. Durbeej, *Cell Tissue Res.*, 2010, **339**, 259–268.
23. J. S. Desgrosellier and D. A. Cheresh, *Nat. Rev. Cancer*, 2010, **10**, 9–22.
24. A. J. Singer and R. Clark, *N. Engl. J. Med.*, 1999, **341**, 738–746.
25. P. Martin, *Science*, 1997, **276**, 75–81.
26. K. H. Kwan, X. Liu, M. K. To, K. W. Yeung, C.-m. Ho and K. K. Wong, *Nanomed.: Nanotechnol., Biol. Med.*, 2011, **7**, 497–504.
27. M. P. Lutolf, G. P. Raeber, A. H. Zisch, N. Tirelli and J. A. Hubbell, *Adv. Mater.*, 2003, **15**, 888–892.
28. J. L. West and J. A. Hubbell, *React. Polym.*, 1995, **25**, 139–147.
29. J. A. Burdick and K. S. Anseth, *Biomaterials*, 2002, **23**, 4315–4323.
30. N. Yamaguchi, L. Zhang, B.-S. Chae, C. S. Palla, E. M. Furst and K. L. Kiick, *J. Am. Chem. Soc.*, 2007, **129**, 3040–3041.
31. R. H. Schmedlen, K. S. Masters and J. L. West, *Biomaterials*, 2002, **23**, 4325–4332.
32. Y. Luo, K. R. Kirker and G. D. Prestwich, *J. Controlled Release*, 2000, **69**, 169–184.
33. J. A. Burdick, C. Chung, X. Jia, M. A. Randolph and R. Langer, *Biomacromolecules*, 2005, **6**, 386–391.
34. J. A. Rowley, G. Madlambayan and D. J. Mooney, *Biomaterials*, 1999, **20**, 45–53.
35. L. E. O'Leary, J. A. Fallas, E. L. Bakota, M. K. Kang and J. D. Hartgerink, *Nat. Chem.*, 2011, **3**, 821–828.

36. H. J. Lee, J.-S. Lee, T. Chansakul, C. Yu, J. H. Elisseeff and S. M. Yu, *Biomaterials*, 2006, **27**, 5268–5276.
37. B. Brodsky, G. Thiagarajan, B. Madhan and K. Kar, *Biopolymers*, 2008, **89**, 345–353.
38. O. D. Krishna and K. L. Kiick, *Biomacromolecules*, 2009, **10**, 2626–2631.
39. A. Bertz, S. Wöhl-Bruhn, S. Miethe, B. Tiersch, J. Koetz, M. Hust, H. Bunjes and H. Menzel, *J. Biotechnol.*, 2012, **163**, 243–249.
40. W. J. Li, C. T. Laurencin, E. J. Caterson, R. S. Tuan and F. K. Ko, *J. Biomed. Mater. Res.*, 2002, **60**, 613–621.
41. J.-Y. Rho, L. Kuhn-Spearing and P. Zioupos, *Med. Eng. Phys.*, 1998, **20**, 92–102.
42. J. P. Winer, P. A. Janmey, M. E. McCormick and M. Funaki, *Tissue Eng., Part A*, 2008, **15**, 147–154.
43. I. Levental, P. C. Georges and P. A. Janmey, *Soft Matter*, 2007, **3**, 299–306.
44. J. M. Mansour, *Kinesiology*, 2004, **2**, 66–79.
45. P. Agache, C. Monneur, J. Leveque and J. De Rigal, *Arch. Dermatol. Res.*, 1980, **269**, 221–232.
46. C. Frantz, K. M. Stewart and V. M. Weaver, *J. Cell Sci.*, 2010, **123**, 4195–4200.
47. F. Liu, J. D. Mih, B. S. Shea, A. T. Kho, A. S. Sharif, A. M. Tager and D. J. Tschumperlin, *J. Cell Biol.*, 2010, **190**, 693–706.
48. D. E. Discher, P. Janmey and Y.-l. Wang, *Science*, 2005, **310**, 1139–1143.
49. A. M. Kloxin, A. M. Kasko, C. N. Salinas and K. S. Anseth, *Science*, 2009, **324**, 59–63.
50. A. Metters and J. Hubbell, *Biomacromolecules*, 2005, **6**, 290–301.
51. P. J. Martens, S. J. Bryant and K. S. Anseth, *Biomacromolecules*, 2003, **4**, 283–292.
52. A. T. Metters, C. N. Bowman and K. S. Anseth, *J. Phys. Chem. B*, 2000, **104**, 7043–7049.
53. P. Bao, A. Kodra, M. Tomic-Canic, M. S. Golinko, H. P. Ehrlich and H. Brem, *J. Surg. Res.*, 2009, **153**, 347–358.
54. R. J. Wordinger and A. F. Clark, Growth Factors and Neurotrophic Factors as Targets, in *Ocular Therapeutics: Eye on New Discoveries,* ed. A. C. Thomas Yorio and M. B. Wax, Academic Press, New York, NY, 1st edn, 2008, pp. 87–88.
55. D. J. Cua and C. M. Tato, *Nat. Rev. Immunol.*, 2010, **10**, 479–489.
56. M. J. Cameron and D. J. Kelvin, *Cytokines, Chemokines and Their Receptors*, Landes Bioscience, Austin, TX, 2000.

57. A. Crawford, J. M. Angelosanto, K. L. Nadwodny, S. D. Blackburn and E. J. Wherry, *PLoS Pathog.*, 2011, **7**, e1002098.
58. A. Leask and D. J. Abraham, *FASEB J.*, 2004, **18**, 816–827.
59. S. Sant, M. J. Hancock, J. P. Donnelly, D. Iyer and A. Khademhosseini, *Can. J. Chem. Eng.*, 2010, **88**, 899–911.
60. F. Brandl, F. Sommer and A. Goepferich, *Biomaterials*, 2007, **28**, 134–146.
61. C. A. DeForest and K. S. Anseth, *Angew. Chem.*, 2012, **124**, 1852–1855.
62. A. H. Zisch, M. P. Lutolf and J. A. Hubbell, *Cardiovasc. Pathol.*, 2003, **12**, 295–310.
63. J. J. Rice, M. M. Martino, L. De Laporte, F. Tortelli, P. S. Briquez and J. A. Hubbell, *Adv. Healthcare Mater.*, 2013, **2**, 57–71.
64. E. Ruoslahti, *Annu. Rev. Cell Dev. Biol.*, 1996, **12**, 697–715.
65. S. P. Massia and J. A. Hubbell, *J. Cell Biol.*, 1991, **114**, 1089–1100.
66. D. R. Clemmons, *Mol. Cell. Endocrinol.*, 1998, **140**, 19–24.
67. C. M. Kirschner and K. S. Anseth, *Acta Mater.*, 2013, **61**, 931–944.
68. G. S. Schultz and A. Wysocki, *Wound Repair Regener.*, 2009, **17**, 153–162.
69. N. E. Fedorovich, M. H. Oudshoorn, D. van Geemen, W. E. Hennink, J. Alblas and W. J. Dhert, *Biomaterials*, 2009, **30**, 344–353.
70. S. J. Bryant, C. R. Nuttelman and K. S. Anseth, *J. Biomater. Sci., Polym. Ed.*, 2000, **11**, 439–457.
71. M. Guvendiren and J. A. Burdick, *Nat. Commun.*, 2012, **3**, 792.
72. M. W. Tibbitt, A. M. Kloxin, K. U. Dyamenahalli and K. S. Anseth, *Soft Matter*, 2010, **6**, 5100–5108.
73. B. D. Fairbanks, M. P. Schwartz, C. N. Bowman and K. S. Anseth, *Biomaterials*, 2009, **30**, 6702–6707.
74. C. G. Williams, A. N. Malik, T. K. Kim, P. N. Manson and J. H. Elisseeff, *Biomaterials*, 2005, **26**, 1211–1218.
75. H. Shih and C. C. Lin, *Macromol. Rapid Commun.*, 2012, **34**, 269–273.
76. J. Torgersen, X. H. Qin, Z. Li, A. Ovsianikov, R. Liska and J. Stampfl, *Adv. Funct. Mater.*, 2013, **23**, 4524–4554.
77. B. J. Adzima, Y. Tao, C. J. Kloxin, C. A. DeForest, K. S. Anseth and C. N. Bowman, *Nat. Chem.*, 2011, **3**, 256–259.
78. G. G. Odian, *Principles of Polymerization*, John Wiley & Sons, Inc., Hoboken, NJ, 4th edn, 2007, pp. 216–217.
79. L. M. Johnson, B. D. Fairbanks, K. S. Anseth and C. N. Bowman, *Biomacromolecules*, 2009, **10**, 3114–3121.

80. P. S. Hume, C. N. Bowman and K. S. Anseth, *Biomaterials*, 2011, **32**, 6204–6212.
81. J. S. Temenoff, H. Shin, D. E. Conway, P. S. Engel and A. G. Mikos, *Biomacromolecules*, 2003, **4**, 1605–1613.
82. A. M. Kloxin, M. W. Tibbitt and K. S. Anseth, *Nat. Protoc.*, 2010, **5**, 1867–1887.
83. Y. Luo and M. S. Shoichet, *Nat. Mater.*, 2004, **3**, 249–253.
84. C. E. Hoyle and C. N. Bowman, *Angew. Chem., Int. Ed.*, 2010, **49**, 1540–1573.
85. C. E. Hoyle, A. B. Lowe and C. N. Bowman, *Chem. Soc. Rev.*, 2010, **39**, 1355–1387.
86. M. Lutolf, J. Lauer-Fields, H. Schmoekel, A. Metters, F. Weber, G. Fields and J. Hubbell, *Proc. Natl. Acad. Sci. U. S. A.*, 2003, **100**, 5413–5418.
87. Y. Fu and W. J. Kao, *J. Biomed. Mater. Res., Part A*, 2011, **98**, 201–211.
88. C. N. Salinas and K. S. Anseth, *Macromolecules*, 2008, **41**, 6019–6026.
89. M. S. Hahn, L. J. Taite, J. J. Moon, M. C. Rowland, K. A. Ruffino and J. L. West, *Biomaterials*, 2006, **27**, 2519–2524.
90. J. A. Johnson, J. M. Baskin, C. R. Bertozzi, J. T. Koberstein and N. J. Turro, *Chem. Commun.*, 2008, 3064–3066.
91. J. M. Baskin, J. A. Prescher, S. T. Laughlin, N. J. Agard, P. V. Chang, I. A. Miller, A. Lo, J. A. Codelli and C. R. Bertozzi, *Proc. Natl. Acad. Sci. U. S. A.*, 2007, **104**, 16793–16797.
92. D. L. Alge, M. A. Azagarsamy, D. F. Donohue and K. S. Anseth, *Biomacromolecules*, 2013, **14**, 949–953.
93. M. L. Blackman, M. Royzen and J. M. Fox, *J. Am. Chem. Soc.*, 2008, **130**, 13518–13519.
94. T. Nie, A. Baldwin, N. Yamaguchi and K. L. Kiick, *J. Controlled Release*, 2007, **122**, 287–296.
95. E. A. Phelps, N. O. Enemchukwu, V. F. Fiore, J. C. Sy, N. Murthy, T. A. Sulchek, T. H. Barker and A. J. García, *Adv. Mater.*, 2012, **24**, 64–70.
96. A. D. Baldwin and K. L. Kiick, *Bioconjugate Chem.*, 2011, **22**, 1946–1953.
97. B. D. Fairbanks, S. P. Singh, C. N. Bowman and K. S. Anseth, *Macromolecules*, 2011, **44**, 2444–2450.
98. X. Z. Shu, Y. Liu, Y. Luo, M. C. Roberts and G. D. Prestwich, *Biomacromolecules*, 2002, **3**, 1304–1311.
99. X. Zheng Shu, Y. Liu, F. S. Palumbo, Y. Luo and G. D. Prestwich, *Biomaterials*, 2004, **25**, 1339–1348.

100. H. Nandivada, X. Jiang and J. Lahann, *Adv. Mater.*, 2007, **19**, 2197–2208.
101. K. C. Koehler, K. S. Anseth and C. N. Bowman, *Biomacromolecules*, 2013, **14**, 538–547.
102. J. L. West and J. A. Hubbell, *Macromolecules*, 1999, **32**, 241–244.
103. Y. Dong, S. Liao, M. Ngiam, C. K. Chan and S. Ramakrishna, *Tissue Eng., Part B*, 2009, **15**, 333–351.
104. J. Patterson and J. A. Hubbell, *Biomaterials*, 2010, **31**, 7836–7845.
105. A. A. Aimetti, M. W. Tibbitt and K. S. Anseth, *Biomacromolecules*, 2009, **10**, 1484–1489.
106. H. Zhao, E. S. Sterner, E. B. Coughlin and P. Theato, *Macromolecules*, 2012, **45**, 1723–1736.
107. C. P. Holmes, *J. Org. Chem.*, 1997, **62**, 2370–2380.
108. D. R. Griffin and A. M. Kasko, *ACS Macro Lett.*, 2012, **1**, 1330–1334.
109. D. R. Griffin and A. M. Kasko, *J. Am. Chem. Soc.*, 2012, **134**, 13103–13107.
110. D. R. Griffin, J. L. Schlosser, S. F. Lam, T. H. Nguyen, H. D. Maynard and A. M. Kasko, *Biomacromolecules*, 2013, **14**, 1199–1207.
111. Y. Zhao, Q. Zheng, K. Dakin, K. Xu, M. L. Martinez and W.-H. Li, *J. Am. Chem. Soc.*, 2004, **126**, 4653–4663.
112. J. H. Wosnick and M. S. Shoichet, *Chem. Mater.*, 2007, **20**, 55–60.
113. R. G. Wylie and M. S. Shoichet, *J. Mater. Chem.*, 2008, **18**, 2716–2721.
114. D. Y. Wong, D. R. Griffin, J. Reed and A. M. Kasko, *Macromolecules*, 2010, **43**, 2824–2831.
115. V. San Miguel, C. G. Bochet and A. del Campo, *J. Am. Chem. Soc.*, 2011, **133**, 5380–5388.
116. A. P. Pelliccioli and J. Wirz, *Photochem. Photobiol. Sci.*, 2002, **1**, 441–458.
117. R. S. Givens, M. Rubina and J. Wirz, *Photochem. Photobiol. Sci.*, 2012, **11**, 472–488.
118. A. M. Kloxin, M. W. Tibbitt, A. M. Kasko, J. A. Fairbairn and K. S. Anseth, *Adv. Mater.*, 2010, **22**, 61–66.
119. M. W. Tibbitt, A. M. Kloxin and K. S. Anseth, *J. Polym. Sci., Part A: Polym. Chem.*, 2013, **51**, 1899–1911.
120. M. W. Tibbitt, A. M. Kloxin, L. A. Sawicki and K. S. Anseth, *Macromolecules*, 2013, **46**, 2785–2792.
121. Y. S. Jo, J. Gantz, J. A. Hubbell and M. P. Lutolf, *Soft Matter*, 2009, **5**, 440–446.

122. R. A. Larson and E. J. Weber, *Reaction mechanisms in environmental organic chemistry*, CRC press, 1994.

123. P. J. Bracher, P. W. Snyder, B. R. Bohall and G. M. Whitesides, *Origins Life Evol. Biospheres*, 2011, **41**, 399–412.

124. S. J. Bryant and K. S. Anseth, *Scaffolding In Tissue Engineering*, CRC Press, 2005, pp. 71–90.

125. A. T. Metters, K. S. Anseth and C. N. Bowman, *J. Phys. Chem. B*, 2001, **105**, 8069–8076.

126. M. A. Cejas, W. A. Kinney, C. Chen, J. G. Vinter, H. R. Almond, K. M. Balss, C. A. Maryanoff, U. Schmidt, M. Breslav and A. Mahan, *Proc. Natl. Acad. Sci. U. S. A.*, 2008, **105**, 8513–8518.

127. C. J. Bowerman and B. L. Nilsson, *J. Am. Chem. Soc.*, 2010, **132**, 9526–9527.

128. W. A. Petka, J. L. Harden, K. P. McGrath, D. Wirtz and D. A. Tirrell, *Science*, 1998, **281**, 389–392.

129. J. D. Hartgerink, E. Beniash and S. I. Stupp, *Science*, 2001, **294**, 1684–1688.

130. J. P. Schneider, D. J. Pochan, B. Ozbas, K. Rajagopal, L. Pakstis and J. Kretsinger, *J. Am. Chem. Soc.*, 2002, **124**, 15030–15037.

131. A. P. Nowak, V. Breedveld, L. Pakstis, B. Ozbas, D. J. Pine, D. Pochan and T. J. Deming, *Nature*, 2002, **417**, 424–428.

132. C. Guo and T. S. Bailey, *Soft Matter*, 2010, **6**, 4807–4818.

133. J. N. Hunt, K. E. Feldman, N. A. Lynd, J. Deek, L. M. Campos, J. M. Spruell, B. M. Hernandez, E. J. Kramer and C. J. Hawker, *Adv. Mater.*, 2011, **23**, 2327–2331.

134. C. LoPresti, M. Massignani, C. Fernyhough, A. Blanazs, A. J. Ryan, J. Madsen, N. J. Warren, S. P. Armes, A. L. Lewis and S. Chirasatitsin, *ACS nano*, 2011, **5**, 1775–1784.

135. M. Krogsgaard, M. A. Behrens, J. S. Pedersen and H. Birkedal, *Biomacromolecules*, 2013, **14**, 297–301.

136. D. B. Pike, S. Cai, K. R. Pomraning, M. A. Firpo, R. J. Fisher, X. Z. Shu, G. D. Prestwich and R. A. Peattie, *Biomaterials*, 2006, **27**, 5242–5251.

137. J. D. McCall, C.-C. Lin and K. S. Anseth, *Biomacromolecules*, 2011, **12**, 1051–1057.

138. C. J. Wilson, R. E. Clegg, D. I. Leavesley and M. J. Pearcy, *Tissue Eng.*, 2005, **11**, 1–18.

139. Z. Sui, W. J. King and W. L. Murphy, *Adv. Mater.*, 2007, **19**, 3377–3380.

140. A. W. York, C. W. Scales, F. Huang and C. L. McCormick, *Biomacromolecules*, 2007, **8**, 2337–2341.

141. C.-C. Lin, A. T. Metters and K. S. Anseth, *Biomaterials*, 2009, **30**, 4907–4914.
142. S.-H. Lee, J. J. Moon, J. S. Miller and J. L. West, *Biomaterials*, 2007, **28**, 3163–3170.
143. K. M. Schultz, A. D. Baldwin, K. L. Kiick and E. M. Furst, *ACS Macro Lett.*, 2012, **1**, 706–708.
144. K. M. Schultz and K. S. Anseth, *Soft Matter*, 2013, **9**, 1570–1579.
145. J. Tian, S. Alimperti, P. Lei and S. T. Andreadis, *Lab Chip*, 2010, **10**, 1967–1975.
146. M. H. Zaman, L. M. Trapani, A. L. Sieminski, D. MacKellar, H. Gong, R. D. Kamm, A. Wells, D. A. Lauffenburger and P. Matsudaira, *Proc. Natl. Acad. Sci. U. S. A.*, 2006, **103**, 10889–10894.
147. A. E. Carpenter, T. R. Jones, M. R. Lamprecht, C. Clarke, I. H. Kang, O. Friman, D. A. Guertin, J. H. Chang, R. A. Lindquist and J. Moffat, *Genome Biol.*, 2006, **7**, R100.
148. K. Ohki, S. Chung, Y. H. Ch'ng, P. Kara and R. C. Reid, *Nature*, 2005, **433**, 597–603.
149. A. M. Kloxin, J. A. Benton and K. S. Anseth, *Biomaterials*, 2010, **31**, 1–8.
150. H. Wang, S. M. Haeger, A. M. Kloxin, L. A. Leinwand and K. S. Anseth, *PloS One*, 2012, **7**, e39969.
151. C. A. DeForest and K. S. Anseth, *Nat. Chem.*, 2011, **3**, 925–931.
152. S. P. Zustiak and J. B. Leach, *Biomacromolecules*, 2010, **11**, 1348–1357.
153. O. Jeon, K. H. Bouhadir, J. M. Mansour and E. Alsberg, *Biomaterials*, 2009, **30**, 2724–2734.
154. C. M. Kirschner and K. S. Anseth, *Small*, 2012, **9**, 578–584.
155. M. Guvendiren and J. A. Burdick, *Adv. Healthcare Mater.*, 2013, **2**, 155–164.
156. K. A. Mosiewicz, L. Kolb, A. J. van der Vlies, M. M. Martino, P. S. Lienemann, J. A. Hubbell, M. Ehrbar and M. P. Lutolf, *Nat. Mater.*, 2013, **12**, 1072–1078.
157. M. Kim, J. Y. Lee, C. N. Jones, A. Revzin and G. Tae, *Biomaterials*, 2010, **31**, 3596–3603.
158. C. C. Lin and K. S. Anseth, *Adv. Funct. Mater.*, 2009, **19**, 2325–2331.

CHAPTER 10

Biomaterials: Supramolecular Artificial Extracellular Matrices

GREGORY A. HUDALLA*[a] AND JOEL H. COLLIER*[b]

[a]J. Crayton Pruitt Family Department of Biomedical Engineering, University of Florida, 1275 Center Drive, Gainesville, FL 32611, USA; [b]Department of Surgery, University of Chicago, 5841 S. Maryland Ave, Chicago, IL 60637, USA
*E-mail: ghudalla@bme.ufl.edu, collier@uchicago.edu

10.1 ARTIFICIAL EXTRACELLULAR MATRICES BASED ON NATURALLY DERIVED BIOMOLECULES: MULTIFUNCTIONAL BUT NOT ALWAYS MODULAR

Many of the biomolecules that impart natural extracellular matrices (ECMs) with cell-instructive and cell-responsive features, such as collagen, elastin, fibrin, hyaluronic acid, and heparin, can be extracted from mammalian tissues and fabricated into artificial ECMs (aECMs) having functional properties related to the integrated molecule.[1-12] However, because the extraction and purification processes are intentionally designed to isolate one molecule from a complex, multicomponent mixture, they inherently disrupt the hierarchy and organization of biomolecules that underlie natural ECM function, as discussed in detail in Chapter 4, by Hamill *et al.* In addition, aECMs based on a single naturally derived biomolecule, such as collagen or fibrin, will lack the diversity of functional features inherent to native

Mimicking the Extracellular Matrix: The Intersection of Matrix Biology and Biomaterials
Edited by Gregory A. Hudalla and William L. Murphy
© The Royal Society of Chemistry 2020
Published by the Royal Society of Chemistry, www.rsc.org

ECMs, whose composition is considerably more complex. For example, although collagenous aECMs will likely enable cell adhesion *via* collagen-binding integrins, such as $\alpha_1\beta_1$ and $\alpha_2\beta_1$ integrins, these materials will likely be unable to mediate cell adhesion *via* integrin receptors, such as $\alpha_v\beta_3$, which preferentially bind to laminin or fibronectin domains that are often present in close association with collagen in natural ECMs.[13]

Multicomponent aECMs are therefore an interesting alternative for approaching the diversity of functional features found within natural ECMs. However, combinatorial effects resulting from simply mixing different isolated biomolecules into a multicomponent aECM often cannot be predicted *a priori*. For example, in a survey of aECMs consisting of collagen I, fibronectin, laminin, and hyaluronic acid for neurite extension, Schmidt and co-workers determined that laminin had a strong, dose-dependent effect on neurite length and outgrowth, while collagen I and hyaluronic acid concentration had no significant effect on neurite extension.[14] Alternatively, ECMs produced by cells or harvested directly from mammalian tissues are receiving significant attention as the basis for aECMs, having physiologically relevant combinations of cell-instructive and cell-responsive features. For example, Matrigel™ is a basement membrane-like extract isolated from Engelbreth-Holm-Swarm sarcoma cells that is rich in the ECM biomolecules laminin, collagen IV, heparin proteoglycans, and entactin.[15] Upon warming from refrigerated temperatures to room temperature, Matrigel undergoes a liquid-to-solid transition, forming a complex biologically active aECM that can guide a range of cellular behaviors, including migration, proliferation, and differentiation. Owing to its useful thermal gelation properties and its ability to support these cellular processes, Matrigel is widely used for three-dimensional cell culture and the transplantation of xenogeneic cells into experimental animals.[16,17]

Although Matrigel provides a closer approximation of natural basement membrane composition, the processing conditions required to render this solid ECM soluble and re-form the solid network in the presence of cells are unlikely to maintain the molecular hierarchy of native tissue. An alternative method that can maintain both natural ECM composition and hierarchical structure involves creating an "empty" or "ghost" matrix by decellularizing whole tissue using various chemical or physical treatments.[18] Decellularized ECMs can subsequently be repopulated with cells that engraft within the matrix, migrate to physiologically appropriate locations, and organize into multi-cellular structures similar to those observed within native tissues. In a few striking examples, such an approach has generated bioengineered hearts, lungs, kidneys, and livers with basic physiological function.[19-23]

Together, these examples suggest an important correlation between ECM composition and hierarchical organization for directing cell and tissue formation, which should be considered in the design of aECMs. Despite providing advantages over single-component aECMs, however, naturally derived multicomponent aECMs are not without their limitations. As discussed in Chapter 4, Matrigel and decellularized ECM preparations are subject to batch-to-batch variability that is often dependent on tissue processing conditions, which complicates efforts to determine the composition of cell-instructive and cell-responsive features inherent in these aECMs. For example, Matrigel consists of hundreds of different proteins,[24] and although not explicitly characterized yet, "empty" or "ghost" decellularized ECMs are likely to be similarly complex. This introduces significant challenges with regard to elucidating the role of different features in guiding cell behavior, as experimental control is sacrificed in the face of such complexity. In addition, it is difficult to systematically vary the type, ratio, or organization of cell-instructive or cell-responsive features within Matrigel or decellularized ECMs to optimize cell responses, since specific biomolecules must be selectively altered or removed without perturbing the structure or function of other biomolecules contained therein.

In light of the practical challenges associated with engineering aECMs comprised solely of naturally derived biomolecules, there is growing interest in aECMs based on synthetic biomaterials that provide a "blank slate" that can be modified with precise combinations of cell-instructive or cell-responsive features according to the needs of the intended application. Many such aECMs were discussed at length in the preceding chapters, and therefore will not be discussed in detail here. Additional discussion of "hybrid" or "composite" aECMs created from mixtures of synthetic polymers and ECM-derived biomolecules is provided in many excellent articles and chapters on this topic.[25-33] Instead, the remainder of this chapter will focus on a particular class of multifunctional aECMs that provide modular design and finely tunable control of composition, structure, and function: *self-assembled peptide nanofibers.*

10.2 HARNESSING MOLECULAR SELF-ASSEMBLY TO CREATE aECMs

Protein self-assembly into fibrous structures underlies the formation of natural ECMs.[34] For example, the ECM "basement membrane" is formed *via* the self-assembly of laminin with collagen IV and entactin, as described in detail in Chapter 4. Collagen fibers, which are the primary structural component of connective tissue ECMs, self-assemble from triple helical pro-collagen molecules that are synthesized inside a cell and are proteolytically cleaved by cell surface enzymes once secreted into the extracellular space.[34] Elastin fibers that provide recoil properties to mechanically compliant ECMs are created in a sequential process in which cell-secreted tropoelastin monomers first self-assemble *via* their hydrophobic domains—a process often referred to as coacervation—followed by the association of coacervate droplets with fibrillin-rich microfibrils.[34] However, the biological processes integral to natural ECM self-assembly are complicated, relying as they do on protein synthesis, processing in the endoplasmic reticulum, the action of chaperones, highly regulated and spatially controlled secretion pathways, and the positioning and phenotypes of the secreting cells. Thus, self-assembly strategies for building aECMs based on less complicated, more technically achievable processes represent promising and practical approaches for engineering biomaterials with the compositional diversity and hierarchical organization of their natural counterparts.

10.2.1 Peptide Self-Assembly into Fibrillar Peptide Networks

Synthetic peptides can be engineered to self-assemble into fibrillar architectures, giving rise to supramolecular networks having morphological similarities to natural ECMs.[35–43] In general, peptides engineered to self-assemble into fibrillar architectures can be subdivided into two classes: (1) those designed to assemble *via* common natural secondary structure motifs, such as coiled-coils and β-sheets, by exploiting rules relating amino acid primary sequence to protein folding; and (2) those leveraging folding and assembly designs not found in native ECMs, such as "peptide amphiphiles" having appended hydrophobic tails, or those modified with aromatic moieties that assemble *via* π–π stacking. In the following sections, we will discuss molecular design principles and highlight examples of fibrillar peptide networks based on peptides from each of these classes.

10.2.1.1 β-Sheet-Forming Peptides. Important theoretical observations by Pauling and Corey in the 1950s,[44,45] in conjunction with pioneering experimental work by Walton and colleagues in the 1970s,[46,47] and Meredith and Osterman in the 1980s,[48,49] established the general guideline that peptides with alternating polar and non-polar amino acids can self-assemble into β-sheets *via* H-bonds formed between amide backbone groups of extended β-strands. Of pivotal importance to the use of β-sheet fibrillizing peptides as aECMs, however, was the demonstration by Zhang and colleagues in 1993 that an alternating polar/non-polar peptide having ionic self-complementarity, Ac-(AE-AEAKAK)$_2$-NH$_2$ (EAK16-II), self-assembled into β-sheet fibrils in water and further assembled into an insoluble macroscopic membrane in the presence of salt.[50] Initially, it was proposed that membrane assembly was mediated by formation of staggered pleated β-sheets *via* charge–charge interactions between Glu and Lys residues and hydrophobic interactions between Ala residues. However, later studies determined that the membrane was largely composed of entagled "fibrils",[51] in which β-sheets are associated into bilayers or higher-ordered laminate assemblies *via* interactions between hydrophobic faces or hydrophilic faces. Since these early demonstrations, there has remained a growing interest in understanding correlations between peptide sequence, β-sheet fibrillization, and hydrogel formation. Importantly, Zhang and colleagues demonstrated that replacing Glu and Lys in EAK16-II with Arg and Asp provided a peptide, Ac-(RADARADA)$_2$-NH$_2$ (RAD16-I), which self-assembled into β-sheet-rich hydrogels under physiological conditions (Figure 10.1A).[52] This peptide is commercially available as PuraMatrix™, and its use as an aECM is discussed at length in later sections.

An additional significant contribution is the MAX family of peptides designed by Schneider and colleagues, which assemble into β-hairpins under specified pH, ionic strength, and temperature conditions. For example, H$_2$N-VKVKVKVK-V^{D}PPT-KVKVKVKV-NH$_2$ (MAX1) assembles into nanofiber networks above pH = 9,[53] MAX2 and MAX3 (V16T and V7T/V16T mutants of MAX1, respectively) have reduced self-assembly and hydrogelation kinetics that decrease proportionally with the number of V/T mutations,[54] while MAX6 (V16E) is unable to assemble at pH 4, due electrostatic repulsion of deprotonated Glu residues.[55]

In 1997, Aggeli and Boden demonstrated that peptides derived from β-sheet domains of two proteins, IsK and hen egg white lysozyme, can self-assemble into tapes that entangle into gels at appropriate volume fractions.[56] Later, they developed the synthetic polyglutamine P$_{11}$ peptides, which assemble into antiparallel β-sheets *via* complementary interactions between cationic and anionic residues inserted near the N- and C-termini of the peptide, respectively.[57] Notably, Ac-QQRQQQQQEQQ-NH$_2$ (P$_{11}$-1) self-assembled into fibrils ~300 nm in length, whereas Ac-QQRFQWQFEQQ-NH$_2$ (P$_{11}$-2), with an alternating hydrophobic core, self-assembled into fibers that exceed 1 μm in length. In addition, replacing Glu with Gln (Ac-QQRFQWQFQQQ-NH$_2$, P$_{11}$-3) led to precipitate formation above pH 10, whereas exchanging

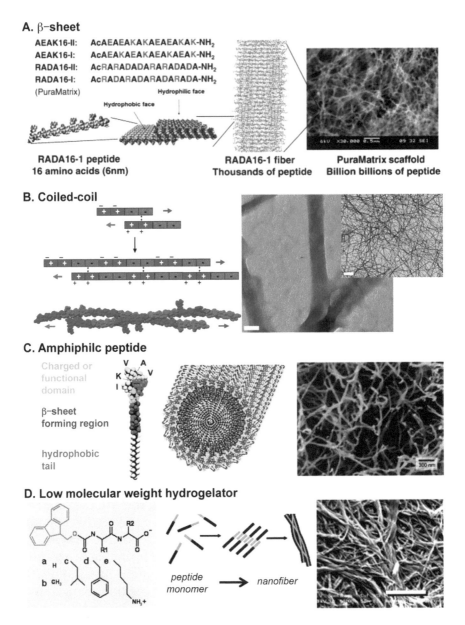

A. β-sheet

AEAK16-II: AcAEAEAKAKAEAEAKAK-NH₂
AEAK16-I: AcAEAKAEAKAEAKAEAK-NH₂
RADA16-II: AcRARADADARARADADA-NH₂
RADA16-I: AcRADARADARADARADA-NH₂
(PuraMatrix)

Hydrophobic face
Hydrophilic face

RADA16-1 peptide
16 amino acids (6nm)

RADA16-1 fiber
Thousands of peptide

PuraMatrix scaffold
Billion billions of peptide

B. Coiled-coil

C. Amphiphilc peptide

Charged or
functional
domain

β–sheet
forming region

hydrophobic
tail

D. Low molecular weight hydrogelator

peptide
monomer → nanofiber

Figure 10.1 Examples of synthetic peptides that self-assemble into nanofibers. (A) Peptides, such as *RADA16* developed by Zhang and colleagues, can self-assemble into β-sheet nanofibers *via* hydrogen-bonding between neighboring β-strands. Adapted with permission from Loo *et al.*, *Biotechnol. Adv.*, 2012.[36] (B) Offset coiled-coils, such as the *SAF peptides* developed by Woolfson and colleagues, can self-assemble into elongated fibers. Adapted with permission from Papapostolou *et al.*, *Proc. Natl. Acad. Sci. U. S. A.*, 2007.[284] (C) Peptides having a hydrophilic headgroup, a β-sheet-forming domain, and a hydrophobic tail, such as the *Peptide Amphiphiles* developed by Stupp and colleagues, can self-assemble into cylindrical nanofibers. Adapted with permission from Cui *et al.*, *Biopolymers*, 2010.[40] (D) Short peptides having an appended aromatic moiety, such as the *Fmoc-dipeptides* developed by Ulijn and colleagues, can self-assemble into nanofibers *via* π–π and hydrophobic interactions. Adapted from Jayawarna *et al.*, *Adv. Mater.*, 2006.[87]

Glu for Gln (Ac-QQRFEWEFEQQ-NH$_2$, P$_{11}$-4) hindered self-assembly at neutral and basic pH.[58] Collier and Messersmith developed a variant of P$_{11}$-2, QQKFQFQFEQQ (Q11),[59] which has been widely used for creating multifunctional aECMs and will be discussed in more detail below. For additional information on correlations between peptide sequence and β-sheet fibrillization, we direct the reader to a detailed review.[38]

10.2.1.2 Coiled-Coil-Forming Peptides. Theoretical observations by Pauling and Crick in the 1950s,[60,61] followed by identification of the leucine zipper motif,[62] and significant experimental work by DeGrado, Kim, Parry, and others,[63–66] established general "design rules" governing the self-assembly of "α-helical coiled-coils", entwined bundles of α-helices. In particular, due to the strict 3.6 residue per turn architecture of the α-helix, peptides with a regular repeated sequence of (hpphppp) (often referred to as the "heptad repeat"; h = hydrophobic, and p = polar), fold into amphipathic α-helices that can preferentially bundle *via* their hydrophobic faces under aqueous conditions.[67] One of the earliest examples of harnessing coiled-coil formation to create a 3-D network was provided by Tirrell and colleagues, who designed an artificial protein that self-assembled in near-neutral aqueous solutions *via* terminal leucine zipper domains flanking a central, water-soluble polyelectrolyte segment.[68] Around this same time, Kojima and colleagues demonstrated that a synthetic peptide, (LETLAKA)$_3$, forms fibrils that are 5–10 nm thick and micrometers long,[69] likely due to slipping of hydrophobic faces into staggered assemblies. Inspired by the latter, in 2000, Woolfson and colleagues rationally designed a heterodimeric staggered coiled-coil having "sticky ends" that directed peptide self-assembly into fibrillar networks,[70] which they refer to as the "self-assembling fiber" (SAF) system. Their design relied on peptides having two positively charged heptads at the N-terminus and two negatively charged heptads at the C-terminus, as well as an asparagine residue in heptad 4 or heptad 1, which specifies co-assembly of the peptides into a heterodimeric coiled-coil having electrostatically complementary overhangs (Figure 10.1B). Since then, other staggered coiled-coil designs, such as the sequence-αFFP pentamer by Kajava *et al.*,[71] and the YZ1 dimer by Conticello *et al.*,[72] as well as an "out of phase" hydrophobic face design by Fairman *et al.*,[73] have been developed to create SAFs based on α-helical peptides. The use of SAFs as aECMs will be discussed later in this chapter.

10.2.1.3 Amphiphilic Peptides. As their name suggests, amphiphilic peptides are peptide-based molecules having a hydrophobic tail and a hydrophilic head group. In general, amphiphilic peptides can be divided into two general types: (1) "peptide amphiphiles" (PAs) consisting of a hydrophilic peptide attached to hydrophobic alkyl tails; and (2) "surfactant-like peptides" (SLPs) having a tail of multiple hydrophobic amino acids, such as glycine, alanine, leucine, or valine, and a hydrophilic amino acid head group.

Self-assembly of PAs is enabled by their architecture (Figure 10.1C), which consists of: (1) a hydrophobic tail that assembles under aqueous conditions above a critical aggregation concentration, similar to surfactants and lipids; (2) a peptide sequence that is able to form intermolecular H-bonds; (3) a charged amino acid to promote solubility; and (4) a functional peptide domain,[40,74] as discussed in greater detail in subsequent sections related to their use as aECMs. The earliest examples of PAs were demonstrated by Fields and Tirrell in 1996, and consisted of a collagen-derived peptide domain linked to a dialkyl tail that self-assembled into a triple-helical polyproline II-like structure under aqueous conditions.[75] The dialkyl tail enabled adsorption of PAs onto hydrophobic substrates, which mediated cell adhesion and intracellular signaling *via* recognition of the alpha 1(IV) 1263–1277 sequence of collagen type IV.[76] The use of PAs as the basis for aECMs has expanded greatly since the early 2000s through pioneering work by Stupp and colleagues, following their initial demonstration that a PA consisting of a 16 carbon atom alkyl tail linked to an ionic peptide self-assembles into a fibrous scaffold with nanoscale features.[77] Subsequent work demonstrated that the alkyl tail length is an important determinant of self-assembly and gelation, with PAs having C_{10}, C_{16}, and C_{22} tails forming self-supporting hydrogels at a pH that neutralized peptide electrostatic repulsion, while those with a C_6 tail remained soluble.[78] In addition, fibrillization of peptide domains into β-sheets *via* intermolecular H-bonding was important for favoring self-assembly of PAs into nanofibers rather than cylindrical micelles.[79] The use of PAs as aECMs is discussed in more detail below, but for additional depth in this specific area we direct the reader to excellent recent review articles.[80–82]

SLPs pioneered by Zhang and colleagues primarily form nanotubes and vesicles,[74] and have not received much attention as aECMs to date. Nonetheless, similarly to PAs, SLP hydrophobic tails are important determinants of nanostructural properties. In general, alanine-rich tails provide more stable structures due to more favorable intertail hydrophobic interactions.[83–85] In addition, increasing hydrophobic tail length or decreasing the number of hydrophilic amino acids favors assembly of nanotubes or nanoribbons, instead of nanovesicles.[86] For additional details on amphiphilic peptide assembly, we direct the reader to a detailed review.[41]

10.2.1.4   Low Molecular Weight Peptide Hydrogelators.   Short peptides conjugated to aromatic moieties provide another class of molecules that can self-assemble into fibrillar networks *via* a design not found in nature. In the mid-2000s, Ulijn and colleagues pioneered the use of "Fmoc dipeptides", which assemble into self-supporting hydrogels *via* the Fmoc moiety commonly employed as an N-terminal protecting group in solid-phase peptide synthesis (Figure 10.1D).[87] In particular, the aromatic Fmoc moiety drives self-assembly under aqueous conditions *via* hydrophobic and π–π interactions, with a secondary role for hydrogen bonding between peptide domains.[88] Fmoc dipeptides and their analogs have received much attention as aECMs, and are discussed in more detail in later sections.

Using a similar design, Xu and colleagues demonstrated that naphthalene conjugates of β-amino acids,[89,90] dipeptides,[91] and D-amino acid dipeptides,[92] can self-assemble into hydrogels, while the Fmoc moiety, naphthalene, or pyrene can also mediate self-assembly of pentapeptides into hydrogels.[93] Many of these molecular designs have not been explored as the basis for aECMs, however, and are included here primarily to highlight the breadth of this concept. More recently, though, Xu and colleagues demonstrated that molecules consisting of a nucleobase, an amino acid, and a glycoside can self-assemble into hydrogels,[94] and similar molecular designs are currently being explored as the basis for aECMs that support cell adhesion, viability, and proliferation.[95,96] For additional details on low molecular weight hydrogelators, we direct the reader to a more detailed review.[97]

10.2.2   Design Flexibility Inherent in Self-Assembling Peptides

One clear benefit afforded by engineered peptides, when compared to naturally derived biomolecules, is that their primary sequence can be precisely and systematically varied to alter their assembly into supramolecular structures.[38,40,67,98,99] For example, for peptides having a general amino acid sequence of *hydrophobic-anionic-hydrophobic-cationic*, increasing side-chain hydrophobicity decreased the critical salt concentration required for gelation, while increasing the number of repeats provided a biphasic dependence on salt concentration.[100] Notably, these features could be used to rationally design a peptide that was viscous in water at pH 7, yet underwent gelation in the presence of physiological salt concentrations.[101] More recently, Bowerman *et al.* demonstrated that aromaticity and hydrophobicity influence ionic strength dependence of charged peptide assembly, as well as the mechanical properties of fibrillar networks.[102] The relative hydrophobicity of the non-polar amino acids can also be tuned to provide temperature-mediated control of fibril assembly,[54,103] while ionizable amino acids can provide pH-controlled or ion-mediated fiber formation and entanglement.[104–106] In addition, fibril length and width can be tuned by varying sequence hydrophobicity or charge,[107–109] and sequence interchange can induce structure-switching from anisotropic to branched fibers.[110]

Engineered peptides can also be designed to assemble from non-assembling precursor molecules *via* a specific enzymatic reaction,[111–115] analogous to enzyme-mediated collagen and elastin assembly described above. In an early example, Xu and colleagues engineered a phosphorylated peptide that cannot self-assemble due to electrostatic repulsion, yet rapidly self-assembles into a fibrillar gel network in the presence of alkaline phosphatase, an enzyme that dephosphorylates the peptide and renders it electrostatically neutral.[114] Similarly, Stupp and colleagues demonstrated that phosphorylation/dephosphorylation *via* enzymes can provide reversible assembly of peptide amphiphiles.[116] On the flipside, Ulijn and colleagues demonstrated that non-assembling precursors can self-assemble when covalently joined *via* enzyme-catalyzed "reverse hydrolysis".[115] The self-assembly process is reversible in the presence of additional enzyme, and thus these non-assembling precursors can undergo sol–gel–sol transition.[117] Coupling self-assembly with enzyme-mediated assembly–disassembly therefore provided a dynamic component library that selects for the most stable self-assembling components.[118,119] Notably, fibrillar network properties may also be dependent on enzymatic reaction conditions, as exemplified recently by the increased shear modulus with increasing enzyme concentration of hydrogels formed *via* reverse hydrolysis of Phe-Glu-Phe-Lys (FEFK) fragments *via* thermolysine.[120]

Together, the examples discussed above provide an overview of different molecular designs that can be used to create fibrillar networks *via* engineered peptide assembly. In light of their morphological similarities with natural ECMs, fibrillar peptide networks have been widely explored as aECMs. In the following sections, we highlight examples of fibrillar peptide networks as aECMs, and their evolution from basic structural analogs to multifunctional entities that can approach the hierarchy and diversity of activity of natural ECMs. Throughout the remaining sections, we will use the term, "fibrillar peptide networks" as a general descriptor of supramolecular assemblies based on β-sheets, α-helical coiled-coils, peptide amphiphiles, or low molecular weight hydrogelators.

10.3   FIBRILLAR PEPTIDE NETWORKS AS aECMs

Shortly after demonstrating that β-sheet fibrillizing peptides can assemble into stable fibrillar networks,[50] Zhang and colleagues demonstrated that these same networks can provide aECMs capable of supporting mammalian cell attachment.[121] Since then, fibrillar peptide networks have been explored as aECMs for a wide variety of cell types and tissue regeneration applications. For example, Ellis-Behnke *et al.* demonstrated that fibrillar peptide networks can support axon regeneration at an acute injury site, leading to axonal reconnection to target tissues and restored vision.[122] Davis *et al.* demonstrated that empty fibrillar peptide networks injected into myocardium can recruit progenitor cells expressing endothelial markers, followed by vascular smooth muscle cells that form functional vascular structures, while neonatal cardiomyocytes injected with a fibrillar peptide network survived and enhanced endothelial cell recruitment.[123] Misawa *et al.* demonstrated that fibrillar peptide networks injected into calvarial bone defects upregulated expression of bone-related genes, including alkaline phosphatase, Runx2, and osterix, promoted formation of bony bridges observable *via* X-ray radiography, and induced formation of mature cortical bone having medullary cavities.[124] Quintana *et al.* demonstrated that mouse embryonic fibroblasts cultured in fibrillar peptide networks proliferated and migrated into bilateral structures, upregulated expression of mesodermal genes, and spontaneously differentiated into cartilage-like tissues.[125] Malinen *et al.* demonstrated that fibrillar peptide networks promoted HepG2 liver cell structural polarity (*i.e.* filamentous actin accumulation and large tubular bile canalicular structure formation consistent with apicobasal polarity) and functional polarity (*i.e.* multidrug-resistant protein-1 and -2 expression and transport of fluorescent probes into bile canalicular structures).[126] In addition, fibrillar peptide networks have been investigated as aECMs for neural,[127–129] osteo-chondral,[130–135] cardiovascular,[136,137] hepatic,[138,139] connective tissue,[140,141] adrenal gland,[142] pancreatic,[143,144] and stem cells.[145–148] Important here is that fibrillar peptide networks often have low cytotoxicity and are biocompatible *in vivo*, typically eliciting negligible inflammatory or adaptive immune responses.[122,123,149–151]

10.3.1   Sequence-Structure-Function Relationships of aECMs Based on Fibrillar Peptide Networks

As discussed above in Section 10.2.2, the ability to precisely vary the primary sequence of a self-assembling peptide provides opportunities to tune network chemical or physical properties. In the context of aECMs, such tunability can be harnessed to optimize cell responses, such as adhesion, proliferation, migration, or differentiation. For example, gels comprised of RADA16-based β-sheet peptide nanofibers promoted human umbilical vein endothelial cell adhesion, elongation, and formation of interconnected capillary-like networks, whereas those composed of KFE8 or KLD12 favored a rounded cell morphology, ultimately leading to cell clustering instead of interconnected network formation.[152] A Lys-Ala-Ser-Glu-Ala (KASEA) peptide variant bearing an additional cationic lysine residue assembled into nanofibers that promoted fibroblast elongation into a spindle-like morphology, whereas a KASEA variant bearing an additional anionic glutamate residue promoted clustering of cells lacking extended cytoplasmic processes, despite each peptide forming nanofibrillar networks with similar morphologies.[153] Self-assembling Fmoc dipeptides bearing a terminal serine residue supported viability of mouse and human fibroblasts, as well as chondrocytes, whereas cationic, anionic, or hydrophobic termini induced death of one or more of these cell types.[154] Sulfated, negatively charged nanofibers favored formation of a greater number of prechondrocyte aggregates having a smaller diameter when compared to neutral and non-sulfated, negatively charged and non-sulfated, or neutral and sulfated nanofibers, and this morphology correlated with increased chondrocyte gene expression.[155]

Fibrillar network mechanics can be tuned by varying self-assembling peptide primary sequence or chemical composition,[100,108,156] peptide concentration,[157,158] or fibrillization conditions,[159,160] which in turn can influence cell behavior. For example, decreasing stiffness of RADA16-based networks correlated with increased human umbilical vein endothelial elongation and capillary morphogenesis.[161] RADA16-based gels with low stiffness also enhanced migration and cell–cell contact of mouse preosteoblasts, leading to upregulated expression of bone-related proteins, including collagen type I, bone sialoprotein, and osteocalcin.[162] RADA16-based networks with low elastic moduli favored co-expression of BMP-4 and its antagonist, Noggin, by mouse embryonic fibroblasts, leading to their spontaneous chondrogenic commitment, whereas Noggin expression was suppressed at higher elastic modulus values.[163] Engineering the self-assembling peptide Q11 to undergo intermolecular native chemical ligation increased fibrillar Q11 network stiffness, which enhanced human umbilical vein endothelial cell proliferation and expression of CD31.[164] In addition to proliferation and differentiation, Stupp and colleagues recently demonstrated that intermolecular bond cohesion within self-assembled nanofibers is an important determinant of cell viability, with nanofibers having weak intermolecular cohesive forces promoting cell death *via* disruption of lipid membranes.[165] Notably, fibrillar peptide networks can also be designed to flow under applied stress yet recover when stress is removed, which is referred to as "shear-thinning", and peptide primary sequence can be systematically altered to provide optimal shear-thinning conditions for minimally invasive *in vivo* cell delivery.[166,167]

It is worthwhile to note that the mechanical properties of naturally derived Matrigel networks can similarly be tuned by simply varying molecular concentration in the pre-gelled state, leading to differences in cell migration through the network.[168] The ill-defined composition and inherent complexity of Matrigel, however, complicate understanding of mechanistic details because multiple different experimental variables are changed concurrently. In addition, the inability to directly tailor Matrigel molecular composition necessitates use of cell surface receptor inhibitors, such as antibodies or small molecules, to indirectly probe influence of network chemistry on cell behavior. Thus, peptide nanofiber networks provide a useful platform for probing the influence of ECM features on cell behavior because their molecular components can be engineered at the amino acid level and their chemical composition can be precisely defined.

10.3.2 Comparison of Cell Response to Self-Assembled Nanofiber aECMs *Versus* Naturally Derived aECMs

Despite their greatly reduced molecular composition, fibrillar peptide networks often perform as well as or better than aECMs based on naturally derived components for supporting cell adhesion, viability, differentiation, and physiological function. For example, RADA16-based networks enhanced calvarial bone regeneration when compared to Matrigel.[124] Similar networks also maintained explanted chick cochleae hair cell viability, density, morphology, and physiological function to a greater extent than collagen type 1, collagen type 1 plus chondroitin-6-sulfate, or Matrigel networks.[169] In addition, fibrillar peptide networks also promoted more rapid functional recovery of isolated rat hepatocytes when compared to a conventional collagen sandwich.[139] RADA16-based networks also had low cytotoxicity for human neural stem cells, and maintained their proliferation and neuronal differentiation to a greater extent than Matrigel, which was unable to support neuronal differentiation.[170] In addition, similar networks provided improved long-term neural stem cell survival when compared to Matrigel or collagen type 1.[171] Finally, RADA16-based networks promoted greater neovascular network formation *in vivo* when compared to collagen or fibrin networks.[136] Thus, when taken together, these examples demonstrate that bare fibrillar peptide networks with morphological properties similar to natural ECMs provide aECMs with potential for a wide variety of cell delivery and tissue regeneration applications.

10.4 CREATING FUNCTIONAL aECMs BY INTEGRATING CELL-INSTRUCTIVE OR CELL-RESPONSIVE MOLECULES INTO SELF-ASSEMBLED NANOFIBERS

Recently, it has been suggested that the β-sheet fibrillizing peptide RADA16, which is commonly used for creating aECMs, may promote cell adhesion by mediating low-affinity, non-specific integrin interactions.[172] For other fibrillizing peptides, however, cell adhesion is likely mediated by non-specific adsorption of proteins onto the fibrillar network, similar to conventional polymeric culture substrates and biomaterials. Although advantageous for promoting many basic cell functions, non-specific protein adsorption is ill-defined and difficult to control, and thus bare fibrillar peptide networks lack the defined composition and hierarchical organization of integrated cell-instructive and cell-responsive features inherent to natural ECMs. The following sections will highlight advances in integrating cell-instructive and cell-responsive features into fibrillar peptide networks toward the development of multifunctional, hierarchically organized aECMs.

10.4.1 Extracting Cell-Instructive or Cell-Responsive Features from ECM-Derived Biomolecules

Many cell-instructive or cell-responsive properties of ECM proteins can be mimicked by contiguous stretches of amino acids isolated from the protein of interest and re-created as a synthetic peptide. For example, peptides as simple as the arginine-glycine-aspartic acid (RGD) triad, initially isolated from fibronectin,[173] but also found in vitronectin, laminin, collagen, and others,[174] can bind to cell surface integrin receptors analogous to their parent proteins.[175] Since the initial determination of RGD as a minimum motif for binding to certain integrin heterodimers,[176] various other integrin-binding peptide sequences have been identified within ECM proteins.[177] Similarly, contiguous stretches of amino acids within an ECM protein that are cleaved by an enzyme are often also sensitive to cleavage by the same enzyme when re-created as synthetic peptides. An early example demonstrated that collagenase degradation of Gly_{775}–Ile_{776} in the α1(I) collagen chain can be harnessed to render aECMs susceptible to collagenase-mediated degradation *via* a synthetic peptide crosslinker, Ala-Gly-Ile-Pro, while the synthetic peptide crosslinker, Val-Arg-Asn, rendered aECMs susceptible to plasmin, an enzyme that preferentially cleaves fibrin at various locations C-terminal to Arg or Lys residues.[178] Richer discussion on this topic was covered in preceding chapters and will not be covered in more detail here. Nonetheless, the precision afforded by conventional peptide synthesis techniques allows for peptide primary sequence to be systematically altered to identify optimal functional properties (*e.g.* enzymatic cleavage specificity),[179] which can be used to tailor aECM cell-instructive or cell-responsive features.[180] Additionally, peptides having non-natural chemical moities, such as non-canonical amino acids, expand the number of covalent reaction mechanisms available for integrating ligands into biomaterials.[181,182] Finally, peptides can be designed to mimic folding of the parent protein that they are derived from, which is often essential for recapitulating ECM protein activity. For example, high-affinity cyclic RGD ligands that more closely mimic RGD conformation in native fibronectin enhanced neurite outgrowth at significantly lower concentrations than those required for a flexible, low-affinity analog,[183] promoted formation of a greater number of smaller focal adhesions by fibroblasts when compared to a flexible linear analog,[184] enhanced myoblast proliferation,[185] promoted endothelial cell adhesion and spreading at significantly lower ligand densities than flexible analogs,[186] and enhanced osteogenic differentiation of mesenchymal stem cells in 2-D and 3-D.[187,188] The following sections will highlight the use of ECM-derived peptides to create fibrillar peptide networks with integrated cell-instructive and cell-responsive properties.

10.4.2 Integrating Cell-Instructive or Cell-Responsive Features into Peptide Nanofibers Post-Assembly

Toward the development of functional aECMs, bare fibrillar peptide networks can be functionalized with biomolecular ligands post-assembly. For example, cell adhesive RGD peptides can be covalently conjugated onto glutamine-rich fibrillar peptide assemblies by the enzyme tissue transglutaminase.[59] Proteins can also be conjugated onto fibrillar peptide assemblies engineered to present "capture ligands" that form a specific covalent bond with a target enzyme.[189] "Click" reactions, such as Cu(I)-catalyzed azide-alkyne cycloaddition, photoinitiated thiol–ene reactions, and native chemical ligation, can be used to conjugate bioactive ligands onto fibrillar peptide networks.[190,191] Ligands can also be immobilized onto fibrillar peptide networks *via* specific, non-covalent interactions. For example, fibril-binding peptides linked to a biomolecular ligand mediate ligand immobilization onto fibrillar networks.[192–196] Fibrillar peptide networks can also be designed to present a non-covalent "capture ligand", such as biotin, which can bind to functional ligands conjugated to avidin or streptavidin.[149,197–199] Similar approaches are commonly used to introduce biomolecular ligands into polymeric biomaterials, as discussed in Chapter 5, and pragmatic challenges or limitations noted for biomolecule conjugation to polymeric biomaterials will likely also exist for fibrillar peptide networks.

10.4.3 Engineering Self-Assembling Building Blocks to Directly Integrate Cell-Instructive or Cell-Responsive Features into Peptide Nanofibers

Cell-instructive or cell-responsive peptide ligands can either be integrated into or appended onto a fibril-forming, "assembly domain" in the pre-assembled state using conventional peptide synthesis or protein expression methods. The ligand is then integrated into the fibrillar peptide network during the course of assembly, resulting in an aECM having activity related to the ligand. In the following sections, we highlight examples of using this approach to incorporate various ECM functional features into fibrillar peptide networks, such as cell adhesion, modulation of soluble factor signaling, degradation, and mineralization.

10.4.3.1 Cell Adhesion. Self-assembling domains modified with ECM-derived integrin-binding peptides have been widely used to fabricate fibrillar peptide networks with cell adhesive properties.[43,200–220] However, other non-integrin-binding peptides have also been explored.[221–225] Notably, integrating ECM-derived adhesion domains into fibrillar peptide networks can induce cell responses that are not observed with bare networks alone. For example, our group demonstrated that human umbilical vein endothelial cell proliferation was increased by appending an RGD motif onto fibrillar Q11 networks having tunable stiffness conferred by native chemical ligation.[164] Zhang and colleagues demonstrated that osteoblast proliferation, differentiation, and migration were modestly enhanced by RADA16 networks modified with an osteogenic growth peptide or osteopontin cell adhesion motif, while a two-unit RGD motif further enhanced these responses.[213] Similarly, RADA16 variants modified with the same two-unit RGD motif, as well as a laminin-derived cell adhesion motif, promoted periodontal ligament fibroblast adhesion, proliferation, and secretion of type I and type III collagens.[210] In addition, RADA16 variants functionalized with bone marrow homing peptides enhanced murine adult neural stem cell differentiation into neurons and glial cells without requiring additional soluble signaling molecules.[221] Stupp and colleagues demonstrated that PAs modified with the laminin-derived peptide, IKVAV, favored neural progenitor cell differentiation into neurons, while suppressing their differentiation into astrocytes.[202] More recent work demonstrated efficacy of these materials for enhancing functional recovery following spinal cord injury.[203,226] In addition, Stupp and colleagues demonstrated that RGD-modified PA nanofibers injected into enamel organ epithelia of mouse incisors enhanced enamel organ epithelial cell proliferation and differentiation into ameloblasts,[207] in part by activating focal adhesion kinase to increase phosphorylation of JNK and c-Jun.[227] Ulijn and colleagues demonstrated that fibrillar networks fabricated *via* co-assembly of Fmoc-RGD and Fmoc-FF supported encapsulated fibroblast spreading in an RGD-dependent manner, suggesting specific integrin engagement by these aECMs.[215] Finally, Woolfson and colleagues recently demonstrated that SAF nanofibers decorated with Arg-Gly-Asp-Ser increased neural stem cell proliferation, migration, differentiation, formation of larger neurospheres, and electrophysiological activity, suggesting cell differentiation toward a phenotype with mature neuron-like behavior.[228]

Adhesion ligand structure and display are important variables, as discussed in relation to other biomaterials in Section 10.4.1 above, which should be considered in the design of fibrillar peptide networks as cell-instructive aECMs. For example, a longer glycine spacer between RADA16 and the functional motif PFSSTKT provided more stable nanostructures and increased availability of the functional motif, as demonstrated by more effective neural stem cell adhesion, viability, and differentiation on networks with a Gly–Gly–Gly–Gly spacer when compared to Gly–Gly or no spacer.[229] Similarly, Stupp and colleagues recently demonstrated that spacing between a peptide nanofiber and an integrin-binding ligand influences extent of cell spreading, actin bundling, and morphology.[230] A constrained, cyclic RGD presented on PA nanofibers enhanced fibroblast adhesion and spreading when compared to an unstructured, linear RGD.[231] Mechanics of PA networks modified with the laminin-derived KDI sequence influenced hippocampal neuron function, with softer substrates enhancing neuronal polarity that is important for axon differentiation.[232] Finally, appended ligands can influence fibrillar peptide network assembly, as demonstrated by the bundling of IKVAV-modified PA nanofibers, which decreased ligand activity.[233] Notably, incorporating charged amino acids into IKVAV-PAs reduced bundling *via* electrostatic repulsion, and in turn enhanced neurite outgrowth.

10.4.3.2   Soluble Factor Signaling.   Chapter 2 of this book highlighted the importance of soluble factor signaling, non-covalent sequestration of soluble factors by the ECM, and soluble factor gradient formation, for modulating cell behavior in both space and time. Chapter 7 presented examples of polymeric aECMs that can mimic soluble factor sequestering and gradient formation *via* presentation of ligands that bind to heparin glycosaminoglycans, and in turn various soluble heparin-binding growth factors. Fibrillar peptide networks derivatized with heparin-binding ligands can also modulate soluble signal transport to guide cell behavior. For example, Stupp and colleagues developed PA variants that assembled in the presence of heparin, bound angiogenic growth factors, including vascular endothelial growth factor (VEGF) and fibroblast growth factor-2 (FGF-2), and stimulated angiogenesis *in vivo*.[234] Localized delivery of VEGF and FGF-2 from this same PA network also enhanced pancreatic islet engraftment by stimulating angiogenesis, leading to more effective restoration of normoglycemia in diabetic mice.[235] Shao and colleagues demonstrated that RADA16 variants terminated with a heparin-binding peptide provided sustained release of VEGF, which improved cardiac function following infarction by reducing scar size and collagen deposition, as well as promoting cardiac cell survival.[236] Similarly, Cui and colleagues demonstrated that RADA16 variants modified with heparin-binding peptides increased human adipose stem cell proliferation, migration, and secretion of angiogenic growth factors that upregulated endothelial cell growth when compared to bare RADA16 networks.[223]

Fibrillar peptide networks can also be engineered to mimic the activity of heparin. For example, Guler, Tekinay, and colleagues demonstrated that a "heparin-mimetic PA", having integrated sulfonate, carboxylate, and hydroxyl groups enhanced endothelial cell tubule formation *in vitro* in the absence of growth factor supplements, and corneal angiogenesis when loaded with VEGF.[237] More recently, this same group demonstrated that heparin-mimetic PA networks can maintain isolated pancreatic islet glucose stimulation index *in vitro*, and restore normoglycemia in diabetic mice more effectively than islets transplanted alone.[238]

Fibrillar peptide networks can also be modified to present ligands that bind directly to growth factors *via* reversible, non-covalent interactions. For example, Stupp and colleagues demonstrated that PA fibrillar networks modified with a peptide that binds to transforming growth factor-β (TGF-β) released TGF-β more slowly than bare PA networks, promoted chondrogenic differentiation of human mesenchymal stem cells, and enhanced regeneration of articular cartilage in a rabbit defect model without the need for exogenous growth factor.[239] Similarly, PA fibrillar networks modified with a peptide that binds to bone morphogenetic protein-2 (BMP-2) were more effective for inducing spinal fusion in a rat model at BMP-2 doses that were 10-fold less than those used clinically in collagen scaffolds.[240] Notably, BMP-2-binding PA networks also induced fusion at an unexpectedly high rate (~42%) without the use of exogenous BMP-2, suggesting that these networks can capture and localize the activity of endogenous BMP-2.

Finally, peptides that mimic the activity of growth factors or other soluble signaling molecules can also be tethered to fibrillar peptide networks to modulate cell behavior. For example, Zhang and colleagues demonstrated that a RADA16 variant terminated with a VEGF-mimetic peptide that activates the VEGF receptor enhanced endothelial cell survival, proliferation, migration, and tubule formation.[211] In a subsequent study, this same RADA16 variant was shown to induce capillary vessel formation within fibrillar peptide networks using the chicken chorioallantoic membrane (CAM) assay.[212] Stupp and colleagues demonstrated that a PA modified with a VEGF-mimetic peptide enhanced endothelial cell proliferation and migration, elicited capillary formation in the CAM model, and increased tissue perfusion and functional recovery in a mouse hind-limb ischemia model.[241] A PA modified with a peptide mimicking the activity of glucagon-like peptide 1 stimulated insulin secretion from rat insulinoma cells to a greater extent than the peptide mimetic alone and a similar extent as a clinically used drug.[242] Thus, peptide fragments having growth factor-like activity can promote cell proliferation, migration and organization, differentiation, and physiological function when covalently tethered to fibrillar peptide networks.

10.4.3.3 Degradation. Cell-responsive features can also be introduced into fibrillar peptide networks by tailoring peptide sequence. For example, peptide sequences that are cleaved by cell-secreted enzymes, such as matrix metalloproteases, can be integrated into fibrillar peptide networks to provide cell-mediated aECM degradation.[243,244] Our group recently demonstrated that a Q11 "depsipeptide", which has a hydrolytically degradable ester bond in place of one backbone amide, renders peptide nanofibers susceptible to hydrolytic degradation at rates that can be tuned by varying the hydrophobicity of the side-chain on the α-carbon adjacent to the ester bond.[245] Network stiffness decreased over time in accordance with the rate of nanofiber hydrolytic degradation. In turn, cells encapsulated in hydrolytically degradable networks spread and proliferated to a much greater extent than cells within non-degradable networks, which remained round and did not proliferate to any appreciable extent. Thus, integrating degradative properties into fibrillar peptide networks provides a dynamic means of modulating cell function by varying aECM mechanics over time, which cannot be achieved *via* the methods to tailor network mechanical properties discussed in Section 10.3.1 above. Importantly, such dynamic properties may ultimately provide aECMs that more closely mimic the temporal turnover of their natural counterparts, as discussed in Chapter 3.

10.4.3.4 Mineralization. As discussed in Chapter 1, calcium-phosphate minerals, such as hydroxyapatite (HA), are important structural and functional components of bony ECMs, playing a role in matrix assembly, molecular organization, and matrix mechanics. Stupp and colleagues demonstrated that a PA with an integrated phosphoserine residue assembled into nanofibers that nucleated HA mineral formation.[77] Notably, the HA crystals aligned with the nanofibers, providing a morphology that resembled HA crystals associated with collagen fibrils in natural bony ECMs. More recently, these nanofibers demonstrated efficacy for enhancing bone formation in a femoral defect.[246] In addition, Guler, Tekinay, and colleagues demonstrated that mineralized PA nanofibers promoted mesenchymal stem cell differentiation into osteoblasts, as indicated by upregulated Runx2 and collagen type-1 expression, as well as redistribution of DMP-1 from the nucleus to the cytoplasm and extracellular space.[247] Similarly, Schneider and colleagues demonstrated that a β-hairpin peptide terminated with a mineral nucleating peptide domain assembled into fibrillar peptide networks that were mineralized *via* alkaline phosphatase and $CaCl_2$, as well as cementoblast cells.[248] Taken together, these examples demonstrate that fibrillar peptide networks can be engineered to promote HA mineral nucleation, thereby providing aECMs with appropriate structural and functional properties for bony tissue regeneration applications.

10.5 CO-INTEGRATION OF CELL-INSTRUCTIVE AND CELL-RESPONSIVE FEATURES INTO SELF-ASSEMBLED PEPTIDE NANOFIBERS

Supramolecular peptide assembly is particularly advantageous for creating aECMs with *multiple different* co-integrated cell-instructive and cell-responsive features. For example, a single self-assembling peptide can be designed to include more than one integrated functional component (*i.e.* a "multidomain peptide"), such as a cell adhesion ligand and its "synergy site", or a cell adhesion ligand and an enzymatically degradable sequence, to fabricate fibrillar networks with functionalities related to each domain.[249–253] One limitation of fibrillar networks created *via* multidomain peptides, however, is that the concentration of each ligand in the aECM cannot be independently varied to elicit optimal cell responses. Thus, although modular, the inability to independently tune fibrillar network content of cell-instructive and cell-responsive features *via* multidomain peptides will likely hinder development of aECMs with optimal functional properties.

10.5.1 Modular, Multifunctional aECMs *via* Engineered Peptide Co-Assembly

Multifunctional aECMs can also be created by mixing peptides having an identical "assembly domain" but different functional ligand domains in the pre-assembled state, and then inducing their co-assembly into multicomponent nanofibers. Using this approach, our group demonstrated that aECMs presenting tunable concentrations of a cell adhesion ligand increased endothelial cell growth,[150] while a combination of cell adhesion ligands further enhanced this response.[254] Combining an adhesion ligand and an enzymatic degradation domain enhanced periodontal ligament cell proliferation and migration.[255] Nanofibers displaying a laminin-derived adhesion ligand and a heparin-binding peptide enhanced neurite outgrowth, even in the presence of chondroitin sulfate proteoglycans that potently suppress neurite extension following central nervous system injury.[256] Finally, an RGD adhesion peptide and a mineral-nucleating domain worked synergistically to enhance mineral deposition and osteogenic differentiation of mesenchymal stem cells.[257]

Notably, the composition of functional ligands integrated into these co-fibrillized peptide networks often correlates precisely with the molar ratio of engineered peptides that are mixed together in the pre-assembled state.[150,254] This provides materials with easily interchangeable ligand composition, in which the concentration of each ligand can be precisely and independently varied to create aECMs that elicit optimal cell responses. Thus, co-assembly of engineered peptides into multi-component fibrillar networks provides an unprecedented strategy to create aECMs that approach both the compositional diversity and hierarchical organization of their natural counterparts.

10.5.2   Molecular Features Enabling Predictable Co-Assembly

The precise composition of multifunctional aECMs based on multicomponent fibrillar peptide networks can be explained by the kinetic and thermodynamic properties of peptide self-assembly. Protein fibrillization is often observed in nature,[258] for example in Alzheimer's and Parkinson's plaques, and occurs *via* a nucleation-elongation mechanism.[259] Fiber nucleation is a stochastic process that can be influenced by peptide concentration, whereas elongation is often less dependent on concentration.[260] Elongation is driven by peptide "monomer" affinity for nuclei and fibril "chain ends", and proceeds until thermodynamic equilibrium between the monomer and fibril is reached.[261] For engineered peptides having a high fibrillization propensity (*i.e.* "high monomer-fiber affinity"), nucleation lag time can be minimized and equilibrium can approach near-quantitative conversion of monomeric peptides into nanofibers. For example, we previously observed rapid fibrillization kinetics of a β-sheet fibrillizing peptide at physiologic pH and millimolar peptide concentrations, with greater than 90% of peptide in solution incorporated into nanofibers.[262]

For multicomponent assemblies, the formulation at equilibrium is determined by the mutual interactions between each component in the system, and sluggish dynamics of component interchange (*e.g.* due to significant differences in affinity of each component for nuclei/fibril chain ends) can lead to non-equilibrium formulations,[263] often referred to as "composition drift" in co-polymer synthesis. In the case of multicomponent fibrillar networks created from variants of a given self-assembling peptide domain (*e.g.* Q11-based co-assemblies demonstrated by our group), however, multicomponent assembly can be approximated as single-component assembly. In particular, peptides having an identical assembly domain and different appended ligands will have comparable affinity for nuclei/fibril chain ends as long as mutual interactions between assembly domains are not perturbed by the presence of the ligand (*e.g. via* steric clashing or electrostatic repulsion). Thus, the ratio of assembly domains in the fibrillar (*i.e.* bound) state to monomer (*i.e.* unbound) state will be independent of the ligand at thermodynamic equilibrium, and governed entirely by the affinity of the assembly domain for the chain end. As a result, an equimolar mixture of self-assembling domain variants terminated with ligand A and ligand B would be predicted to provide a population of nanofibers having an equimolar ratio of A and B, while an excess of ligand A to B would be predicted to provide nanofibers with an equivalent excess of A to B. This ability to reduce system complexity from multi- to single-component can circumvent composition drift, thereby providing exquisite control of fibrillar peptide network molecular composition. As described in the following sections, such precisely tunable composition provides unique opportunities to fabricate aECMs with optimal functional properties.

10.5.3 Scaffold Design Approaches Facilitated by Self-Assembly

The modular nature and minimal compositional drift of engineered peptide self-assembly opens up opportunities to use powerful investigational styles that are not amenable to biologically derived or covalently polymerized scaffolds. For example, in our group we took advantage of the modular nature of β-sheet fibrillizing peptides with appended cell binding peptides, by applying "Design of Experiments (DOE)" optimization processes.[254] Using factorial experiments and response surface methodology, both DOE techniques, we systematically optimized the concentrations and ratios of different cell binding ligands to promote endothelial cell adhesion and growth on soft peptide hydrogels. This process was simple and straightforward because a set of gels with widely varying combinations of different ligands could be easily created in multi-well culture plates by mixing predetermined ratios of the different peptides, then inducing their gelation *via* the assembly of their appended β-sheet fibrillizing domain (the peptide Q11). Because the ratios of peptides mixed in the precursor solutions matched the ratios that ultimately formed the gels (*i.e.* there was no compositional drift), the multi-peptide formulation of the gels could be quickly adjusted with sequential optimization experiments to arrive at an optimum mixture. Owing in part to the difficulty of independently manipulating specific components of biologically sourced materials (such as Matrigel or decellularized tissues) and the fact that synthetic co-polymers can be prone to compositional drift, DOE had been rarely used previously to design synthetic extracellular matrices despite its centrality to many other areas of engineering design. Modular supramolecular aECMs promise to make DOE and other multifactorial design processes more common.

In another study, we illustrated that different ligands could be segregated into separate populations of fibers or mixed uniformly within homogeneous, multi-ligand fibers.[264] This was done by controlling the point in the assembly process at which two fibrillizing peptides with different appended ligands were mixed (Figure 10.2). If the peptides were mixed as dry powders, formed into protofibrils in water, and subsequently formed into mature nanofibers by adding salt-containing buffers, then intermixed fibers were formed (where every nanofiber

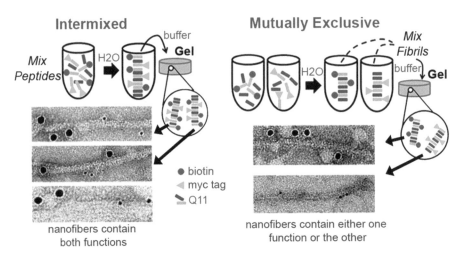

Figure 10.2   Modular supramolecular peptide approaches can be used to specify whether two or more functional components, such as ligands, occupy the same nanofiber (left) or are segregated into mutually exclusive nanofibers, each containing only one type of functional component (right). Adapted with permission from Gasiorowski *et al.*, *Biomacromolecules*, 2011.[264] Blue rectangles represent the self-assembling peptide Q11, red circles represent one model functionality (biotin tag), and green triangles represent the second functionality (myc tag peptide). Mixing the peptides at the very beginning of the assembly process produces intermixed nanofibers (left), whereas keeping the peptides in separate containers until the fibrillar structure has formed results in mutually exclusive nanofibers that can subsequently be mixed into gels without forming intermixed nanofibers (right). Tags in fibers were labeled in the transmission electron microscope images using 15 nm gold nanoparticles specific for the myc tag and 5 nm gold particles specific for biotin. Intermixed nanofibers bind both size particles (left), whereas separately assembled nanofibers only bind one sized particle or the other (right).

contained both functionalities). If, on the other hand, the peptides were taken through the first protofibrillization step in separate containers, and only after this mixed together with buffer to form mature nanofibers, the different functionalities could be maintained in two distinct populations of nanofibers (Figure 10.2).

10.5.4   Biological Impact of Modulating the Composition and Arrangement of Different Molecular Components within Supramolecular Nanomaterials

A tremendous number of biological processes are sensitive to the clustering and co-presentation of ligands, so control over their presentation on nanofibers is useful in a range of applications. For example, it is well known that the clustering of integrin ligands has a strong influence over cell attachment, growth, survival, and migration,[265,266] and peptide nanofibers can be utilized to adjust such clustering.[264] In another consideration of clustering and co-presentation, the presence and spatial arrangement of various different types of immune epitopes on particles, fibers, and antigens also has a powerful effect on the overall immune responses that such materials raise. Because proper engagement of the immune response is central for the design of aECMs that will be used clinically, *i.e.* for tissue engineering, regenerative medicine, or cell delivery, it is important to understand and control these responses. It is also becoming appreciated that some materials similar to those under investigation for aECMs can be used specifically to engage immunological processes for positive, therapeutic effects. One critical aspect for both stimulating and diminishing immune responses against fibrillized peptides is the co-presentation of T cell epitopes and B cell epitopes.[267] T cell epitopes are short peptide sequences or non-peptidic biomolecules that are loaded onto a cell's major histocompatibility complex (MHC) molecules. In a classical response to an extracellular proteinaceous material, antigen-presenting cells such as dendritic cells phagocytose the material, degrade it into short peptides usually around 13–17 amino acids in length, load the peptides into Class-II MHC molecules, and present the peptide/MHC complexes extracellularly for CD4+ (helper) T cells to detect. If additional "danger" signals are also received by the dendritic cell as it is processing the T cell epitope, it becomes "activated" and can signal to T cells to mount a response. Conversely, B cell epitopes operate *via* a completely different mechanism, in which portions of the material's surface bind to a clonal population of B cells specific to that epitope. B cell epitopes can be single peptide strands, or they can be more complicated surfaces generated by two or more adjacent peptide strands comprising a protein's surface. It should also be noted that T cell epitopes and B cell epitopes can be formed by molecules other than peptides. When a B cell epitope on a nanomaterial is bound by a B cell through its B cell receptor, it is internalized and proteolyzed, with any associated T cell epitopes also residing in the material or protein being loaded onto Class-II MHC molecules. In this way, although the precise B cell epitope determines which specific B cells can bind and internalize the material, any associated T cell epitopes also present in the material now become identified with the specific B cells. The CD4+ T cells that have been stimulated to respond to the T cell epitope in the antigen then provide "help" (*i.e.* cytokines that augment and specify the phenotype of the immune response) when they encounter a B cell displaying the T cell epitope. The B cell, after receiving help, then can go on to differentiate into antibody-producing B cell descendants such as plasma cells.

This process, cycling as it does between multiple populations of cells (dendritic cells, T cells, B cells), is highly sensitive to the co-presentation of B cell and T cell epitopes,[267] and using modular peptide materials, we have found that the ratios between these epitopes also modulate the phenotype of the response. Using the peptide co-assembly method described above, we studied how immune responses could be adjusted by controlling the co-fibrillization of a model B cell epitope peptide (E214 from *Staphylococcus aureus* enolase protein) and a model T cell epitope (PADRE, engineered previously to bind to multiple different Class-II MHC molecules, regardless of species or haplotype) (Figure 10.3). Each epitope was synthesized as a fusion peptide with Q11, a short peptide that forms self-assembled nanofibers even when appended with a variety of additional peptide sequences.[150,164,254,267–269] When the two Q11-epitope peptides were "separately assembled" into different nanofibers and co-injected in mice, no detectable antibody responses were raised. However, if the two peptides were co-assembled into the same integrated nanofibers, strong antibody responses were measured against the B cell epitope peptide.[267] The response was modular, in that antibodies were directed only against the B cell epitope, whereas T cell responses were directed only against the T cell epitope (Figure 10.3c). This result suggested a way to either promote or diminish immune responses to peptide nanofibers simply by controlling the intermixing and proximity of different T cell and B cell epitopes. Even more interesting was the finding that different *ratios* of T cell and B cell epitopes in the nanofibers raised different phenotypes of the immune response (Figure 10.3d). Different populations of specific types of CD4+ T cells were preferentially stimulated by different ratios of the T and B cell epitope peptides. In this way, simply controlling the loading of each epitope type within the nanofibers could be used to favor one type of T cell response over the other.[267] Such an example illustrates the powerful effects that simple compositional changes in modular self-assembled biomaterials can have *in vivo*, spanning all the way from a strong antibody response to an absent one, and including subtle phenotypic variations in between. As modular materials and aECMs are increasingly varied with fine control over the amount and relative positioning of different ligands, for example in the context of scaffolds for cells and tissues, it is likely that similarly subtle effects will be increasingly understood and refined.

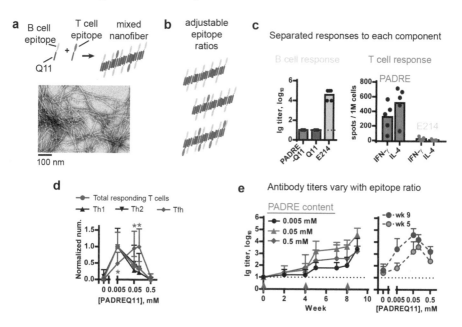

Figure 10.3 When multiple fibrillizing peptides are co-assembled, their types (B or T cell epitope) and ratios have significant influences on the strength and phenotype of the immune responses raised. Adapted with permission from Pompano *et al.*, *Adv. Healthcare Mater.*, 2014.[267] In this way, the modular assembly of the system allows tuning of the immune response elicited. Q11 peptides with an appended B cell epitope (the MRSA epitope peptide E214 from the protein enolase) or a T cell epitope (the pan-DR CD4+ T cell epitope PADRE) can co-assemble into mixed nanofibers (a). Simply mixing them in different ratios can afford nanofibers with varying epitope content (b). When injected into mice, B cell/antibody responses were measured only against the E214 B cell epitope, and T cell responses were measured by ELISPOT against only the PADRE T cell epitope (c), indicating the modularity of the system. When the epitope ratios were varied, different formulations favored different T cell phenotypes. For example, Th1 and Th2 cells were maximally stimulated at a composition of 0.005 mM PADRE/1 mM E214, whereas T follicular helper (Tfh) cells were maximally stimulated at a composition of 0.05 mM PADRE/1 mM E214 (d). Antibody responses against E214 were also maximized at 0.05 mM PADRE, tracking with the Tfh response (e).

10.6   EXTENDING BEYOND PEPTIDES: MODULAR SELF-ASSEMBLED aECMs FROM PROTEINS

Most of the modular, bottom-up aECMs described so far in this chapter have been constructed using short peptides, but there is also tremendous research activity currently that focuses on designing whole expressed proteins as modular aECMs. Beginning with the use of α-helical leucine zipper motifs in the 1990s to assemble hydrogels,[68] the field of engineered proteins as aECMs has blossomed to include matrices assembled using molecular recognition,[270,271] using native structures such as silks,[272] elastins,[273] or resilins,[274] or combining expressed proteins with peptides to create composite modular materials (Figure 10.4).[262]

Proteins, with their complex secondary, tertiary, and quaternary structures, are obviously more capable of installing more specific cell-responsive and cell-instructive functions than short peptides. For example, enzymatic activity and the binding of select subsets of integrins is much more readily achieved with conformationally precise whole proteins or protein domains than correspondingly less specific peptides. Owing to the importance of achieving controlled amounts of various different ligands in biological applications, illustrated above with the example regarding immune responses, we sought to design modular materials in which different expressed proteins could be inserted with exact control over their relative ratios. To accomplish this, we designed a tag, called a "β-tail" that can be expressed within *E. coli* in an unassembled state, to allow production and purification, but then can be induced to assemble and insert into β-sheet fibrillar structures when mixed with additional fibrillizing peptide such as Q11.[262] To illustrate that different combinations of multiple proteins could be precisely targeted, we cloned three different fluorescent proteins into *E. coli* and expressed them with the β-tail tag. Using red–green–blue color mixing to illustrate how a precise combination of different proteins could be targeted, we predicted the ratios of red, green, and blue that would make various hues, then mixed the β-tail proteins and assembled them into Q11 nanofibers to see how closely the actual colors matched those predicted. We found excellent color matching (Figure 10.4, bottom), but only when a functional β-tail was used, as proteins with a mutated β-tail could not achieve precise color matching. Looking towards the future, modular protein-displaying systems such as this one should provide a powerful tool for understanding how precise combinations of various protein ligands can be used as aECMs to control the phenotype of cells that are in contact with them.

10.7 CONCLUDING REMARKS AND FUTURE DIRECTIONS

The ECM is a hierarchical assemblage of various biomolecules, including proteins, proteoglycans, glycosaminoglycans, and glycoproteins, which influences cell behavior through co-presentation of various cell-instructive and cell-responsive cues, including growth factors, cell adhesion sites, enzymatically sensitive domains,

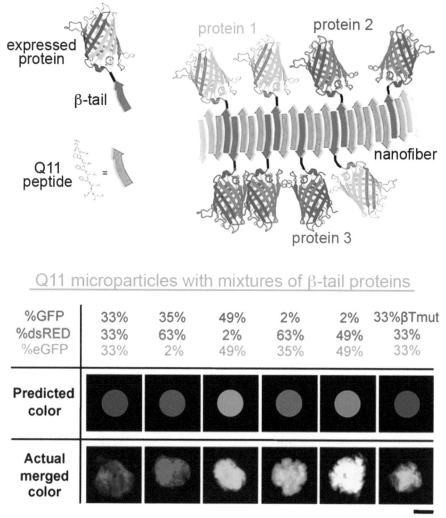

Figure 10.4 The "β-tail" system can be used for integrating whole proteins into supramolecular biomaterials. Adapted with permission from Hudalla *et al.*, *Nat. Mater.*, 2014.[262] Proteins are expressed in *E. coli* with a conditionally assembling tag, which does not self-assemble during expression and purification, but can be later induced to integrate with Q11 and other self-assembling peptides to form multi-protein nanofibers (top). Multiple different proteins can be precisely dosed into the nanofiber matrix, as illustrated here with Q11-based nanofiber microparticles containing three different colors of fluorescent proteins (bottom half of figure). Any predicted composite color could be targeted when the three fluorescent proteins were mixed in the appropriate red:green:blue ratio. Proteins with a mutated β-tail abolished this ability to target a predicted color (bottom right panel). Scale bar = 40 μm.

and mechanical properties that range from compliant to hard and stiff. Due to the essential role of the ECM in developmental, regenerative, and pathological processes, there is enormous interest in creating aECMs that can modulate cell behavior *via* presentation of cell-instructive and cell-responsive features for various biomedical and biotechnological applications. To date, ECM-derived biomolecules, intact cell-free ECMs, natural biopolymers, and synthetic polymers have been explored as aECMs, with many noteworthy successes for each material. However, truly mimicking natural ECM composition, structure, and organization with these materials is often challenged by difficulties associated with replicating ECM hierarchical organization or precisely manipulating composition of cell-instructive and cell-responsive features.

The ECM is formed *via* non-covalent self-assembly of various biomolecules into a "supramolecular" structure, and there is growing interest in developing synthetic peptides and associated derivatives that can similarly self-assemble into supramolecular structures as the basis for aECMs. Since the earliest suggestion that self-assembled peptide nanofibers can provide a structural analog of natural ECM proteins, there have been significant advances in harnessing self-assembly to create aECMs that approximate natural ECM composition, structure, function, and hierarchy. In spite of these advances, however, many challenges remain for creating aECMs that truly mimic their natural counterparts. In particular, the preceding sections highlight numerous examples of aECMs based on fusions of a self-assembling peptide domain linked to a peptide or protein "ligand". aECMs composed solely of proteinaceous ligands contrast starkly with natural ECMs, however, which are composites of both proteinaceous and non-proteinaceous molecules. Thus, there is growing interest in combining fibrillar peptides and other non-proteinaceous molecules into composite biomaterials. Some notable examples to date include hierarchically ordered membranes created *via* co-assembly of hyaluronic acid and peptide amphiphiles by Stupp and colleagues,[275,276] as well as composites of demineralized bone matrix and RADA16 nanofibers developed by Hou *et al.*,[277] hydroxyapatite nanoparticles and cell-adhesive PAs developed by Jun *et al.*,[278,279] Fmoc-dipeptides and polysaccharide hybrids developed by Huang *et al.*,[280] and composites of heparin-binding peptide amphiphiles, heparin, and collagen developed by Lee *et al.*[281] Along these lines, we are also exploring glycopeptides that can self-assemble into β-sheet nanofibers as the basis for aECMs that can modulate the bioactivity of a family of carbohydrate-binding proteins, known as "galectins". In particular, we recently demonstrated that a fusion of a self-assembling peptide domain and a monosaccharide can provide nanofibers that bind to galectins in a carbohydrate-dependent manner.[282] Galectin binding affinity of these nanofibers could be tailored by varying the concentration of carbohydrate ligand integrated into the nanofibers, which was controlled using the same method of mixing components in the pre-assembled state (as discussed above in Section 10.5.1) for creating nanofibers with tunable composition of peptide or protein ligands.[150,254,262] In addition, nanofibers could be exposed to glycosyltransferase enzymes to modify the carbohydrate type, which enabled nanofiber carbohydrate-binding specificity and affinity to be tailored.[282] Moving forward, we anticipate that these glycopeptide nanofibers may provide the basis for aECMs that can more accurately mimic the function of laminin glycoproteins that are prevalent within epithelial basement membranes, as discussed in Chapter 4. Another area of growing interest is in achieving finer control of ligand spatial and temporal presentation within self-assembled aECMs. Although controlled intermixing can be used to enable ligand co-integration or segregation within self-assembled nanofibers due to "kinetic trapping" of peptide molecules in the assembled state, ligand presentation within natural ECMs is often patterned in both time and space, with changes in ECM composition often correlating with specific developmental events. Sur *et al.* recently demonstrated simultaneous temporal and spatial control of RGD ligand presentation within self-assembled aECMs *via* photolabile "caging", which renders RGD ligands inactive until exposure to light.[283] Thus, by combining increasing knowledge of matrix biology with emerging capabilities in molecular self-assembly there will be a wide range of opportunities in the future for creating aECMs that more accurately recapitulate their natural counterparts.

ACKNOWLEDGEMENTS

This work was supported by the National Institutes of Health (NIBIB, 1R01EB009701; NCI, U54 CA151880; NIAID, 1F32AI096769) and the National Science Foundation (DMR-1455201). The content is solely the responsibility of the authors and does not necessarily represent the official views of the National Institute of Biomedical Imaging and BioEngineering, the National Institute of Allergy and Infectious Disease, the National Cancer Institute, the National Institutes of Health, or the National Science Foundation.

REFERENCES

1. L. Gasperini, J. F. Mano and R. L. Reis, Natural polymers for the microencapsulation of cells, *J. R. Soc., Interface*, 2014, **11**, 20140817.
2. J. A. Burdick and G. D. Prestwich, Hyaluronic Acid Hydrogels for Biomedical Applications, *Adv. Mater.*, 2011, **23**, H41–H56.
3. S. F. Badylak, D. O. Freytes and T. W. Gilbert, Extracellular matrix as a biological scaffold material: Structure and function, *Acta. Biomater.*, 2009, **5**, 1–13.
4. B. D. Walters and J. P. Stegemann, Strategies for directing the structure and function of three-dimensional collagen biomaterials across length scales, *Acta. Biomater.*, 2014, **10**, 1488–1501.
5. J. Glowacki and S. Mizuno, Collagen scaffolds for tissue engineering, *Biopolymers*, 2008, **89**, 338–344.
6. W. F. Daamen, J. H. Veerkamp, J. C. van Hest and T. H. van Kuppevelt, Elastin as a biomaterial for tissue engineering, *Biomaterials*, 2007, **28**, 4378–4398.
7. J. F. Almine, D. V. Bax, S. M. Mithieux, L. Nivison-Smith, J. Rnjak, A. Waterhouse, S. G. Wise and A. S. Weiss, Elastin-based materials, *Chem. Soc. Rev.*, 2010, **39**, 3371–3379.
8. S. E. Sakiyama-Elbert, Incorporation of heparin into biomaterials, *Acta. Biomater.*, 2014, **10**, 1581–1587.
9. P. de la Puente and D. Ludena, Cell culture in autologous fibrin scaffolds for applications in tissue engineering, *Exp. Cell Res.*, 2014, **322**, 1–11.
10. T. A. Ahmed, E. V. Dare and M. Hincke, Fibrin: a versatile scaffold for tissue engineering applications, *Tissue Eng., Part B*, 2008, **14**, 199–215.
11. A. N. Renth and M. S. Detamore, Leveraging "raw materials" as building blocks and bioactive signals in regenerative medicine, *Tissue Eng., Part B*, 2012, **18**, 341–362.
12. N. C. Hunt and L. M. Grover, Cell encapsulation using biopolymer gels for regenerative medicine, *Biotechnol. Lett.*, 2010, **32**, 733–742.
13. M. Barczyk, S. Carracedo and D. Gullberg, Integrins, *Cell Tissue Res.*, 2010, **339**, 269–280.
14. C. Deister, S. Aljabari and C. E. Schmidt, Effects of collagen 1, fibronectin, laminin and hyaluronic acid concentration in multi-component gels on neurite extension, *J. Biomater. Sci., Polym. Ed.*, 2007, **18**, 983–997.
15. H. K. Kleinman and G. R. Martin, Matrigel: basement membrane matrix with biological activity, *Semin. Cancer Biol.*, 2005, **15**, 378–386.

16. G. Benton, H. K. Kleinman, J. George and I. Arnaoutova, Multiple uses of basement membrane-like matrix (BME/Matrigel) in vitro and in vivo with cancer cells, *Int. J. Cancer*, 2011, **128**, 1751–1757.

17. A. Albini and D. M. Noonan, The 'chemoinvasion' assay, 25 years and still going strong: the use of reconstituted basement membranes to study cell invasion and angiogenesis, *Curr. Opin. Cell Biol.*, 2010, **22**, 677–689.

18. T. J. Keane, I. T. Swinehart and S. F. Badylak, Methods of tissue decellularization used for preparation of biologic scaffolds and in vivo relevance, *Methods*, 2015, **84**, 25–34.

19. J. J. Song, J. P. Guyette, S. E. Gilpin, G. Gonzalez, J. P. Vacanti and H. C. Ott, Regeneration and experimental orthotopic transplantation of a bioengineered kidney, *Nat. Med.*, 2013, **19**, 646–651.

20. T. H. Petersen, E. A. Calle, L. Zhao, E. J. Lee, L. Gui, M. B. Raredon, K. Gavrilov, T. Yi, Z. W. Zhuang, C. Breuer, E. Herzog and L. E. Niklason, Tissue-engineered lungs for in vivo implantation, *Science*, 2010, **329**, 538–541.

21. L. F. Tapias and H. C. Ott, Decellularized scaffolds as a platform for bioengineered organs, *Curr. Opin. Organ Transplant.*, 2014, **19**, 145–152.

22. H. C. Ott, T. S. Matthiesen, S. K. Goh, L. D. Black, S. M. Kren, T. I. Netoff and D. A. Taylor, Perfusion-decellularized matrix: using nature's platform to engineer a bioartificial heart, *Nat. Med.*, 2008, **14**, 213–221.

23. B. E. Uygun, A. Soto-Gutierrez, H. Yagi, M. L. Izamis, M. A. Guzzardi, C. Shulman, J. Milwid, N. Kobayashi, A. Tilles, F. Berthiaume, M. Hertl, Y. Nahmias, M. L. Yarmush and K. Uygun, Organ reengineering through development of a transplantable recellularized liver graft using decellularized liver matrix, *Nat. Med.*, 2010, **16**, 814–820.

24. C. S. Hughes, L. M. Postovit and G. A. Lajoie, Matrigel: a complex protein mixture required for optimal growth of cell culture, *Proteomics*, 2010, **10**, 1886–1890.

25. H. E. Davis and J. K. Leach, in *Topics in Multifunctional Biomaterials and Devices*, ed. N. Ashammakhi, 2008, vol. 1.

26. H. K. Lau and K. L. Kiick, Opportunities for multicomponent hybrid hydrogels in biomedical applications, *Biomacromolecules*, 2015, **16**, 28–42.

27. G. A. Hudalla and W. L. Murphy, Biomaterials that regulate growth factor activity via bioinspired interactions, *Adv. Funct. Mater.*, 2011, **21**, 1754–1768.

28. D. Seliktar, Designing cell-compatible hydrogels for biomedical applications, *Science*, 2012, **336**, 1124–1128.

29. M. W. Tibbitt and K. S. Anseth, Hydrogels as extracellular matrix mimics for 3D cell culture, *Biotechnol. Bioeng.*, 2009, **103**, 655–663.

30. F. Edalat, I. Sheu, S. Manoucheri and A. Khademhosseini, Material strategies for creating artificial cell-instructive niches, *Curr. Opin. Biotechnol.*, 2012, **23**, 820–825.

31. J. J. Rice, M. M. Martino, L. De Laporte, F. Tortelli, P. S. Briquez and J. A. Hubbell, Engineering the regenerative microenvironment with biomaterials, *Adv. Healthcare Mater.*, 2013, **2**, 57–71.

32. O. D. Krishna and K. L. Kiick, Protein- and peptide-modified synthetic polymeric biomaterials, *Biopolymers*, 2010, **94**, 32–48.

33. M. Guvendiren and J. A. Burdick, Engineering synthetic hydrogel microenvironments to instruct stem cells, *Curr. Opin. Biotechnol.*, 2013, **24**, 841–846.

34. L. D. Muiznieks and F. W. Keeley, Molecular assembly and mechanical properties of the extracellular matrix: A fibrous protein perspective, *Biochim. Biophys. Acta*, 2013, **1832**, 866–875.

35. D. N. Woolfson and Z. N. Mahmoud, More than just bare scaffolds: towards multi-component and decorated fibrous biomaterials, *Chem. Soc. Rev.*, 2010, **39**, 3464–3479.

36. Y. Loo, S. Zhang and C. A. Hauser, From short peptides to nanofibers to macromolecular assemblies in biomedicine, *Biotechnol. Adv.*, 2012, **30**, 593–603.

37. C. A. Hauser and S. Zhang, Designer self-assembling peptide nanofiber biological materials, *Chem. Soc. Rev.*, 2010, **39**, 2780–2790.

38. C. J. Bowerman and B. L. Nilsson, Review self-assembly of amphipathic beta-sheet peptides: Insights and applications, *Biopolymers*, 2012, **98**, 169–184.

39. T. Aida, E. W. Meijer and S. I. Stupp, Functional supramolecular polymers, *Science*, 2012, **335**, 813–817.

40. H. Cui, M. J. Webber and S. I. Stupp, Self-assembly of peptide amphiphiles: from molecules to nanostructures to biomaterials, *Biopolymers*, 2010, **94**, 1–18.

41. A. Dehsorkhi, V. Castelletto and I. W. Hamley, Self-assembling amphiphilic peptides, *J. Pept. Sci.*, 2014, **20**, 453–467.

42. R. V. Ulijn and A. M. Smith, Designing peptide based nanomaterials, *Chem. Soc. Rev.*, 2008, **37**, 664–675.

43. R. Orbach, L. Adler-Abramovich, S. Zigerson, I. Mironi-Harpaz, D. Seliktar and E. Gazit, Self-assembled Fmoc-peptides as

a platform for the formation of nanostructures and hydrogels, *Biomacromolecules*, 2009, **10**, 2646–2651.

44. L. Pauling and R. B. Corey, Configurations of Polypeptide Chains With Favored Orientations Around Single Bonds: Two New Pleated Sheets, *Proc. Natl. Acad. Sci. U. S. A.*, 1951, **37**, 729–740.

45. L. Pauling and R. B. Corey, Two Rippled-Sheet Configurations of Polypeptide Chains, and a Note about the Pleated Sheets, *Proc. Natl. Acad. Sci. U. S. A.*, 1953, **39**, 253–256.

46. W. B. Rippon, H. H. Chen and A. G. Walton, Spectroscopic characterization of poly(Glu-Ala), *J. Mol. Biol.*, 1973, **75**, 369–375.

47. J. M. Anderson, H. H. Chen, W. B. Rippon and A. G. Walton, Characterization of three sequential polydipeptides containing glycine, *J. Mol. Biol.*, 1972, **67**, 459–468.

48. D. Osterman, R. Mora, F. J. Kezdy, E. T. Kaiser and S. C. A. Meredith, Synthetic Amphiphilic Beta-Strand Tridecapeptide - a Model for Apolipoprotein-B, *J. Am. Chem. Soc.*, 1984, **106**, 6845–6847.

49. D. G. Osterman and E. T. Kaiser, Design and Characterization of Peptides with Amphiphilic Beta-Strand Structures, *J. Cell. Biochem.*, 1985, **29**, 57–72.

50. S. Zhang, T. Holmes, C. Lockshin and A. Rich, Spontaneous assembly of a self-complementary oligopeptide to form a stable macroscopic membrane, *Proc. Natl. Acad. Sci. U. S. A.*, 1993, **90**, 3334–3338.

51. S. Zhang, C. Lockshin, R. Cook and A. Rich, Unusually stable beta-sheet formation in an ionic self-complementary oligopeptide, *Biopolymers*, 1994, **34**, 663–672.

52. S. Zhang and A. Rich, Direct conversion of an oligopeptide from a beta-sheet to an alpha-helix: a model for amyloid formation, *Proc. Natl. Acad. Sci. U. S. A.*, 1997, **94**, 23–28.

53. J. P. Schneider, D. J. Pochan, B. Ozbas, K. Rajagopal, L. Pakstis and J. Kretsinger, Responsive hydrogels from the intramolecular folding and self-assembly of a designed peptide, *J. Am. Chem. Soc.*, 2002, **124**, 15030–15037.

54. D. J. Pochan, J. P. Schneider, J. Kretsinger, B. Ozbas, K. Rajagopal and L. Haines, Thermally reversible hydrogels via intramolecular folding and consequent self-assembly of a de novo designed peptide, *J. Am. Chem. Soc.*, 2003, **125**, 11802–11803.

55. L. A. Haines, K. Rajagopal, B. Ozbas, D. A. Salick, D. J. Pochan and J. P. Schneider, Light-activated hydrogel formation via the triggered folding and self-assembly of a designed peptide, *J. Am. Chem. Soc.*, 2005, **127**, 17025–17029.

56. A. Aggeli, M. Bell, N. Boden, J. N. Keen, P. F. Knowles, T. C. McLeish, M. Pitkeathly and S. E. Radford, Responsive gels formed by the spontaneous self-assembly of peptides into polymeric beta-sheet tapes, *Nature*, 1997, **386**, 259–262.

57. A. Aggeli, I. A. Nyrkova, M. Bell, R. Harding, L. Carrick, T. C. McLeish, A. N. Semenov and N. Boden, Hierarchical self-assembly of chiral rod-like molecules as a model for peptide beta-sheet tapes, ribbons, fibrils, and fibers, *Proc. Natl. Acad. Sci. U. S. A.*, 2001, **98**, 11857–11862.

58. A. Aggeli, M. Bell, L. M. Carrick, C. W. Fishwick, R. Harding, P. J. Mawer, S. E. Radford, A. E. Strong and N. Boden, pH as a trigger of peptide beta-sheet self-assembly and reversible switching between nematic and isotropic phases, *J. Am. Chem. Soc.*, 2003, **125**, 9619–9628.

59. J. H. Collier and P. B. Messersmith, Enzymatic modification of self-assembled peptide structures with tissue transglutaminase, *Bioconjugate Chem.*, 2003, **14**, 748–755.

60. F. H. Crick, Is alpha-keratin a coiled coil? *Nature*, 1952, **170**, 882–883.

61. L. Pauling, R. B. Corey and H. R. Branson, The structure of proteins; two hydrogen-bonded helical configurations of the polypeptide chain, *Proc. Natl. Acad. Sci. U. S. A.*, 1951, **37**, 205–211.

62. W. H. Landschulz, P. F. Johnson and S. L. McKnight, The leucine zipper: a hypothetical structure common to a new class of DNA binding proteins, *Science*, 1988, **240**, 1759–1764.

63. C. Cohen and D. A. Parry, Alpha-helical coiled coils and bundles: how to design an alpha-helical protein, *Proteins*, 1990, **7**, 1–15.

64. C. Cohen and D. A. Parry, Alpha-helical coiled coils: more facts and better predictions, *Science*, 1994, **263**, 488–489.

65. W. D. Kohn and R. S. Hodges, De novo design of alpha-helical coiled coils and bundles: models for the development of protein-design principles, *Trends Biotechnol.*, 1998, **16**, 379–389.

66. J. P. Schneider, A. Lombardi and W. F. DeGrado, Analysis and design of three-stranded coiled coils and three-helix bundles, *Folding Des.*, 1998, **3**, R29–R40.

67. D. N. Woolfson, The design of coiled-coil structures and assemblies, *Adv. Protein Chem.*, 2005, **70**, 79–112.

68. W. A. Petka, J. L. Harden, K. P. McGrath, D. Wirtz and D. A. Tirrell, Reversible hydrogels from self-assembling artificial proteins, *Science*, 1998, **281**, 389–392.

69. S. Kojima, Y. Kuriki, T. Yoshida, K. Yazaki and K. Miura, Fibril formation by an amphipathic alpha-helix-forming polypeptide produced by gene engineering, *Proc. Jpn. Acad., Ser. B*, 1997, **73**, 7–11.

70. M. J. Pandya, G. M. Spooner, M. Sunde, J. R. Thorpe, A. Rodger and D. N. Woolfson, Sticky-end assembly of a designed peptide fiber provides insight into protein fibrillogenesis, *Biochemistry*, 2000, **39**, 8728–8734.

71. S. A. Potekhin, T. N. Melnik, V. Popov, N. F. Lanina, A. A. Vazina, P. Rigler, A. S. Verdini, G. Corradin and A. V. Kajava, De novo design of fibrils made of short alpha-helical coiled coil peptides, *Chem. Biol.*, 2001, **8**, 1025–1032.

72. Y. Zimenkov, V. P. Conticello, L. Guo and P. Thiyagarajan, Rational design of a nanoscale helical scaffold derived from self-assembly of a dimeric coiled coil motif, *Tetrahedron*, 2004, **60**, 7237–7246.

73. D. E. Wagner, C. L. Phillips, W. M. Ali, G. E. Nybakken, E. D. Crawford, A. D. Schwab, W. F. Smith and R. Fairman, Toward the development of peptide nanofilaments and nanoropes as smart materials, *Proc. Natl. Acad. Sci. U. S. A.*, 2005, **102**, 12656–12661.

74. S. Vauthey, S. Santoso, H. Gong, N. Watson and S. Zhang, Molecular self-assembly of surfactant-like peptides to form nanotubes and nanovesicles, *Proc. Natl. Acad. Sci. U. S. A.*, 2002, **99**, 5355–5360.

75. Y. C. Yu, P. Berndt, M. Tirrell and G. B. Fields, Self-assembling amphiphiles for construction of protein molecular architecture, *J. Am. Chem. Soc.*, 1996, **118**, 12515–12520.

76. G. B. Fields, J. L. Lauer, Y. Dori, P. Forns, Y. C. Yu and M. Tirrell, Protein-like molecular architecture: biomaterial applications for inducing cellular receptor binding and signal transduction, *Biopolymers*, 1998, **47**, 143–151.

77. J. D. Hartgerink, E. Beniash and S. I. Stupp, Self-assembly and mineralization of peptide-amphiphile nanofibers, *Science*, 2001, **294**, 1684–1688.

78. J. D. Hartgerink, E. Beniash and S. I. Stupp, Peptide-amphiphile nanofibers: a versatile scaffold for the preparation of self-assembling materials, *Proc. Natl. Acad. Sci. U. S. A.*, 2002, **99**, 5133–5138.

79. S. E. Paramonov, H. W. Jun and J. D. Hartgerink, Self-assembly of peptide-amphiphile nanofibers: the roles of hydrogen bonding and amphiphilic packing, *J. Am. Chem. Soc.*, 2006, **128**, 7291–7298.

80. J. Boekhoven and S. I. Stupp, 25th anniversary article: supramolecular materials for regenerative medicine, *Adv. Mater.*, 2014, **26**, 1642–1659.

81. J. B. Matson and S. I. Stupp, Self-assembling peptide scaffolds for regenerative medicine, *Chem. Commun.*, 2012, **48**, 26–33.

82. M. J. Webber, J. A. Kessler and S. I. Stupp, Emerging peptide nanomedicine to regenerate tissues and organs, *J. Intern. Med.*, 2010, **267**, 71–88.

83. C. K. Tang, F. Qiu and X. J. Zhao, Molecular Design and Applications of Self-Assembling Surfactant-Like Peptides, *J. Nanomater.*, 2013, DOI: 10.1155/2013/469261.

84. S. Scanlon and A. Aggeli, Self-assembling peptide nanotubes, *Nano Today*, 2008, **3**, 22–30.

85. X. J. Zhao, Design of self-assembling surfactant-like peptides and their applications, *Curr. Opin. Colloid Interface Sci.*, 2009, **14**, 340–348.

86. Q. Meng, Y. Kou, X. Ma, Y. Liang, L. Guo, C. Ni and K. Liu, Tunable self-assembled peptide amphiphile nanostructures, *Langmuir*, 2012, **28**, 5017–5022.

87. V. Jayawarna, M. Ali, T. A. Jowitt, A. E. Miller, A. Saiani, J. E. Gough and R. V. Ulijn, Nanostructured hydrogels for three-dimensional cell culture through self-assembly of fluorenylmethoxycarbonyl-dipeptides, *Adv. Mater.*, 2006, **18**, 611–614.

88. C. Tang, R. V. Ulijn and A. Saiani, Effect of glycine substitution on Fmoc-diphenylalanine self-assembly and gelation properties, *Langmuir*, 2011, **27**, 14438–14449.

89. Z. M. Yang, G. L. Liang and B. Xu, Supramolecular hydrogels based on beta-amino acid derivatives, *Chem. Commun.*, 2006, 738–740.

90. Z. M. Yang, G. L. Liang and B. Xu, Supramolecular hydrogels based on beta-amino acid derivatives, *Chem. Commun.*, 2006, **3**, 738.

91. Z. M. Yang, G. L. Liang, M. L. Ma, Y. Gao and B. Xu, Conjugates of naphthalene and dipeptides produce molecular hydrogelators with high efficiency of hydrogelation and superhelical nanofibers, *J. Mater. Chem.*, 2007, **17**, 850–854.

92. G. Liang, Z. Yang, R. Zhang, L. Li, Y. Fan, Y. Kuang, Y. Gao, T. Wang, W. W. Lu and B. Xu, Supramolecular hydrogel of a D-amino acid dipeptide for controlled drug release in vivo, *Langmuir*, 2009, **25**, 8419–8422.

93. M. Ma, Y. Kuang, Y. Gao, Y. Zhang, P. Gao and B. Xu, Aromatic-aromatic interactions induce the self-assembly of pentapeptidic derivatives in water to form nanofibers and supramolecular hydrogels, *J. Am. Chem. Soc.*, 2010, **132**, 2719–2728.

94. X. Li, Y. Kuang, J. Shi, Y. Gao, H. C. Lin and B. Xu, Multifunctional, biocompatible supramolecular hydrogelators consist only of nucleobase, amino acid, and glycoside, *J. Am. Chem. Soc.*, 2011, **133**, 17513–17518.

95. X. W. Du, J. Zhou, O. Guvench, F. O. Sangiorgi, X. M. Li, N. Zhou and B. Xu, Supramolecular Assemblies of a Conjugate of Nucleobase, Amino Acids, and Saccharide Act as Agonists for Proliferation of Embryonic Stem Cells and Development of Zygotes, *Bioconjugate Chem.*, 2014, **25**, 1031–1035.

96. X. M. Li, X. W. Du, Y. Gao, J. F. Shi, Y. Kuang and B. Xu, Supramolecular hydrogels formed by the conjugates of nucleobases, Arg-Gly-Asp (RGD) peptides, and glucosamine, *Soft Matter*, 2012, **8**, 7402–7407.

97. S. Fleming and R. V. Ulijn, Design of nanostructures based on aromatic peptide amphiphiles, *Chem. Soc. Rev.*, 2014, **43**, 8150–8177.

98. R. Orbach, I. Mironi-Harpaz, L. Adler-Abramovich, E. Mossou, E. P. Mitchell, V. T. Forsyth, E. Gazit and D. Seliktar, The rheological and structural properties of Fmoc-peptide-based hydrogels: the effect of aromatic molecular architecture on self-assembly and physical characteristics, *Langmuir*, 2012, **28**, 2015–2022.

99. C. Tang, R. V. Ulijn and A. Saiani, Self-assembly and gelation properties of glycine/leucine Fmoc-dipeptides, *Eur. Phys. J. E*, 2013, **36**, 111–121.

100. M. R. Caplan, E. M. Schwartzfarb, S. Zhang, R. D. Kamm and D. A. Lauffenburger, Effects of systematic variation of amino acid sequence on the mechanical properties of a self-assembling, oligopeptide biomaterial, *J. Biomater. Sci., Polym. Ed.*, 2002, **13**, 225–236.

101. M. R. Caplan, E. M. Schwartzfarb, S. Zhang, R. D. Kamm and D. A. Lauffenburger, Control of self-assembling oligopeptide matrix formation through systematic variation of amino acid sequence, *Biomaterials*, 2002, **23**, 219–227.

102. C. J. Bowerman, W. Liyanage, A. J. Federation and B. L. Nilsson, Tuning beta-sheet peptide self-assembly and hydrogelation behavior by modification of sequence hydrophobicity and aromaticity, *Biomacromolecules*, 2011, **12**, 2735–2745.

103. M. M. Nguyen, K. M. Eckes and L. J. Suggs, Charge and sequence effects on the self-assembly and subsequent hydrogelation of Fmoc-depsipeptides, *Soft Matter*, 2014, **10**, 2693–2702.

104. K. Rajagopal, M. S. Lamm, L. A. Haines-Butterick, D. J. Pochan and J. P. Schneider, Tuning the pH responsiveness of beta-hairpin peptide folding, self-assembly, and hydrogel material formation, *Biomacromolecules*, 2009, **10**, 2619–2625.

105. J. H. Collier, B. H. Hu, J. W. Ruberti, J. Zhang, P. Shum, D. H. Thompson and P. B. Messersmith, Thermally and photochemically triggered self-assembly of peptide hydrogels, *J. Am. Chem. Soc.*, 2001, **123**, 9463–9464.

106. B. Ozbas, J. Kretsinger, K. Rajagopal, J. P. Schneider and D. J. Pochan, Salt-triggered peptide folding and consequent self-assembly into hydrogels with tunable modulus, *Macromolecules*, 2004, **37**, 7331–7337.

107. C. J. Bowerman, D. M. Ryan, D. A. Nissan and B. L. Nilsson, The effect of increasing hydrophobicity on the self-assembly of amphipathic beta-sheet peptides, *Mol. Biosyst.*, 2009, **5**, 1058–1069.

108. L. Aulisa, H. Dong and J. D. Hartgerink, Self-assembly of multidomain peptides: sequence variation allows control over cross-linking and viscoelasticity, *Biomacromolecules*, 2009, **10**, 2694–2698.

109. H. Dong, S. E. Paramonov, L. Aulisa, E. L. Bakota and J. D. Hartgerink, Self-assembly of multidomain peptides: balancing molecular frustration controls conformation and nanostructure, *J. Am. Chem. Soc.*, 2007, **129**, 12468–12472.

110. G. Cheng, V. Castelletto, C. M. Moulton, G. E. Newby and I. W. Hamley, Hydrogelation and self-assembly of Fmoc-tripeptides: unexpected influence of sequence on self-assembled fibril structure, and hydrogel modulus and anisotropy, *Langmuir*, 2010, **26**, 4990–4998.

111. Y. Gao, Z. M. Yang, Y. Kuang, M. L. Ma, J. Y. Li, F. Zhao and B. Xu, Enzyme-Instructed Self-Assembly of Peptide Derivatives to Form Nanofibers and Hydrogels, *Biopolymers*, 2010, **94**, 19–31.

112. M. E. Hahn and N. C. Gianneschi, Enzyme-directed assembly and manipulation of organic nanomaterials, *Chem. Commun.*, 2011, **47**, 11814–11821.

113. R. J. Williams, R. J. Mart and R. V. Ulijn, Exploiting biocatalysis in peptide self-assembly, *Biopolymers*, 2010, **94**, 107–117.

114. Z. M. Yang, H. W. Gu, D. G. Fu, P. Gao, J. K. Lam and B. Xu, Enzymatic formation of supramolecular hydrogels, *Adv. Mater.*, 2004, **16**, 1440–1444.

115. S. Toledano, R. J. Williams, V. Jayawarna and R. V. Ulijn, Enzyme-triggered self-assembly of peptide hydrogels via reversed hydrolysis, *J. Am. Chem. Soc.*, 2006, **128**, 1070–1071.

116. M. J. Webber, C. J. Newcomb, R. Bitton and S. I. Stupp, Switching of Self-Assembly in a Peptide Nanostructure with a Specific Enzyme, *Soft Matter*, 2011, **7**, 9665–9672.

117. A. K. Das, R. Collins and R. V. Ulijn, Exploiting enzymatic (reversed) hydrolysis in directed setf-assembly of peptide nanostructures, *Small*, 2008, **4**, 279–287.

118. R. J. Williams, A. M. Smith, R. Collins, N. Hodson, A. K. Das and R. V. Ulijn, Enzyme-assisted self-assembly under thermodynamic control, *Nat. Nanotechnol.*, 2009, **4**, 19–24.

119. A. K. Das, A. R. Hirst and R. V. Ulijn, Evolving nanomaterials using enzyme-driven dynamic peptide libraries (eDPL), *Faraday Discuss.*, 2009, **143**, 293–303.

120. J. B. Guilbaud, C. Rochas, A. F. Miller and A. Saiani, Effect of enzyme concentration of the morphology and properties of enzymatically triggered peptide hydrogels, *Biomacromolecules*, 2013, **14**, 1403–1411.

121. S. Zhang, T. C. Holmes, C. M. DiPersio, R. O. Hynes, X. Su and A. Rich, Self-complementary oligopeptide matrices support mammalian cell attachment, *Biomaterials*, 1995, **16**, 1385–1393.

122. R. G. Ellis-Behnke, Y. X. Liang, S. W. You, D. K. Tay, S. Zhang, K. F. So and G. E. Schneider, Nano neuro knitting: peptide nanofiber scaffold for brain repair and axon regeneration with functional return of vision, *Proc. Natl. Acad. Sci. U. S. A.*, 2006, **103**, 5054–5059.

123. M. E. Davis, J. P. Motion, D. A. Narmoneva, T. Takahashi, D. Hakuno, R. D. Kamm, S. Zhang and R. T. Lee, Injectable self-assembling peptide nanofibers create intramyocardial microenvironments for endothelial cells, *Circulation*, 2005, **111**, 442–450.

124. H. Misawa, N. Kobayashi, A. Soto-Gutierrez, Y. Chen, A. Yoshida, J. D. Rivas-Carrillo, N. Navarro-Alvarez, K. Tanaka, A. Miki, J. Takei, T. Ueda, M. Tanaka, H. Endo, N. Tanaka and T. Ozaki, PuraMatrix facilitates bone regeneration in bone defects of calvaria in mice, *Cell Transplant.*, 2006, **15**, 903–910.

125. L. Quintana, T. F. Muinos, E. Genove, M. Del Mar Olmos, S. Borros and C. E. Semino, Early tissue patterning recreated by mouse embryonic fibroblasts in a three-dimensional environment, *Tissue Eng., Part A*, 2009, **15**, 45–54.

126. M. M. Malinen, H. Palokangas, M. Yliperttula and A. Urtti, Peptide nanofiber hydrogel induces formation of bile canaliculi structures in three-dimensional hepatic cell culture, *Tissue Eng., Part A*, 2012, **18**, 2418–2425.

127. T. C. Holmes, S. de Lacalle, X. Su, G. Liu, A. Rich and S. Zhang, Extensive neurite outgrowth and active synapse formation on self-assembling peptide scaffolds, *Proc. Natl. Acad. Sci. U. S. A.*, 2000, **97**, 6728–6733.

128. C. E. Semino, J. Kasahara, Y. Hayashi and S. Zhang, Entrapment of migrating hippocampal neural cells in three-dimensional peptide nanofiber scaffold, *Tissue Eng.*, 2004, **10**, 643–655.

129. T. Hou, Y. Wu, L. Wang, Y. Liu, L. Zeng, M. Li, Z. Long, H. Chen, Y. Li and Z. Wang, Cellular prostheses fabricated with motor neurons seeded in self-assembling peptide promotes partial functional recovery after spinal cord injury in rats, *Tissue Eng., Part A*, 2012, **18**, 974–985.

130. R. A. Hule, R. P. Nagarkar, A. Altunbas, H. R. Ramay, M. C. Branco, J. P. Schneider and D. J. Pochan, Correlations between structure, material properties and bioproperties in self-assembled beta-hairpin peptide hydrogels, *Faraday Discuss.*, 2008, **139**, 251–264, discussion 309–225, 419–220.

131. J. Kisiday, M. Jin, B. Kurz, H. Hung, C. Semino, S. Zhang and A. J. Grodzinsky, Self-assembling peptide hydrogel fosters chondrocyte extracellular matrix production and cell division: implications for cartilage tissue repair, *Proc. Natl. Acad. Sci. U. S. A.*, 2002, **99**, 9996–10001.

132. J. Liu, H. Song, L. Zhang, H. Xu and X. Zhao, Self-assembly-peptide hydrogels as tissue-engineering scaffolds for three-dimensional culture of chondrocytes in vitro, *Macromol. Biosci.*, 2010, **10**, 1164–1170.

133. A. Mujeeb, A. F. Miller, A. Saiani and J. E. Gough, Self-assembled octapeptide scaffolds for in vitro chondrocyte culture, *Acta. Biomater.*, 2013, **9**, 4609–4617.

134. J. H. Sun, Q. X. Zheng, Y. C. Wu, Y. D. Liu, X. D. Guo and W. G. Wu, Culture of nucleus pulposus cells from intervertebral disc on self-assembling KLD-12 peptide hydrogel scaffold, *Mater. Sci. Eng., C*, 2010, **30**, 975–980.

135. J. H. Sun, Q. X. Zheng, Y. C. Wu, Y. D. Liu, X. D. Guo and W. G. Wu, Biocompatibility of KLD-12 Peptide Hydrogel as a Scaffold in Tissue Engineering of Intervertebral Discs in Rabbits, *J. Huazhong Univ. Sci. Technol., Med. Sci.*, 2010, **30**, 173–177.

136. P. Allen, J. Melero-Martin and J. Bischoff, Type I collagen, fibrin and PuraMatrix matrices provide permissive environments for human endothelial and mesenchymal progenitor cells to form neovascular networks, *J. Tissue Eng. Regener. Med.*, 2011, **5**, e74–e86.

137. D. A. Narmoneva, O. Oni, A. L. Sieminski, S. Zhang, J. P. Gertler, R. D. Kamm and R. T. Lee, Self-assembling short oligopeptides and the promotion of angiogenesis, *Biomaterials*, 2005, **26**, 4837–4846.

138. C. E. Semino, J. R. Merok, G. G. Crane, G. Panagiotakos and S. Zhang, Functional differentiation of hepatocyte-like spheroid structures from putative liver progenitor cells in three-dimensional peptide scaffolds, *Differentiation*, 2003, **71**, 262–270.

139. S. Wang, D. Nagrath, P. C. Chen, F. Berthiaume and M. L. Yarmush, Three-dimensional primary hepatocyte culture in synthetic self-assembling peptide hydrogel, *Tissue Eng., Part A*, 2008, **14**, 227–236.

140. J. K. Kretsinger, L. A. Haines, B. Ozbas, D. J. Pochan and J. P. Schneider, Cytocompatibility of self-assembled beta-hairpin peptide hydrogel surfaces, *Biomaterials*, 2005, **26**, 5177–5186.

141. M. A. Serban, Y. Liu and G. D. Prestwich, Effects of extracellular matrix analogues on primary human fibroblast behavior, *Acta. Biomater.*, 2008, **4**, 67–75.

142. E. F. Banwell, E. S. Abelardo, D. J. Adams, M. A. Birchall, A. Corrigan, A. M. Donald, M. Kirkland, L. C. Serpell, M. F. Butler and D. N. Woolfson, Rational design and application of responsive alpha-helical peptide hydrogels, *Nat. Mater.*, 2009, **8**, 596–600.

143. Y. Yuan, C. Cong, J. Zhang, L. Wei, S. Li, Y. Chen, W. Tan, J. Cheng, Y. Li, X. Zhao and Y. Lu, Self-assembling peptide nanofiber as potential substrates in islet transplantation, *Transplant. Proc.*, 2008, **40**, 2571–2574.

144. M. Zhao, C. Song, W. Zhang, Y. Hou, R. Huang, Y. Song, W. Xie and Y. Shi, The three-dimensional nanofiber scaffold culture condition improves viability and function of islets, *J. Biomed. Mater. Res., Part A*, 2010, **94**, 667–672.

145. K. Hamada, M. Hirose, T. Yamashita and H. Ohgushi, Spatial distribution of mineralized bone matrix produced by marrow mesenchymal stem cells in self-assembling peptide hydrogel scaffold, *J. Biomed. Mater. Res., Part A*, 2008, **84**, 128–136.

146. N. Ni, Y. Hu, H. Ren, C. Luo, P. Li, J. B. Wan and H. Su, Self-assembling peptide nanofiber scaffolds enhance dopaminergic differentiation of mouse pluripotent stem cells in 3-dimensional culture, *PLoS One*, 2013, **8**, e84504.

147. M. Ozeki, S. Kuroda, K. Kon and S. Kasugai, Differentiation of bone marrow stromal cells into osteoblasts in a self-assembling peptide hydrogel: in vitro and in vivo studies, *J. Biomater. Appl.*, 2011, **25**, 663–684.

148. B. N. Cavalcanti, B. D. Zeitlin and J. E. Nor, A hydrogel scaffold that maintains viability and supports differentiation of dental pulp stem cells, *Dent. Biomater.*, 2013, **29**, 97–102.

149. M. E. Davis, P. C. Hsieh, T. Takahashi, Q. Song, S. Zhang, R. D. Kamm, A. J. Grodzinsky, P. Anversa and R. T. Lee, Local myocardial insulin-like growth factor 1 (IGF-1) delivery with biotinylated peptide nanofibers improves cell therapy for myocardial infarction, *Proc. Natl. Acad. Sci. U. S. A.*, 2006, **103**, 8155–8160.

150. J. P. Jung, A. K. Nagaraj, E. K. Fox, J. S. Rudra, J. M. Devgun and J. H. Collier, Co-assembling peptides as defined matrices for endothelial cells, *Biomaterials*, 2009, **30**, 2400–2410.
151. J. Chen, R. R. Pompano, F. W. Santiago, L. Maillat, R. Sciammas, T. Sun, H. Han, D. J. Topham, A. S. Chong and J. H. Collier, The use of self-adjuvanting nanofiber vaccines to elicit high-affinity B cell responses to peptide antigens without inflammation, *Biomaterials*, 2013, **34**, 8776–8785.
152. A. L. Sieminski, C. E. Semino, H. Gong and R. D. Kamm, Primary sequence of ionic self-assembling peptide gels affects endothelial cell adhesion and capillary morphogenesis, *J. Biomed. Mater. Res., Part A*, 2008, **87**, 494–504.
153. A. Nagayasu, H. Yokoi, J. A. Minaguchi, Y. Z. Hosaka, H. Ueda and K. Takehana, Efficacy of self-assembled hydrogels composed of positively or negatively charged peptides as scaffolds for cell culture, *J. Biomater. Appl.*, 2012, **26**, 651–665.
154. V. Jayawarna, S. M. Richardson, A. R. Hirst, N. W. Hodson, A. Saiani, J. E. Gough and R. V. Ulijn, Introducing chemical functionality in Fmoc-peptide gels for cell culture, *Acta. Biomater.*, 2009, **5**, 934–943.
155. S. Ustun, A. Tombuloglu, M. Kilinc, M. O. Guler and A. B. Tekinay, Growth and differentiation of prechondrogenic cells on bioactive self-assembled peptide nanofibers, *Biomacromolecules*, 2013, **14**, 17–26.
156. E. T. Pashuck, H. Cui and S. I. Stupp, Tuning supramolecular rigidity of peptide fibers through molecular structure, *J. Am. Chem. Soc.*, 2010, **132**, 6041–6046.
157. E. J. Leon, N. Verma, S. Zhang, D. A. Lauffenburger and R. D. Kamm, Mechanical properties of a self-assembling oligopeptide matrix, *J. Biomater. Sci., Polym. Ed.*, 1998, **9**, 297–312.
158. N. A. Dudukovic and C. F. Zukoski, Mechanical properties of self-assembled Fmoc-diphenylalanine molecular gels, *Langmuir*, 2014, **30**, 4493–4500.
159. J. Raeburn, A. Zamith Cardoso and D. J. Adams, The importance of the self-assembly process to control mechanical properties of low molecular weight hydrogels, *Chem. Soc. Rev.*, 2013, **42**, 5143–5156.
160. M. A. Greenfield, J. R. Hoffman, M. O. de la Cruz and S. I. Stupp, Tunable mechanics of peptide nanofiber gels, *Langmuir*, 2010, **26**, 3641–3647.
161. A. L. Sieminski, A. S. Was, G. Kim, H. Gong and R. D. Kamm, The stiffness of three-dimensional ionic self-assembling peptide gels affects the extent of capillary-like network formation, *Cell Biochem. Biophys.*, 2007, **49**, 73–83.

162. N. Mari-Buye, T. Luque, D. Navajas and C. E. Semino, Development of a three-dimensional bone-like construct in a soft self-assembling peptide matrix, *Tissue Eng., Part A*, 2013, **19**, 870–881.

163. T. Fernandez-Muinos, M. Suarez-Munoz, M. Sanmarti-Espinal and C. E. Semino, Matrix dimensions, stiffness, and structural properties modulate spontaneous chondrogenic commitment of mouse embryonic fibroblasts, *Tissue Eng., Part A*, 2014, **20**, 1145–1155.

164. J. P. Jung, J. L. Jones, S. A. Cronier and J. H. Collier, Modulating the mechanical properties of self-assembled peptide hydrogels via native chemical ligation, *Biomaterials*, 2008, **29**, 2143–2151.

165. C. J. Newcomb, S. Sur, J. H. Ortony, O. S. Lee, J. B. Matson, J. Boekhoven, J. M. Yu, G. C. Schatz and S. I. Stupp, Cell death versus cell survival instructed by supramolecular cohesion of nanostructures, *Nat. Commun.*, 2014, **5**, 3321.

166. C. Yan, M. E. Mackay, K. Czymmek, R. P. Nagarkar, J. P. Schneider and D. J. Pochan, Injectable solid peptide hydrogel as a cell carrier: effects of shear flow on hydrogels and cell payload, *Langmuir*, 2012, **28**, 6076–6087.

167. L. Haines-Butterick, K. Rajagopal, M. Branco, D. Salick, R. Rughani, M. Pilarz, M. S. Lamm, D. J. Pochan and J. P. Schneider, Controlling hydrogelation kinetics by peptide design for three-dimensional encapsulation and injectable delivery of cells, *Proc. Natl. Acad. Sci. U. S. A.*, 2007, **104**, 7791–7796.

168. M. H. Zaman, L. M. Trapani, A. L. Sieminski, D. Mackellar, H. Gong, R. D. Kamm, A. Wells, D. A. Lauffenburger and P. Matsudaira, Migration of tumor cells in 3D matrices is governed by matrix stiffness along with cell-matrix adhesion and proteolysis, *Proc. Natl. Acad. Sci. U. S. A.*, 2006, **103**, 10889–10894.

169. N. J. Spencer, D. A. Cotanche and C. M. Klapperich, Peptide- and collagen-based hydrogel substrates for in vitro culture of chick cochleae, *Biomaterials*, 2008, **29**, 1028–1042.

170. J. R. Thonhoff, D. I. Lou, P. M. Jordan, X. Zhao and P. Wu, Compatibility of human fetal neural stem cells with hydrogel biomaterials in vitro, *Brain Res.*, 2008, **1187**, 42–51.

171. S. Koutsopoulos and S. Zhang, Long-term three-dimensional neural tissue cultures in functionalized self-assembling peptide hydrogels, matrigel and collagen I, *Acta. Biomater.*, 2013, **9**, 5162–5169.

172. H. Cho, S. Balaji, A. Q. Sheikh, J. R. Hurley, Y. F. Tian, J. H. Collier, T. M. Crombleholme and D. A. Narmoneva, Regulation of endothelial cell activation and angiogenesis by injectable peptide nanofibers, *Acta. Biomater.*, 2012, **8**, 154–164.

173. M. D. Pierschbacher and E. Ruoslahti, Cell attachment activity of fibronectin can be duplicated by small synthetic fragments of the molecule, *Nature*, 1984, **309**, 30–33.

174. E. Ruoslahti, RGD and other recognition sequences for integrins, *Annu. Rev. Cell Dev. Biol.*, 1996, **12**, 697–715.

175. R. Pytela, M. D. Pierschbacher and E. Ruoslahti, Identification and isolation of a 140 kd cell surface glycoprotein with properties expected of a fibronectin receptor, *Cell*, 1985, **40**, 191–198.

176. E. Ruoslahti and M. D. Pierschbacher, Arg-Gly-Asp: a versatile cell recognition signal, *Cell*, 1986, **44**, 517–518.

177. E. F. Plow, T. A. Haas, L. Zhang, J. Loftus and J. W. Smith, Ligand binding to integrins, *J. Biol. Chem.*, 2000, **275**, 21785–21788.

178. J. L. West and J. A. Hubbell, Polymeric biomaterials with degradation sites for proteases involved in cell migration, *Macromolecules*, 1999, **32**, 241–244.

179. H. Nagase, in *Cancer Drug Discovery and Development: Matrix Metalloprotease Inhibitors in Cancer Therapy*, ed. N. J. Clendeninn and K. Appelt, Humana Press, 2001, pp. 39–66.

180. J. Patterson and J. A. Hubbell, Enhanced proteolytic degradation of molecularly engineered PEG hydrogels in response to MMP-1 and MMP-2, *Biomaterials*, 2010, **31**, 7836–7845.

181. G. A. Hudalla and W. L. Murphy, Using "click" chemistry to prepare SAM substrates to study stem cell adhesion, *Langmuir*, 2009, **25**, 5737–5746.

182. G. A. Hudalla and W. L. Murphy, Immobilization of peptides with distinct biological activities onto stem cell culture substrates using orthogonal chemistries, *Langmuir*, 2010, **26**, 6449–6456.

183. J. C. Schense and J. A. Hubbell, Three-dimensional migration of neurites is mediated by adhesion site density and affinity, *J. Biol. Chem.*, 2000, **275**, 6813–6818.

184. M. Kato and M. Mrksich, Using model substrates to study the dependence of focal adhesion formation on the affinity of integrin-ligand complexes, *Biochemistry*, 2004, **43**, 2699–2707.

185. T. Boontheekul, H. J. Kong, S. X. Hsiong, Y. C. Huang, L. Mahadevan, H. Vandenburgh and D. J. Mooney, Quantifying the relation between bond number and myoblast proliferation, *Faraday Discuss.*, 2008, **139**, 53–70, discussion 105–128, 419–120.

186. P. R. Patel, R. C. Kiser, Y. Y. Lu, E. Fong, W. C. Ho, D. A. Tirrell and R. H. Grubbs, Synthesis and cell adhesive properties of linear and cyclic RGD functionalized polynorbornene thin films, *Biomacromolecules*, 2012, **13**, 2546–2553.

187. K. A. Kilian and M. Mrksich, Directing stem cell fate by controlling the affinity and density of ligand-receptor interactions at the biomaterials interface, *Angew. Chem., Int. Ed. Engl.*, 2012, **51**, 4891–4895.

188. S. X. Hsiong, T. Boontheekul, N. Huebsch and D. J. Mooney, Cyclic arginine-glycine-aspartate peptides enhance three-dimensional stem cell osteogenic differentiation, *Tissue Eng., Part A*, 2009, **15**, 263–272.

189. G. A. Hudalla, J. A. Modica, Y. F. Tian, J. S. Rudra, A. S. Chong, T. Sun, M. Mrksich and J. H. Collier, A self-adjuvanting supramolecular vaccine carrying a folded protein antigen, *Adv. Healthcare Mater.*, 2013, **2**, 1114–1119.

190. Z. N. Mahmoud, S. B. Gunnoo, A. R. Thomson, J. M. Fletcher and D. N. Woolfson, Bioorthogonal dual functionalization of self-assembling peptide fibers, *Biomaterials*, 2011, **32**, 3712–3720.

191. S. Khan, S. Sur, P. Y. Dankers, R. M. da Silva, J. Boekhoven, T. A. Poor and S. I. Stupp, Post-assembly functionalization of supramolecular nanostructures with bioactive peptides and fluorescent proteins by native chemical ligation, *Bioconjugate Chem.*, 2014, **25**, 707–717.

192. S. Sangiambut, K. Channon, N. M. Thomson, S. Sato, T. Tsuge, Y. Doi and E. Sivaniah, A robust route to enzymatically functional, hierarchically self-assembled peptide frameworks, *Adv. Mater.*, 2013, **25**, 2661–2665.

193. T. Sawada, T. Takahashi and H. Mihara, Affinity-based screening of peptides recognizing assembly states of self-assembling peptide nanomaterials, *J. Am. Chem. Soc.*, 2009, **131**, 14434–14441.

194. A. Miyachi, T. Takahashi, S. Matsumura and H. Mihara, Peptide nanofibers modified with a protein by using designed anchor molecules bearing hydrophobic and functional moieties, *Chemistry*, 2010, **16**, 6644–6650.

195. T. Sawada and H. Mihara, Dense surface functionalization using peptides that recognize differences in organized structures of self-assembling nanomaterials, *Mol. Biosyst.*, 2012, **8**, 1264–1274.

196. N. Mehrban, E. Abelardo, A. Wasmuth, K. L. Hudson, L. M. Mullen, A. R. Thomson, M. A. Birchall and D. N. Woolfson, Assessing cellular response to functionalized alpha-helical peptide hydrogels, *Adv. Healthcare Mater.*, 2014, **3**, 1387–1391.

197. M. O. Guler, S. Soukasene, J. F. Hulvat and S. I. Stupp, Presentation and recognition of biotin on nanofibers formed by branched peptide amphiphiles, *Nano Lett.*, 2005, **5**, 249–252.

198. M. Reches and E. Gazit, Biological and chemical decoration of peptide nanostructures via biotin-avidin interactions, *J. Nanosci. Nanotechnol.*, 2007, **7**, 2239–2245.

199. S. Matsumura, S. Uemura and H. Mihara, Construction of biotinylated peptide nanotubes for arranging proteins, *Mol. Biosyst.*, 2005, **1**, 146–148.

200. Q. Li, K. L. Chow and Y. Chau, Three-dimensional self-assembling peptide matrix enhances the formation of embryoid bodies and their neuronal differentiation, *J. Biomed. Mater. Res., Part A*, 2014, **102**, 1991–2000.

201. Q. Li and Y. Chau, Neural differentiation directed by self-assembling peptide scaffolds presenting laminin-derived epitopes, *J. Biomed. Mater. Res., Part A*, 2010, **94**, 688–699.

202. G. A. Silva, C. Czeisler, K. L. Niece, E. Beniash, D. A. Harrington, J. A. Kessler and S. I. Stupp, Selective differentiation of neural progenitor cells by high-epitope density nanofibers, *Science*, 2004, **303**, 1352–1355.

203. V. M. Tysseling-Mattiace, V. Sahni, K. L. Niece, D. Birch, C. Czeisler, M. G. Fehlings, S. I. Stupp and J. A. Kessler, Self-assembling nanofibers inhibit glial scar formation and promote axon elongation after spinal cord injury, *J. Neurosci.*, 2008, **28**, 3814–3823.

204. H. Hosseinkhani, M. Hosseinkhani, F. Tian, H. Kobayashi and Y. Tabata, Osteogenic differentiation of mesenchymal stem cells in self-assembled peptide-amphiphile nanofibers, *Biomaterials*, 2006, **27**, 4079–4086.

205. H. Hosseinkhani, M. Hosseinkhani and H. Kobayashi, Proliferation and differentiation of mesenchymal stem cells using self-assembled peptide amphiphile nanofibers, *Biomed. Mater.*, 2006, **1**, 8–15.

206. F. Zhang, G. S. Shi, L. F. Ren, F. Q. Hu, S. L. Li and Z. J. Xie, Designer self-assembling peptide scaffold stimulates pre-osteoblast attachment, spreading and proliferation, *J. Mater. Sci.: Mater. Med.*, 2009, **20**, 1475–1481.

207. Z. Huang, T. D. Sargeant, J. F. Hulvat, A. Mata, P. Bringas, Jr., C. Y. Koh, S. I. Stupp and M. L. Snead, Bioactive nanofibers instruct cells to proliferate and differentiate during enamel regeneration, *J. Bone Miner. Res.*, 2008, **23**, 1995–2006.

208. M. O. Guler, L. Hsu, S. Soukasene, D. A. Harrington, J. F. Hulvat and S. I. Stupp, Presentation of RGDS epitopes on self-assembled nanofibers of branched peptide amphiphiles, *Biomacromolecules*, 2006, **7**, 1855–1863.

209. Y. F. Tian, J. M. Devgun and J. H. Collier, Fibrillized peptide microgels for cell encapsulation and 3D cell culture, *Soft Matter*, 2011, **7**, 6005–6011.

210. Y. Kumada and S. Zhang, Significant type I and type III collagen production from human periodontal ligament fibroblasts in 3D peptide scaffolds without extra growth factors, *PLoS One*, 2010, **5**, e10305.

211. X. M. Wang, A. Horii and S. G. Zhang, Designer functionalized self-assembling peptide nanofiber scaffolds for growth, migration, and tubulogenesis of human umbilical vein endothelial cells, *Soft Matter*, 2008, **4**, 2388–2395.

212. X. Liu, X. Wang, A. Horii, L. Qiao, S. Zhang and F. Z. Cui, In vivo studies on angiogenic activity of two designer self-assembling peptide scaffold hydrogels in the chicken embryo chorioallantoic membrane, *Nanoscale*, 2012, **4**, 2720–2727.

213. A. Horii, X. Wang, F. Gelain and S. Zhang, Biological designer self-assembling peptide nanofiber scaffolds significantly enhance osteoblast proliferation, differentiation and 3-D migration, *PLoS One*, 2007, **2**, e190.

214. E. Genove, C. Shen, S. Zhang and C. E. Semino, The effect of functionalized self-assembling peptide scaffolds on human aortic endothelial cell function, *Biomaterials*, 2005, **26**, 3341–3351.

215. M. Zhou, A. M. Smith, A. K. Das, N. W. Hodson, R. F. Collins, R. V. Ulijn and J. E. Gough, Self-assembled peptide-based hydrogels as scaffolds for anchorage-dependent cells, *Biomaterials*, 2009, **30**, 2523–2530.

216. M. Zhou, R. V. Ulijn and J. E. Gough, Extracellular matrix formation in self-assembled minimalistic bioactive hydrogels based on aromatic peptide amphiphiles, *J. Tissue Eng.*, 2014, **5**, DOI: 2041731414531593.

217. G. Uzunalli, Z. Soran, T. S. Erkal, Y. S. Dagdas, E. Dinc, A. M. Hondur, K. Bilgihan, B. Aydin, M. O. Guler and A. B. Tekinay, Bioactive self-assembled peptide nanofibers for corneal stroma regeneration, *Acta. Biomater.*, 2014, **10**, 1156–1166.

218. J. M. Anderson, M. Kushwaha, A. Tambralli, S. L. Bellis, R. P. Camata and H. W. Jun, Osteogenic differentiation of human mesenchymal stem cells directed by extracellular matrix-mimicking ligands in a biomimetic self-assembled peptide amphiphile nanomatrix, *Biomacromolecules*, 2009, **10**, 2935–2944.

219. D. J. Lim, S. V. Antipenko, J. B. Vines, A. Andukuri, P. T. Hwang, N. T. Hadley, S. M. Rahman, J. A. Corbett and H. W. Jun, Improved MIN6 beta-cell function on self-assembled peptide amphiphile nanomatrix inscribed with extracellular matrix-derived cell adhesive ligands, *Macromol. Biosci.*, 2013, **13**, 1404–1412.

220. J. Luo and Y. W. Tong, Self-assembly of collagen-mimetic peptide amphiphiles into biofunctional nanofiber, *ACS Nano*, 2011, **5**, 7739–7747.
221. F. Gelain, D. Bottai, A. Vescovi and S. Zhang, Designer self-assembling peptide nanofiber scaffolds for adult mouse neural stem cell 3-dimensional cultures, *PLoS One*, 2006, **1**, e119.
222. C. Cunha, S. Panseri, O. Villa, D. Silva and F. Gelain, 3D culture of adult mouse neural stem cells within functionalized self-assembling peptide scaffolds, *Int. J. Nanomed.*, 2011, **6**, 943–955.
223. X. Liu, X. Wang, H. Ren, J. He, L. Qiao and F. Z. Cui, Functionalized self-assembling peptide nanofiber hydrogels mimic stem cell niche to control human adipose stem cell behavior in vitro, *Acta. Biomater.*, 2013, **9**, 6798–6805.
224. Z. Zou, Q. Zheng, Y. Wu, X. Guo, S. Yang, J. Li and H. Pan, Biocompatibility and bioactivity of designer self-assembling nanofiber scaffold containing FGL motif for rat dorsal root ganglion neurons, *J. Biomed. Mater. Res., Part A*, 2010, **95**, 1125–1131.
225. Z. Zou, T. Liu, J. Li, P. Li, Q. Ding, G. Peng, Q. Zheng, X. Zeng, Y. Wu and X. Guo, Biocompatibility of functionalized designer self-assembling nanofiber scaffolds containing FRM motif for neural stem cells, *J. Biomed. Mater. Res., Part A*, 2014, **102**, 1286–1293.
226. V. M. Tysseling, V. Sahni, E. T. Pashuck, D. Birch, A. Hebert, C. Czeisler, S. I. Stupp and J. A. Kessler, Self-assembling peptide amphiphile promotes plasticity of serotonergic fibers following spinal cord injury, *J. Neurosci. Res.*, 2010, **88**, 3161–3170.
227. Z. Huang, C. J. Newcomb, Y. Zhou, Y. P. Lei, P. Bringas, Jr., S. I. Stupp and M. L. Snead, The role of bioactive nanofibers in enamel regeneration mediated through integrin signals acting upon C/EBPalpha and c-Jun, *Biomaterials*, 2013, **34**, 3303–3314.
228. N. Mehrban, B. Zhu, F. Tamagnini, F. I. Young, A. Wasmuth, K. L. Hudson, A. R. Thomson, M. A. Birchall, A. D. Randall, B. Song and D. N. Woolfson, Functionalized alpha-helical peptide hydrogels for neural tissue engineering, *ACS Biomater. Sci. Eng.*, 2015, 431–439. DOI: 10.1021/acsbiomaterials.5b00051.
229. F. Taraballi, A. Natalello, M. Campione, O. Villa, S. M. Doglia, A. Paleari and F. Gelain, Glycine-spacers influence functional motifs exposure and self-assembling propensity of functionalized substrates tailored for neural stem cell cultures, *Front. Neuroeng.*, 2010, **3**, 1.
230. S. Sur, F. Tantakitti, J. B. Matson and S. I. Stupp, Epitope topography controls bioactivity in supramolecular nanofibers, *Biomater. Sci.*, 2015, **3**, 530–532.

231. H. Storrie, M. O. Guler, S. N. Abu-Amara, T. Volberg, M. Rao, B. Geiger and S. I. Stupp, Supramolecular crafting of cell adhesion, *Biomaterials*, 2007, **28**, 4608–4618.

232. S. Sur, C. J. Newcomb, M. J. Webber and S. I. Stupp, Tuning supramolecular mechanics to guide neuron development, *Biomaterials*, 2013, **34**, 4749–4757.

233. J. E. Goldberger, E. J. Berns, R. Bitton, C. J. Newcomb and S. I. Stupp, Electrostatic Control of Bioactivity, *Angew. Chem., Int. Ed.*, 2011, **50**, 6292–6295.

234. K. Rajangam, H. A. Behanna, M. J. Hui, X. Han, J. F. Hulvat, J. W. Lomasney and S. I. Stupp, Heparin binding nanostructures to promote growth of blood vessels, *Nano Lett.*, 2006, **6**, 2086–2090.

235. J. C. Stendahl, L. J. Wang, L. W. Chow, D. B. Kaufman and S. I. Stupp, Growth factor delivery from self-assembling nanofibers to facilitate islet transplantation, *Transplantation*, 2008, **86**, 478–481.

236. H. D. Guo, G. H. Cui, J. J. Yang, C. Wang, J. Zhu, L. S. Zhang, J. Jiang and S. J. Shao, Sustained delivery of VEGF from designer self-assembling peptides improves cardiac function after myocardial infarction, *Biochem. Biophys. Res. Commun.*, 2012, **424**, 105–111.

237. R. Mammadov, B. Mammadov, S. Toksoz, B. Aydin, R. Yagci, A. B. Tekinay and M. O. Guler, Heparin mimetic peptide nanofibers promote angiogenesis, *Biomacromolecules*, 2011, **12**, 3508–3519.

238. G. Uzunalli, Y. Tumtas, T. Delibasi, O. Yasa, S. Mercan, M. O. Guler and A. B. Tekinay, Improving pancreatic islet in vitro functionality and transplantation efficiency by using heparin mimetic peptide nanofiber gels, *Acta. Biomater.*, 2015, **22**, 8–18.

239. R. N. Shah, N. A. Shah, M. M. Del Rosario Lim, C. Hsieh, G. Nuber and S. I. Stupp, Supramolecular design of self-assembling nanofibers for cartilage regeneration, *Proc. Natl. Acad. Sci. U. S. A.*, 2010, **107**, 3293–3298.

240. S. S. Lee, E. L. Hsu, M. Mendoza, J. Ghodasra, M. S. Nickoli, A. Ashtekar, M. Polavarapu, J. Babu, R. M. Riaz, J. D. Nicolas, D. Nelson, S. Z. Hashmi, S. R. Kaltz, J. S. Earhart, B. R. Merk, J. S. McKee, S. F. Bairstow, R. N. Shah, W. K. Hsu and S. I. Stupp, Gel scaffolds of BMP-2-binding peptide amphiphile nanofibers for spinal arthrodesis, *Adv. Healthcare Mater.*, 2015, **4**, 131–141.

241. M. J. Webber, J. Tongers, C. J. Newcomb, K. T. Marquardt, J. Bauersachs, D. W. Losordo and S. I. Stupp, Supramolecular nanostructures that mimic VEGF as a strategy for ischemic tissue repair, *Proc. Natl. Acad. Sci. U. S. A.*, 2011, **108**, 13438–13443.

242. S. Khan, S. Sur, C. J. Newcomb, E. A. Appelt and S. I. Stupp, Self-assembling glucagon-like peptide 1-mimetic peptide amphiphiles for enhanced activity and proliferation of insulin-secreting cells, *Acta. Biomater.*, 2012, **8**, 1685–1692.

243. Y. Chau, Y. Luo, A. C. Cheung, Y. Nagai, S. Zhang, J. B. Kobler, S. M. Zeitels and R. Langer, Incorporation of a matrix metalloproteinase-sensitive substrate into self-assembling peptides - a model for biofunctional scaffolds, *Biomaterials*, 2008, **29**, 1713–1719.

244. M. C. Giano, D. J. Pochan and J. P. Schneider, Controlled biodegradation of self-assembling beta-hairpin peptide hydrogels by proteolysis with matrix metalloproteinase-13, *Biomaterials*, 2011, **32**, 6471–6477.

245. Y. F. Tian, G. A. Hudalla, H. Han and J. H. Collier, Controllably degradable beta-sheet nanofibers and gels from self-assembling depsipeptides, *Biomater. Sci.*, 2013, **1**, 1037–1045.

246. A. Mata, Y. Geng, K. J. Henrikson, C. Aparicio, S. R. Stock, R. L. Satcher and S. I. Stupp, Bone regeneration mediated by biomimetic mineralization of a nanofiber matrix, *Biomaterials*, 2010, **31**, 6004–6012.

247. H. Ceylan, S. Kocabey, H. Unal Gulsuner, O. S. Balcik, M. O. Guler and A. B. Tekinay, Bone-like mineral nucleating peptide nanofibers induce differentiation of human mesenchymal stem cells into mature osteoblasts, *Biomacromolecules*, 2014, **15**, 2407–2418.

248. M. Gungormus, M. Branco, H. Fong, J. P. Schneider, C. Tamerler and M. Sarikaya, Self assembled bi-functional peptide hydrogels with biomineralization-directing peptides, *Biomaterials*, 2010, **31**, 7266–7274.

249. M. K. Kang, J. S. Colombo, R. N. D'Souza and J. D. Hartgerink, Sequence effects of self-assembling multidomain peptide hydrogels on encapsulated SHED cells, *Biomacromolecules*, 2014, **15**, 2004–2011.

250. K. M. Galler, L. Aulisa, K. R. Regan, R. N. D'Souza and J. D. Hartgerink, Self-assembling multidomain peptide hydrogels: designed susceptibility to enzymatic cleavage allows enhanced cell migration and spreading, *J. Am. Chem. Soc.*, 2010, **132**, 3217–3223.

251. H. W. Jun, V. Yuwono, S. E. Paramonov and J. D. Hartgerink, Enzyme-mediated degradation of peptide-amphiphile nanofiber networks, *Adv. Mater.*, 2005, **17**, 2612–2617.

252. K. Shroff, E. L. Rexeisen, M. A. Arunagirinathan and E. Kokkoli, Fibronectin-mimetic peptide-amphiphile nanofiber gels support increased cell adhesion and promote ECM production, *Soft Matter*, 2010, **6**, 5064–5072.

253. E. L. Rexeisen, W. Fan, T. O. Pangburn, R. R. Taribagil, F. S. Bates, T. P. Lodge, M. Tsapatsis and E. Kokkoli, Self-assembly of fibronectin mimetic peptide-amphiphile nanofibers, *Langmuir*, 2010, **26**, 1953–1959.

254. J. P. Jung, J. V. Moyano and J. H. Collier, Multifactorial optimization of endothelial cell growth using modular synthetic extracellular matrices, *Integr. Biol.*, 2011, **3**, 185–196.

255. Y. Kumada, N. A. Hammond and S. G. Zhang, Functionalized scaffolds of shorter self-assembling peptides containing MMP-2 cleavable motif promote fibroblast proliferation and significantly accelerate 3-D cell migration independent of scaffold stiffness, *Soft Matter*, 2010, **6**, 5073–5079.

256. B. Mammadov, R. Mammadov, M. O. Guler and A. B. Tekinay, Cooperative effect of heparan sulfate and laminin mimetic peptide nanofibers on the promotion of neurite outgrowth, *Acta. Biomater.*, 2012, **8**, 2077–2086.

257. T. D. Sargeant, C. Aparicio, J. E. Goldberger, H. Cui and S. I. Stupp, Mineralization of peptide amphiphile nanofibers and its effect on the differentiation of human mesenchymal stem cells, *Acta. Biomater.*, 2012, **8**, 2456–2465.

258. F. Chiti and C. M. Dobson, Protein misfolding, functional amyloid, and human disease, *Annu. Rev. Biochem.*, 2006, **75**, 333–366.

259. J. D. Harper and P. T. Lansbury, Jr., Models of amyloid seeding in Alzheimer's disease and scrapie: mechanistic truths and physiological consequences of the time-dependent solubility of amyloid proteins, *Annu. Rev. Biochem.*, 1997, **66**, 385–407.

260. P. Hortschansky, V. Schroeckh, T. Christopeit, G. Zandomeneghi and M. Fandrich, The aggregation kinetics of Alzheimer's beta-amyloid peptide is controlled by stochastic nucleation, *Protein Sci.*, 2005, **14**, 1753–1759.

261. J. T. Jarrett and P. T. Lansbury, Jr., Seeding "one-dimensional crystallization" of amyloid: a pathogenic mechanism in Alzheimer's disease and scrapie? *Cell*, 1993, **73**, 1055–1058.

262. G. A. Hudalla, T. Sun, J. Z. Gasiorowski, H. Han, Y. F. Tian, A. S. Chong and J. H. Collier, Gradated assembly of multiple proteins into supramolecular nanomaterials, *Nat. Mater.*, 2014, **13**, 829–836.

263. S. Whitelam, L. O. Hedges and J. D. Schmit, Self-Assembly at a Nonequilibrium Critical Point, *Phys. Rev. Lett.*, 2014, **112**, 155504.

264. J. Z. Gasiorowski and J. H. Collier, Directed intermixing in multicomponent self-assembling biomaterials, *Biomacromolecules*, 2011, **12**, 3549–3558.

265. D. J. Irvine, A. V. Ruzette, A. M. Mayes and L. G. Griffith, Nanoscale clustering of RGD peptides at surfaces using comb polymers. 2. Surface segregation of comb polymers in polylactide, *Biomacromolecules*, 2001, **2**, 545–556.

266. D. J. Irvine, A. M. Mayes and L. G. Griffith, Nanoscale clustering of RGD peptides at surfaces using Comb polymers. 1. Synthesis and characterization of Comb thin films, *Biomacromolecules*, 2001, **2**, 85–94.

267. R. R. Pompano, J. Chen, E. A. Verbus, H. Han, A. Fridman, T. McNeely, J. H. Collier and A. S. Chong, Titrating T-cell epitopes within self-assembled vaccines optimizes CD4+ helper T cell and antibody outputs, *Adv. Healthcare Mater.*, 2014, **3**, 1898–1908.

268. J. S. Rudra, S. Mishra, A. S. Chong, R. A. Mitchell, E. H. Nardin, V. Nussenzweig and J. H. Collier, Self-assembled peptide nanofibers raising durable antibody responses against a malaria epitope, *Biomaterials*, 2012, **33**, 6476–6484.

269. J. S. Rudra, Y. F. Tian, J. P. Jung and J. H. Collier, A self-assembling peptide acting as an immune adjuvant, *Proc. Natl. Acad. Sci. U. S. A.*, 2010, **107**, 622–627.

270. C. T. Wong Po Foo, J. S. Lee, W. Mulyasasmita, A. Parisi-Amon and S. C. Heilshorn, Two-component protein-engineered physical hydrogels for cell encapsulation, *Proc. Natl. Acad. Sci. U. S. A.*, 2009, **106**, 22067–22072.

271. L. Cai and S. C. Heilshorn, Designing ECM-mimetic materials using protein engineering, *Acta. Biomater.*, 2014, **10**, 1751–1760.

272. R. C. Preda, G. Leisk, F. Omenetto and D. L. Kaplan, Bioengineered silk proteins to control cell and tissue functions, *Methods Mol Biol*, 2013, **996**, 19–41.

273. D. L. Nettles, A. Chilkoti and L. A. Setton, Applications of elastin-like polypeptides in tissue engineering, *Adv. Drug Delivery Rev.*, 2010, **62**, 1479–1485.

274. C. M. Elvin, A. G. Carr, M. G. Huson, J. M. Maxwell, R. D. Pearson, T. Vuocolo, N. E. Liyou, D. C. Wong, D. J. Merritt and N. E. Dixon, Synthesis and properties of crosslinked recombinant pro-resilin, *Nature*, 2005, **437**, 999–1002.

275. R. M. Capito, H. S. Azevedo, Y. S. Velichko, A. Mata and S. I. Stupp, Self-assembly of large and small molecules into hierarchically ordered sacs and membranes, *Science*, 2008, **319**, 1812–1816.

276. L. W. Chow, R. Bitton, M. J. Webber, D. Carvajal, K. R. Shull, A. K. Sharma and S. I. Stupp, A bioactive self-assembled membrane to promote angiogenesis, *Biomaterials*, 2011, **32**, 1574–1582.

277. T. Hou, Z. Li, F. Luo, Z. Xie, X. Wu, J. Xing, S. Dong and J. Xu, A composite demineralized bone matrix–self assembling peptide scaffold for enhancing cell and growth factor activity in bone marrow, *Biomaterials*, 2014, **35**, 5689–5699.

278. J. B. Vines, D. J. Lim, J. M. Anderson and H. W. Jun, Hydroxyapatite nanoparticle reinforced peptide amphiphile nanomatrix enhances the osteogenic differentiation of mesenchymal stem cells by compositional ratios, *Acta. Biomater.*, 2012, **8**, 4053–4063.

279. J. M. Anderson, J. L. Patterson, J. B. Vines, A. Javed, S. R. Gilbert and H. W. Jun, Biphasic peptide amphiphile nanomatrix embedded with hydroxyapatite nanoparticles for stimulated osteoinductive response, *ACS Nano*, 2011, **5**, 9463–9479.

280. R. L. Huang, W. Qi, L. B. Feng, R. X. Su and Z. M. He, Self-assembling peptide-polysaccharide hybrid hydrogel as a potential carrier for drug delivery, *Soft Matter*, 2011, **7**, 6222–6230.

281. S. S. Lee, B. J. Huang, S. R. Kaltz, S. Sur, C. J. Newcomb, S. R. Stock, R. N. Shah and S. I. Stupp, Bone regeneration with low dose BMP-2 amplified by biomimetic supramolecular nanofibers within collagen scaffolds, *Biomaterials*, 2013, **34**, 452–459.

282. A. Restuccia, Y. F. Tian, J. H. Collier and G. A. Hudalla, Self-assembled glycopeptide nanofibers as modulators of galectin-1 bioactivity, *Cell. Mol. Bioeng.*, 2015, **8**, 471–487.

283. S. Sur, J. B. Matson, M. J. Webber, C. J. Newcomb and S. I. Stupp, Photodynamic control of bioactivity in a nanofiber matrix, *ACS Nano*, 2012, **6**, 10776–10785.

284. D. Papapostolou, A. M. Smith, E. D. Atkins, S. J. Oliver, M. G. Ryadnov, L. C. Serpell and D. N. Woolfson, Engineering nanoscale order into a designed protein fiber, *Proc. Natl. Acad. Sci. U. S. A.*, 2007, **104**, 10853–10858.

Subject Index